Series in PURE and APPLIED PHYSICS

T0203567

MOLECULAR AND CELLULAR BIOPHYSICS

CRC SERIES *in* PURE *and* APPLIED PHYSICS

Dipak Basu
Editor-in-Chief

PUBLISHED TITLES

Handbook of Particle Physics
M. K. Sundaresan

High-Field Electrodynamics
Frederic V. Hartemann

Fundamentals and Applications of Ultrasonic Waves
J. David N. Cheeke

Introduction to Molecular Biophysics
Jack A. Tuszynski
Michal Kurzynski

Practical Quantum Electrodynamics
Douglas M. Gingrich

Molecular and Cellular Biophysics
Jack A. Tuszynski

Series in PURE and APPLIED PHYSICS

MOLECULAR AND CELLULAR BIOPHYSICS

Jack A. Tuszynski

CRC Press
Taylor & Francis Group
Boca Raton London New York

CRC Press is an imprint of the
Taylor & Francis Group, an **informa** business

A CHAPMAN & HALL BOOK

Cover image provided by Tyler Luchko.

CRC Press
Taylor & Francis Group
6000 Broken Sound Parkway NW, Suite 300
Boca Raton, FL 33487-2742

First issued in paperback 2019

ISBN-13: 978-1-58488-675-4 (hbk)
ISBN-13: 978-0-367-38848-5 (pbk)

Library of Congress Cataloging-in-Publication Data

Tuszynski, J. A.
   Molecular and cellular biophysics / Jack A. Tuszynski.
      p. ; cm. -- (CRC series in pure and applied physics)
   Includes bibliographical references and index.
   ISBN-13: 978-1-58488-675-4 (hardcover : alk. paper)
   ISBN-10: 1-58488-675-7 (hardcover : alk. paper) 1. Biophysics. 2. Molecular biology. I. Title. II. Series.
   [DNLM: 1. Biophysics. 2. Molecular Biology. QH 506 T965m 2007]

QH505.T88 2007
572--dc22
                                          2007019223

Visit the Taylor & Francis Web site at
http://www.taylorandfrancis.com

and the CRC Press Web site at
http://www.crcpress.com

Acknowledgments

This book grew out of an earlier collaboration project with Dr. M. Kurzynski. I am very thankful to Dr. Kurzynski for his many insights.

The author also wishes to thank his students, staff and collaborators for their immense help in various stages of work leading to this book. In particular, the assistance of the following individuals is gratefully acknowledged: Eric Carpenter, Torin Huzil, Tyler Luchko, Hannes Bolterauer, John Dixon, Eberhard Unger, and Avner Priel.

Special thanks are expressed for Michelle Hanlon's tireless work on editing this book and organizing book material.

List of Tables

List of Figures

Contents

Chapter 1

What Is Life?

1.1   Hierarchical Organization of Knowledge

Every branch of science is more than a collection of facts and relations. It is also a philosophy within which empirical facts and observations are organized into a unified conceptual framework providing a more or less coherent concept of reality. Since biology is the study of life and living systems, it is simultaneously the study of human beings and as such can become biased by our philosophical and religious beliefs.

Understandably, bio-philosophy has been the battleground for the two most antagonistic and long-lived scientific controversies between mechanism and vitalism. Mechanism holds that life is basically no different from non-life, both being subject to the same physical and chemical laws, with the living material being simply more complex than the non-living matter. The mechanists firmly believe that ultimately life will be totally explicable in physical and chemical terms. The vitalists, on the other hand, fervently argue that life is much more than a complex ensemble of physically reducible parts and that there are some life processes that are not subject to the normal physical and chemical laws. Consequently, life will never be completely explained on a physiochemical basis alone. Central to the vitalists' doctrine is the concept of a "life force", 'vis vitalis' or 'elan vital', a non-material entity, which is not subject to the usual laws of physics and chemistry. This life force is seen to be animating the complex assembly of biomolecules into an organism and making it "alive". This concept is both ancient and virtually universal, having appeared in some form in all cultures and providing the basis for the religious beliefs of most of them.

It is to the early Greek philosopher-physicians such as Hippocrates that we owe the first organized concepts of the nature of life. These concepts developed within the framework of the medicine of that time and were based upon a modest amount of clinical observation and much conjecture. All functions of living things were the result of "humors", i.e., liquids of mystical properties flowing within the body. Several centuries later, Galen, virtually single-handedly, founded the sciences of anatomy and physiology. He produced a complete, complex system based upon his anatomical observations and an expanded concept of Hippocrates' humors. Galen's ideas became readily accepted and rapidly assumed the status of dogma, remaining unchallenged for more than a thousand years.

In the mid-sixteenth century, Andreas Vesalius questioned the validity of Galen's anatomical concepts and performed his own dissections upon the human body publishing his findings in a book, entitled in Latin *De humanis corporus fabricus* in 1543, which was the first anatomical text based upon actual human dissection. In 1628 William Harvey published the first real series of physiological experiments, describing the circulation of blood as a closed circuit, with the heart as the pumping agent. Vitalism, however, was still the only acceptable concept and Harvey naturally located the "vital spirit" in the blood.

At mid-century, René Descartes, the great French mathematician, attempted to unify biological concepts of structure, function and mind within a framework of mathematical physics. In Descartes' view all life was mechanical with all functions being directed by the brain and the nerves. To him we owe the beginning of the mechanistic concept of living machines-complex, but fully understandable in terms of physics and chemistry. Even Descartes did not break completely with tradition in that he believed that an "animating force" was still necessary to give the machine life "like a wind or a subtle flame" which he located within the nervous system. At about the same time Malphighi, an Italian physician and naturalist, using the new compound microscope to study living organisms, revealed a wealth of detail and complexity in living things.

Continued progress in the biological sciences has pushed the vitalistic viewpoint further and further to the fringes of reputable science. The universe is estimated to be 14-16 billion years old, the solar system is roughly 4.6 billion years old while life on Earth is believed to have emerged 3.5-4 billion years ago. Life is a process commonly described in terms of its properties and functions including self-organization, metabolism (energy utilization), adaptive behaviors, reproduction and evolution. As mentioned earlier, the two main approaches historically developed to understand the nature of the living state are: (i) mechanism or functionalism and (ii) vitalism. Functionalism implies that life is independent of its material substrate.

For example, certain types of self-organizing computer programs (lattice animals, Conway's game of life, etc.) exhibit life-like functions, a so-called "artificial life". Furthermore, all the components of living matter are in turn composed of "ordinary" atoms and molecules. This apparently demystifies life that is viewed as an emergent property of biochemical processes and functional activities. The failure of functionalism can be seen in its inability to consider the "unitary oneness" of all living systems. In 19th century biology (Lamarck) the latter characteristic of living systems was called the life force, elan vital, life energy and was assumed to be of electromagnetic nature. Molecular biology has systematically pushed vitalists ("animists") out of the spotlight viewing electromagnetic effects associated with life as just that: effects but not causes.

An interesting aspect of the organization of science has been very lucidly explained by A.C. Scott [31] in his book "Stairway to the Mind". In a nutshell, each branch of science exists almost autonomously on the broad scientific landscape developing its own set of elements and rules governing their interactions. For example, for all intents and purposes condensed matter physics, other than using electrons and protons as its main building blocks for solid state systems, exists completely apart from elementary particle physics. By extension, biology, as long as it refrains from violating the principles of physics, should have very little in common with physics. This hierarchical organization of knowledge where there is only a tiny set of intersections between the hierarchies involves

- human society

- culture

- consciousness

- physiology of an organism

- autonomous organs

- assembly of neuronal assemblies (brain)

- assembly of neurons (or other types of cells)

- the multiplex neuron (or any other type of cell)

- the axon-dendrite-synapse system

- mitochondria-nucleus-cytoskeleton

- protein-membrane-nucleic acid systems

- phospholipid-ATP-amino acid systems

- inorganic chemistry

- atomic physics-molecular physics

- nuclear physics

- elementary particle physics

This concept is illustrated schematically in Fig. 1.1.

The concept of a combinational barrier explains the rationale for the above, historically developed compartmentalization of the sciences. Mathematician Elsasser [16] coined the term an Immense Number defined as: $I = 10^{110}$ The reason for the choice of the power in the exponential is that: I = atomic weight of the Universe measured in proton's mass (daltons) times the age of the Universe in picoseconds (10^{-12} s)

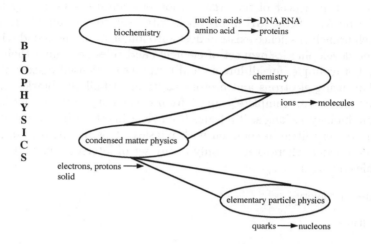

FIGURE 1.1:   Hierarchical organization of science.

Since no conceivable computer (even as big and as old as the whole universe) could store a list of 'I' objects, and even if it could, there would be no time to inspect it, an immense set of objects defines a category that is a virtually inexhaustible arena for intellectual pursuits with no danger of running out of interesting relationships between its elements. Examples of immense sets are: chess games, possible chemical molecules (chemistry), possible proteins (biochemistry), possible nerve cells (neurophysiology), possible ideas (culture), possible tunes (music) and possible types of personalities (psychology). Consequently, both biology and physics are legitimate areas of scientific exploration in their own right that could happily coexist with minimal overlap. What makes this somewhat simplistic separation less applicable to the biology-physics divide is the existence of so-called emergent phenomena. Due to an organizing principle in this hierarchical picture, we can better understand areas on a higher plane by knowing the organization rules for the elements whose roots are in the lower level of the hierarchy. Knowing the interaction principles between electrons and protons certainly helps develop solid state physics. The knowledge of protein-protein interactions should give us a glimpse into the functioning of a dynamical cell. In general, the whole is more than the sum of its parts, so each level of the hierarchy adds new rules of behavior to the structure that emerges.

1.2 General Characteristics of Living Systems

General Characteristics of a Cell

Life is the ultimate example of a complex dynamical system. A living organism develops through a sequence of interlocking transformations involving an immense number of components which are themselves made up of molecular subsystems. Yet when they are combined into a larger functioning unit (e.g., a cell), then so-called emergent properties arise.

For the past several decades, biologists have greatly advanced the understanding of how living systems work by focusing on the structure and function of constituent molecules such as DNA. Understanding what the parts of a complex machine are made of, however, does not explain how the whole system works. Scientific analysis of living systems has posed an enormous challenge and today we are prepared better than ever to tackle this enormous task. Conceptual advances in physics, vast improvements in the experimental techniques of molecular and cell biology (electron microscopy, NMR, AFM, etc.), and exponential progress in computational techniques have brought us to a unique point in the history of science when the expertise of researchers representing many areas of science can be brought to bear on the main unsolved puzzle of life, namely how cells live, divide and how they eventually die.

Cells are the key building blocks of living systems. Some of them are self-sufficient while others co-operate in multi-cellular organisms. The human body is composed of approximately 10^{13} cells of some 200 different types. A typical size of a cell is on the order of 10 microns and its dry weight amounts to about 7×10^{-16} kg. In its natural state, 70% of the contents are water molecules. The fluid contents of a cell are known as the cytoplasm. The cytoplasm is the liquid medium bound within a cell, while the cytoskeleton is the lattice of filaments formed throughout the cytoplasm [11]. Fig. 1.2 is a grossly simplified representation of an animal cell in cross-section.

Two major types of cells are:

- a) prokaryotic: simple cells with no nucleus and no membrane bound compartments. Bacteria, e.g., *E.Coli*, and blue-green algae belong to this group.

- b) eukaryotic: cells with a nucleus and a differentiated structure including compartmentalized organelles as well as a filamentous cytoskeleton. Examples here include higher developed animal and plant cells, green algae and fungi. Eukaryotic cells emerged about 1.5 billion years ago.

Bacteria have linear dimensions in the 1-10 micrometer range while the sizes of eukaryotic cells range between 10 and 100 micrometers. An interesting observation may be made that diffusion constraints can be used for ions

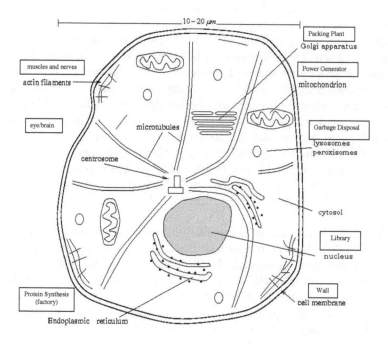

FIGURE 1.2:   General animal cell during interphase. Note major cellular organelles including nucleus, centrosome, and radiating microtubules. The endoplasmic reticulum is dotted with ribosomes.

to show that these are the upper limits on cell size. Moreover, the interior of a bacterium may be under considerable pressure reaching up to several atmospheres. These pressures are maintained due to the presence of a membrane which, except for Archaebacteria, is composed of layers of peptidoglycan sandwiched between two lipid bilayers the inner of which is a plasma membrane.

Plant cells have linear dimensions that vary between 10 and 100 micrometers. They are bounded by a cell wall whose thickness ranges between 0.1 and 10 microns and is composed of cellulose. Among its organelles, plant cells have a nucleus, endoplasmic reticulum, Golgi apparatus and mitochondria. Unique to plant cells is the presence of chloroplasts and vacuoles. Unlike bacteria, plant cells possess a cytoskeletal network adding to their mechanical strength. Animal cells tend to be smaller than plant cells since they do not have liquid-filled vacuoles. The organizing centre for their cytoskeleton appears to be a cylindrical organelle called a centriole that is approximately 0.4 microns long. Instead of chloroplasts, which are sites of photosynthesis, animal cells have mitochondria that produce energy in the form of ATP molecules obtained from reactions involving oxygen and food molecules (e.g., glucose). Mitochondria are shaped like a cylinder with rounded ends.

Eukaryotic cells also possess membrane-bound organelles, some of which are listed below [83]. *Mitochondria* produce energy; a *Golgi* apparatus (where various macromolecules are modified, sorted, and packaged for secretion from the cell for distribution to other organelles) is shaped like a stack of disks. The *endoplasmic reticulum* surrounds the nucleus and is the principial site of protein synthesis. Its volume is small compared to the surface area. A *nucleus* is the residence of chromosomes and the site of DNA replication and transcription. Most of the cell's DNA is housed within the nucleus (an additional site of DNA localization are the mitochondria), which is protected by the nuclear envelope. Within the nucleus is the nucleolus which functions at the site of ribosomal-RNA synthesis. The diameter of a nucleus ranges between 3 and 10 microns. All of the material within the cell excluding the nucleus is defined as the cytoplasm whose liquid components are referred to as the cytosol while the solid protein-based structures that float in it are called the cytoskeleton [11]. The main component of the cytosol is water. Most of the organelles are bound within their own membranes. Despite many differences, both animal and plant cells have striking similarities which are compared and contrasted in Fig. 1.3.

1.3 Artificial Life

Artificial life, or *a-life*, is a discipline within the general area of computer science devoted to the creation and study of lifelike organisms and systems

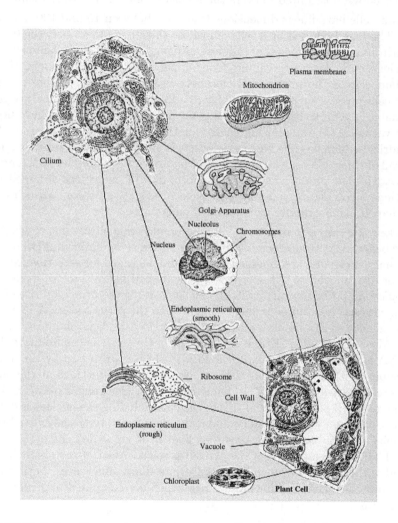

FIGURE 1.3:   Comparison of structures of typical animal and plant cells.

built by humans [30]. The stuff of artificial life is, in contrast to real life, inorganic matter, and its essence is information. Computers are the breeding grounds from which these new organisms emerge. Just as medical scientists and biologists have been studying biological life's mechanisms in vitro, the physicists, mathematicians and computer scientists of a-life hope to create life *in silico*.

The origins of the science of artificial life can be traced to Norber Wiener [40] and to John Von Neumann [38], the brilliant mathematician and physicist who, among his other accomplishments, invented game theory and the stored program computer. Von Neumann also made seminal contributions to quantum mechanics. Shortly before his death in 1955, Von Neumann conceived of an abstract mathematical machine, a cellular automaton or collection of checkerboard squares, each of which could switch between various states. Following a simple set of rules, this automaton was capable of generating a copy of itself that in turn contained the blueprint for generating another copy of itself, ad infinitum. The Von Neumann cellular automaton could not only reproduce itself but also was a version of the "universal computer" first envisioned by Alan Turing. In other words, it was capable of mimicking the operations of any other conceivable computing machine. Then, in the late 1960s, a British mathematician named John Conway created a vastly simplified version of the Von Neumann cellular automaton, now known as the Game of Life. To read more about the fascinating field of cellular automata the reader is referred to Wolfram [41].

The degree to which artificial life resembles real, "wet" life varies; many experimenters admit freely that their laboratory creations are simply simulations of aspects of life. Practitioners of the related science of artificial intelligence (AI) distinguish between two views of what they do: the so-called "weak AI," which deals with simulating intelligence in computers, and "strong AI," in which the work is intended to actually create intelligent computers (machines). Similarly, one can draw a distinction between weak and strong artificial life. The goal of the practitioners of "weak" a-life is to illuminate, model and understand more clearly the life that exists on earth and possibly elsewhere [43]. The boldest practitioners of the science of artificial life engage in so-called "strong" a- life. They look toward the long-term development of actual living organisms whose essence is information and information processing and not matter and energy. These fictitious creatures may be embodied in corporeal form, as a-life robots, or they may live within a computer. Whichever the case may be, these creations, it is insisted, are intended to live under every reasonable definition of the word life, as much as bacteria, plants, animals, and human beings do.

In general, artificial life may be further divided into subfields such as:

- Adaptive Behavior,

- Social Behavior,

- Evolutionary Biology,

- Complex Systems,

- Neutral Evolution,

- Morphogenesis and Development and

- Learning [39].

The Game of Life invented by the Cambridge mathematician John Conway originally began as an experiment to determine if a simple set of rules could create a universal computer. The concept "universal computer", on the other hand, was invented by Alan Turing and denotes a machine that is capable of emulating any kind of information processing by implementing a small set of simple operations. The game of Life is a simple 2-D analog of basic processes in living systems. The game consists in tracking changes through time in the patterns formed by sets of "living" cells arranged in a 2-dimensional grid. Any cell in the grid may be in either of the two states: "alive" or "dead". The state of each cell changes from one generation to the next depending on the state of its immediate neighbors. The rules governing these changes are designed to mimic population change:

- A living cell with only 0 or 1 living neighbors dies from isolation.

- A living cell with 4 or more living neighbors dies from overcrowding.

- A dead cell with exactly 3 living neighbors becomes alive.

- All other cells remain unchanged.

Astonishingly complex patterns are created by these simple rules. Conway's prediction that the system was capable of computation was confirmed. By using "gliders", a pattern that moves diagonally over the board, to represent bit streams one can create a universal computer. The behavior of "Life" is typical of the way in which many cellular automata reproduce features of living systems. That is, regularities in the model tend to produce order: starting from an arbitrary initial configuration, order usually emerges fairly quickly. Ultimately most configurations either disappear entirely or break up into isolated patterns that are either static or else they cycle between several different forms with a fixed period.

Even more fascinating than this are the similarities the cell patterns have with organic life. Another goal of Conway was to find self-reproducing organisms within the "Life" system, agglomerations of cells that with the passage of time split up and form several new organisms, identical to the original. However, no such cell patterns have been found yet. One could imagine such organisms developing through evolutionary mechanisms evolving into steadily fitter and more complex "life forms".

"Life" in the game is in the border area between chaos and order, just as organic life. Any change of the rules results either in a static universe, where cells either die out or fill the entire board, or in a chaotic universe where no structures or patterns can be distinguished.

There are several periodic or static organisms that the initial configurations tend to evolve into, but it is a chaotic system in that it is very sensitive to changes in the environment; organisms are very unstable when subjected to small changes. What is interesting to observe in "Life" is that it tends to spontaneously develop organisms consisting of several cells, in the same way that natural laws in some unknown way seem to lead to steadily more complex organisms. Admittedly, in the game single cells are excluded by rule a). Only multicelled units can survive. Higher life would never have developed in a universe with natural laws like the "Life" rules leading to chaos. The parallels to the development of life are intriguing.

In 1982 Stephen Wolfram set out to create an even simpler, one-dimensional system. The main advantage of a one-dimensional automaton is that changes over time can be illustrated in a single, two-dimensional image and that each cell has only two neighbors. In all, for a CA based on a 3-cell local rule, if the state of a cell is dependent on its own and its two neighbors states in the previous generation, there are 8 configurations for which a result must be specified which makes for a total of 256 rules which makes it possible to examine all the results of all rule sets. However, those rules are usually reduced to a total of 32 "legal" rules. The legal rules come from the fact that the CA were initially used for modeling biological processes such as reproduction. The question was if even such a simplified system could display complex behavior. Astonishingly, the answer is "yes".

All legal rules can be divided into four types according to their behavior when seeded with a single cell. These types of behavior are:

- type 1: disappear over time,

- type 2: merely copy the non-zero cell forever,

- type 3: yield a completely uniform type of rows or uniform pairs of rows,

- type 4: develop nontrivial patterns.

There are 8 rules in each class. In the same way as in the game of "Life" there are rules that yield chaos and those which yield static states. The most interesting rules were, of course, those rules that gave results in between, a state vaguely called "complex", the state resembling that of biological life. Thus, even Wolfram's systems proved capable of displaying complex behavior, though in less spectacular forms than Conway's "Life". It is worth noting that gliders abound in most one-dimensional rule systems. These show up as diagonal lines in the image produced. Just about all rule systems create diagonal patterns. Those which produce gliders out of everything are obviously the ones where the state of the next generation only depends on the state of the left / right neighbor in the previous generation. The main reason one-dimensional systems seem so much more trivial and boring than its two-dimensional equivalents is probably that our vision is adapted to two-dimensional pictures, which makes it easier for us to recognize patterns and organisms / agglomerations in 2-D.

In summary, Wolfram's hypothesis that complexity can arise of out of almost arbitrarily simple rules seems to be confirmed. Consequently, an application of algorithmic, cellular automata models to simulate living systems and their processes is warranted. A particularly interesting class of living processes pertains to the functioning of a neuron and hence to the emergence of consciousness at the most elementary level. Below we discuss how modeling of neuronal processes has led to yet a different type of mathematical description.

Neural network models

As complicated as the biological neuron is, it may be simulated by a simple computer model. The active inputs have a weight that determines how much they contribute to the neuron. Each neuron also has a threshold value for firing. If the sum of all the weights of all active inputs is greater than the threshold, then the neuron is active. The nodes in the network are vast simplifications of real neurons - they can only exist in one of two possible "states" - firing or not firing. Every node is connected to every other node with some strength. At any instant of time a node will change its state (i.e., start or stop firing) depending on the inputs it receives from the other nodes.

It can be shown that a single neuron and its input weighting performs the logical expression A and not B when only two inputs are present. However, a biological neuron may have as many as 10,000 different inputs, and may send its output (the presence or absence of a short-duration spike) to many other neurons. Indeed, neurons are wired up in a 3-dimensional pattern.

While von Neumann computing machines are based on the processing/memory abstraction of human information processing, neural networks present a different paradigm for computing and are based on the parallel architecture of animal brains [20]. Real brains, however, are orders of magnitude more complex than any artificial computer network. In a nutshell, neural networks are a form of multiprocessor computer system, with

- simple processing elements
- a high degree of interconnection
- simple scalar messages
- adaptive interaction between elements.

Neural networks can be explicitly programmed to perform a task by creating the topology and then setting the weights of each link and threshold [17]. However, this bypasses one of the unique strengths of neural nets: the ability to program themselves. The most basic method of training a neural network is trial and error. If the network is not behaving the way it should, we can change the weighting of a random link by a random amount. If the accuracy of the network declines, we can undo the change and make a different one.

Unfortunately, the number of possible weightings rises exponentially as one adds new neurons, making large general-purpose neural nets impossible to construct using trial and error methods. In the early 1980s D. Rumelhart and D. Parker independently rediscovered an old calculus-based learning algorithm called back-propagation. The back-propagation algorithm compares the result that was obtained with the result that was expected. It then uses this information to systematically modify the weights throughout the neural network [22]. This type of training takes only a fraction of the time that the trial and error method takes. It can also be reliably used to train networks on only a portion of the data, since it makes inferences. The resulting networks are often correctly configured to answer problems that they have never been specifically trained on.

In its simplest form, pattern recognition uses one analog neuron for each pattern to be recognized. All the neurons share the entire input space. Training pattern recognition networks is simple. We first draw the desired pattern and select the neuron that should learn that pattern. For each active pixel, we add one to the weight of the link between the pixel and the neuron in training. We subtract one from the weight of each link between an inactive pixel and the neuron in training. A more sophisticated method of pattern recognition involves several neural nets working in parallel, each looking for a particular feature such as "a horizontal line at the top". The results of these feature detectors are then fed into another net that would match the best pattern. This is closer to the way humans recognize patterns.

One of the simplest algorithms implemented in neural networks research is the so-called perceptron. In it, the network adapts as follows: change the weight by an amount proportional to the difference between the desired output and the actual output. Mathematically, we put it as follows:

$$\Delta W_i = \eta(D - Y)I_i \tag{1.1}$$

where η is the learning rate, D is the desired output, and Y is the actual output. The above is called the Perceptron Learning Rule or the Delta Rule, and it was introduced in the early 1960's. We expose the net to the patterns as shown in Table 1.1.

We then train the network on examples. Weights alter after each exposure to a complete set of patterns. Since $(D - Y) = 0$ for all patterns at the end of the procedure, the weights cease adapting. At this point the network has finished learning. However, single perceptrons are limited in what they can learn, therefore, more advanced algorithms have been designed.

Back-Propagated Delta Rule Networks (BP) (also known as multi-layer perceptrons or MLPs) and Radial Basis Function Networks (RBF) are both

TABLE 1.1: Pattern Recognition in Networks

I_0	I_1	Desired Output
0	0	0
0	1	1
1	0	1
1	1	1

extensions of the Perceptron Learning Rule. Both can learn arbitrary mappings or classifications. Furthermore, the inputs and outputs can have real number values.

BP is a development from the simple Delta rule in which hidden layers additional to the input and output layers are added. The network topology is constrained to be feed-forward, i.e., loop-free. Generally connections are allowed from the input layer to the first hidden layer; from the first hidden layer to the second,...,etc., and from the last hidden layer to the output layer. The hidden layer learns to recode the inputs. More than one hidden layer can be used. The architecture is more powerful than single-layer networks.

The weight change rule is a development of the Perceptron Learning Rule. Weights are changed by an amount proportional to the error at that unit times the output of the unit feeding into the weight. Running the network consists of:

- Forward pass: the outputs are calculated and the error at the output units calculated.

- Backward pass: the output unit error is used to alter weights on the output units. Then the error at the hidden nodes is calculated by back-propagating the error at the output units through the weights, and the weights on the hidden nodes are altered using these values.

For each data pair to be learned a forward pass and backwards pass is performed. This is repeated over and over again until the error is at a sufficiently low level.

Radial basis function networks are also feed-forward, but have only one hidden layer. Like BP, RBF nets can learn arbitrary mappings: the primary difference is in the hidden layer. RBF hidden layer units have a receptive field which has a center. That means there is a particular input value at which they have a maximal output. Their output tails off as the input moves away from this point. Generally, the hidden unit function is a Gaussian.

RBF networks are trained by: (a) deciding on how many hidden units there should be, (b) deciding on their centers and the sharpness of their Gaussians and (c) training up the output layer. Generally, the centers and standard deviations of the Gaussians are decided on first by examining the training data. The output layer weights are then trained using the Delta rule. Currently, BP is the most widely applied neural network technique. RBFs have the advantage that one can add extra units with centers near parts of the input which are difficult to classify. Both BP and RBFs can also be used for processing time-varying data. Simple Perceptrons, BP, and RBF networks need a teacher to tell the network what the desired output should be. These are, therefore, supervised networks. In an unsupervised net, the network adapts purely in response to its inputs. Such networks can learn to pick out structure in their input. Although learning in these unsupervised nets can be slow, running the trained net is very fast.

The so-called Kohonen clustering algorithm takes a high-dimensional input, and clusters it, but retains some topological ordering of the output. After training, an input causes some of the output units in some area to become active. Such clustering is very useful as a preprocessing stage. Note that neural networks cannot do anything that cannot be done using traditional computing techniques, but they can achieve some things which would otherwise be very difficult. In particular, they can form a model from their training data alone. This is particularly useful with sensory data where there may be an algorithm, but it is not known a priori, or has too many variables. It is easier to let the network learn from examples than make educated guesses about the algorithm at work.

Neural networks are being used for many applications outside neuroscience, for example: in investment analysis, in signature analysis, in process control and to monitor the performance of industrial systems, among other applications.

Memory

One of the most important functions of a conscious system such as the human brain is the build-up and recall of memories. Due to their operational differences one must distinguish both short and long term memory. Our memories function in what is called an associative or content-addressable fashion. That is, a memory does not exist in some isolated region of the brain, located in a particular set of neurons. All memories are in some sense strings of memories. For example, we remember someone or something in a variety of ways, by the color of the object, its shape, the sound produced, or the smell associated with it. Thus memories are stored in association with one another. These different sensory units lie in completely separate parts of the brain, so it is clear that the memory of the object must be distributed throughout the brain somehow. Indeed, PET scans reveal that during memory recall there is a pattern of brain activity in many widely different parts of the brain. Notice also that it is possible to access the full memory by initially remembering just one or two of these characteristic features. We access the memory by its contents not by where it is stored in the neural pathways of the brain. This is a very powerful and effective mechanism. It is also very different from a traditional computer where specific facts are located in specific places in computer memory. If only partial information is available about this location, the fact or memory cannot be recalled at all.

In addition to performing logic operations, neurons are also capable of storing and retrieving data from 'memory'. A neural network can store data in two formats. Permanent data or long term memory may be designed into the weightings of each neuron in a neural network. Temporary data or short term memory can be actively circulated in a loop, until it is needed again. Since the output of the neuron feeds back onto itself, there is a self-sustaining loop that keeps the neuron firing even when one input is no longer active. The stored binary bit is continuously accessible by looking at the output. This configuration is called a latch. While it works perfectly in a neural network model, a biological neuron does not behave quite this way. After firing, a biological neuron has to rest for a millisecond before it can fire again. Thus one would have to link several neurons together in a duty-cycle chain to achieve the same result.

The Hopfield neural network is a simple artificial network which is able to store certain memories or patterns in a manner similar to the brain. The full pattern can be recovered if the network is presented with only partial information. Furthermore, there is a degree of stability in the system. If just a few of the connections between nodes (neurons) are severed, the recalled memory is not too severely corrupted and the network can respond with a "best guess". A similar phenomenon is observed in the human brain. Namely, during a lifetime many neurons die out (in fact as much as 20 percent of the original neurons) without the individual suffering a catastrophic loss of individual memories. The human brain is very robust in this respect. Any general initial pattern of firing and non-firing nodes changes over time. The crucial property of the Hopfield network which renders it useful for simulating memory recall is that it settles down after a long enough time to some fixed pattern. Certain nodes are always "on" and others "off". Furthermore, it is possible to arrange that these stable firing patterns of the network correspond to the memories we wish to store. If we start the network off with a pattern of firing which approximates one of the "stable firing patterns" (memories), it will end up recalling the original perfect memory. In the Hopfield network there exists a simple way of setting up the connections between nodes in such a way that any desired set of patterns can be made into "stable firing patterns". Thus any set of memories can be burned into the network at the beginning. Then if we perturb the network by node activity we are guaranteed that a "memory" will be recalled. The memory that is recalled is the one which is "closest" to the starting pattern. Of course, there is a limit to it and if the input image is sufficiently poor, it may recall the incorrect memory and the network can become "confused". However, overall the network is reasonably robust such that if we change a few connection strengths just a little the recalled images are "approximately right".

For the Hopfield network one needs to add about ten nodes to store one more image. But the new memory is not stored with these ten new nodes. The stored memories are properties of the entire network since they are associated with the strengths of all inter-node connections. It is these connections or artificial neural pathways which determine the activity and hence function of the neural network. Thus the simple Hopfield neural network can perform some of the functions of memory recall in a manner analogous to the way we believe the brain functions. However, there is a major difference since for the artificial network we have to set up the connection strengths in just the right way in order to store a predetermined set of patterns. Once that is done, the network can be left to itself to handle the pattern-recall process. On the other hand, in the brain, there is no "teacher" to tell the neurons how to link up in order to store useful information. This part of the process is also automatic. The system is said to be self-organizing. In all cases, it is important to understand the nature of biological information hence we discuss it in the section that follows.

1.4 Biological Information, Information Processing and Signalling

Biological Information

One of the characteristic features of living systems is their ability to reduce entropy (organize themselves) and another is the information processing and signaling capability. Living systems must communicate with the environment in order to survive. To cast some light on the latter properties we must first address the issue of entropy reduction and information in biological systems even as small as a single cell.

Shannon [33] defined information as negative entropy given by the formula

$$I = k \ln W = -k \Sigma p_i \ln(p_i) \qquad (1.2)$$

where p_i is the probability of state i and W is the number of possible combinations with k denoting the Boltzmann constant. Shannon's definition has enabled the resolution of the long-standing paradox referred to as Maxwell's demon. The problem involved a fictitious creature that operated a small door between two compartments of a container with two types of gas molecules, for example high- and low-energy ones. The end result would be a separation of the gas into hot and cold with no energy expenditure thus contradicting the second law of thermodynamics. Szilard's solution [35] of the problem endowed the demon with information, i.e., Shannon's information which is negative entropy balancing out the changes in the entropy of the gas. In quantitative terms, the energy cost of 1 bit of information at physiological temperature is $ln2 \times kT = 3 \times 10^{-21} J = 18.5 meV$ which should be kept in mind discussing biological functions that are replete with information content.

Information about the structure and composition of a system to be developed, such as protein sequences and folding patterns is of major importance in this context.

While statistical entropy is a probabilistic measure of uncertainty or ignorance, information is a measure of a reduction in that uncertainty or knowledge. Entropy reaches its maximum value if all microscopic states available to the system are equiprobable, that is, if we have no indication whatsoever to assume that one state is more probable than another state. It is also clear that entropy is 0, if and only if the probability of a certain state is 1 (and of all other states 0). In that case we have maximal certainty or complete information about what state the system is in.

This definition was introduced by Shannon in the context of electronic media as a measure of the capacity for information transmission of a communication channel. Indeed, if we obtain some information about the state of the system, then this will reduce our uncertainty about the system's state, by excluding or reducing the probability of a number of states. The information we receive from an observation is equal to the degree to which uncertainty is reduced [4]. Although Shannon came to disavow the use of the term "information" to describe this measure, because it is purely syntactic and ignores the meaning of the signal, his theory came to be known as Information Theory nonetheless. Entropy has been vigorously pursued as a measure for a number of higher-order relational concepts, including complexity and organization.

Note that there are other methods of weighting the state of a system which do not adhere to probability theory's additivity condition that the sum of the probabilities must be 1. These methods, involving concepts from fuzzy systems theory and possibility theory, lead to alternative information theories. Together with probability theory these are called Generalized Information Theory (GIT). While GIT methods are under development, the probabilistic approach to information theory still dominates applications.

Production of DNA takes place even in non-replicating cells. A typical mammalian cell polymerizes approximately 2×10^8 nucleotides of DNA a minute into hnRNA [8] out of which only 5% end up in the cytoplasm coding for protein synthesis [14]. Since there is redundancy in coding of nucleotide triplets for the 20 amino acids, the original 6 bits of information in DNA translate into $log_2(20) = 4.2$ bits in a protein. Consequently, on the order of 0.7×10^6 bits/s are transmitted from the nucleus to the cytoplasm. This is augmented by a small fraction of information due to mitochondrial DNA [5]. This is but a small fraction of the total information production (understood as negative entropy) of a living cell. The vast majority of information is contained in the organized structure of the cell and its components.

Since the Shannon information formula employs probabilities of particular states, there are inherent dangers of selecting these probabilities, especially when this is done purely combinatorially as is often the case, for example in an amino-acid or nuclei acid sequence determination. This is not necessarily a random choice situation akin to tossing a coin. This means that taking

$p = 1/W$ to represent the probability for a single element selection, where W is the number of possible choices, may not be correct giving an excessively improbable (or high information) estimate. This would be the case if the choices of elements in a sequence are not of the same statistical weight, but instead are biased statistically, so that a more appropriate probability value is given by the canonical ensemble Boltzmann distribution formula $p_i = p_0 e^{-E_i/kT}$. Of course, in order to make this estimate, one needs to know the energies E_i of the individual microstates i and hence the Hamiltonian for the system. Therefore, the apparent information estimate of $I = (k \ln N)^n$ where n is the number of members in a string may be significantly larger than the true value of the corresponding negative entropy $-S$ from thermodynamic estimates of a given state - a maximum entropy state for equilibrium and hence a minimum information content. For a string of choices (e.g., an amino acid sequence in a peptide or a nucleic acid sequence in a DNA or RNA), this may lead to "basins of attraction" favoring some combinations strongly over others. There could be evolutionary retention of favored choices and the establishment of hierarchies of order.

An immense number has been defined by Elasser as $I = 10^{110}$ and represents a clear computational barrier even from the point of view of cataloguing such an enormous number of objects. Immense numbers commonly appear in biology: both DNA and protein sequences are immense numbers arising from the sheer numbers of possible combinations in which these macromolecules may be formed. However, in view of the argument above, restricting the phase space by forming basins of attraction due to intra-molecular interactions may result in a hugely reduced number of combinations one would encounter in practice.

Another comment we wish to make is a clear distinction between information and instruction. While the former was introduced on purely statistical grounds as a measure of the number of choices possible when making a selection for a string of elements, instruction implies the existence of a message, a messenger and a reader who would the execute the message. A classic example of this would be the synthesis of amino-acids contained in the triplet of DNA and RNA base pairs. While there is the same information value in every triplet, namely $k \ln(4^3) = 6$ bits, some amino-acids are coded uniquely by a single triplet and some by two, three, four etc. different ones. This is obvious in view of the fact that there are 64 possible triplets of base pairs while only 20 distinct amino-acids, hence the redundancy. A similar difference between information and instruction can be found in the genome where in addition to the coding sequences of DNA, most of which are of vital importance to the very survival of a given organism, one finds so-called junk DNA that has apparently no coding value but represents a vast majority of the DNA sequence. While it is still too early to state with absolute confidence that junk DNA contains no genetic information, the main thing to stress here is that information is often confused with instruction.

DNA and RNA are thought to be messengers of biological information and so are hormones and various signaling molecules. However, it appears that a vast majority of intracellular information content is not instructional in nature. This is akin to simple algorithms like the logistic map or fractal recursive relations that give rise to great mathematical complexity of the results that follow. Similarly, DNA can be viewed as an algorithm that spans an awe-inspiring complexity of living cells. While coding for protein synthesisis contained in the genetic code, it is most improbable that details of structure formation need special genetic coding. They most likely unfold due to self-organization inherent in the dynamics of the synthesized products within the cell.

In view of the discussion presented in this section regarding information content and information processing in a living cell, we wish to postulate the existence of two types of information in biological systems: (a) structural information, i.e., negative entropy and (b) functional information. The former is simply related to tight packing of the various molecules into macromolecules and macromolecules into organelles that comprise the cell. The latter, meanwhile, pertains to the functioning of the cell and hence the rate and amount of chemical reactions taking place. The two forms are somewhat related but not identical. Imagine the construction of a car as an analogy. It may look perfectly fine but if the gas line is cut, it won't function. The same holds true for a living cell. Some key reactions like the synthesis of tubulin, if not properly executed, will lead to the cell death. While structural information should be maximized, implying entropy reduction by the cell, functional information depends on the rate at which information is being exchanged. Therefore, the cell's tendency should be to increase the speed of biochemical reactions and the amount of molecular interactions if possible.

Living processes are cyclical in nature and one is expecting to maximize the rate of information (or entropy) change over time for maximum functionality. In order to optimize both aspects: structure and functionality, a living system such as a cell should strive to maximize the product of the two quantities, i.e., achieve:

$$\text{Max}\{I^2(t)(\frac{dI(t)}{dt})^2\} \tag{1.3}$$

where we have squared the quantities in the product due to the cyclicity of life's processes. Note that some aspects of this distinction between structural information and functional information in biological systems were already emphasized more than three decades ago by H. Froehlich [18] who coined the term biological coherence to draw attention to both the holistic and functional integration of the information flow in living matter.

Biological Signalling

Signalling by varied means is required to regulate the complex behavior of living systems from the simplest bacterium to yeast cells and larger eukaryotes such as humans. The difference between monocellular and multicellular organisms is that communication must necessarily be possible between the different cells of multicellular organisms. For any particular cell of the organism, this means that in addition to intracellular signalling, it must be prepared to both transmit and receive extracellular signals. The signalling mechanisms discovered so far exhibit the complexity of organic chemistry. In order to interpret signals from other cells, a cell requires special membrane receptors that can detect the presence of signal molecules in the extracellular fluid. In addition to these biochemical signaling pathways, Albrect-Bueller has proposed in his 'intelligent cell' model that cells are able to sense electromagnetic radiation (especially in the near infrared range) through the use of centrioles. Since the centrioles are always found with a perpendicular orientation, this would allow the cell to discern directional information about a signal through latitude and longitude measurements. That this would provide an invaluable signal receptor is clear, but the mechanism through which other cells might transmit such signals is not entirely clear. It is possible that mitochondria may generate light signals at infrared frequencies. (A detailed discussion of this process is given in Section 3.9.) Excluding this hypothesis, the cell has several methods of communication through the use of its varied signalling molecules as outlined below. The molecules are first packaged and then expelled from the cell. In the simplest type of messaging, the chemical signals are dumped outside the cell and carried diffusively. This method of communication is effective for only the signalling of nearby cells. Such local signalling is known as paracrine signalling. Synaptic signalling is a refined version of the paracrine model where the signal molecule, a neurotransmitter, is released at a specifically designed interface providing intimate contact between the source and target cells. This allows quick and direct signalling but still relies on diffusion to carry the signal molecules across the narrow junction. The last type of signalling, known as endocrine signalling, may be used when the target cells of the signal are either more distant or more widespread. These molecular signals known as hormones are secreted by the cell into the circulatory system. Thus although diffusion is used yet again, the stream of blood or sap may carry the signal a long distance. Due to the dilution effect of the circulatory fluid, hormones must be effective even at low concentrations such as $10^{-8}M$.

The mechanism behind the workings of these extracellular messengers is of the familiar antigen-antibody type. That is to say that there are specific integral membrane proteins to which a signalling molecule may bind on the target cell's exterior. The general behavior of these signalling systems is that binding of the signal molecule to the receptor induces a conformation change at the opposite end of the protein receptor that lies within the cell's interior. The conformational change may have a direct or indirect response. In the case of a direct response, the cytoplasmic domain of the protein becomes enzymatic and catalyses a specific chemical reaction until the signal molecule at the extracurricular end breaks down or becomes unbound. In the indirect case, the conformational change may release another signalling molecule on the cell's interior known as a G-protein. This G-protein may then bind to one or more other enzymes, either serving to activate or deactivate them. The indirect signal allows for the co-ordination of complementary reaction pathways and is one way in which the cell regulates its processes. Breaking the signalling scheme into many parts also allows for magnification of the signal at each step and makes it possible to ultimately have a large response to a small number of signal molecules. The ultimate sort of response that is possible by ligand binding is the opening of ion channels. In this case, the conformational change of the receptor is such that a hydrophilic channel is opened through the cell membrane and allows passage of a specific charged species such as calcium (Ca^{2+}). The calcium ion is useful in particular because it exists outside the cell in concentrations 4 to 10 times higher than the intracellular concentration. As a result, it diffuses easily and is used as a secondary signal. The mechanism for the operation of these signals is well understood and has been described in more detail in textbooks such as Alberts et al. [5].

The most interesting biophysics problems can be found within the cell where the methods of signal transduction are yet to be fully elucidated. Intracellular signalling includes mechanisms such as the action potential which is electrical in nature and driven by chemical potentials. Sensitivity of individual cells to concentration and potential gradients is necessary if the cell is to respond to gravitational or electric fields. Intracellular signalling co-ordinates the orchestra of cellular processes to ensure that the entire cell works in harmony. In mitosis, chromosome segregation to each pole of the mother cell is mediated by MTs. However, the simultaneity of the separation must be explained and requires some kind of signal to be mediated by the MTs. Treadmilling by free MTs under conditions of dynamic instability should also be explained since the opposite ends of the MT have such coordinated behavior. Recently, Maniotis and Ingber [180] demonstrated how pulling on actin filaments could induce changes within the nucleus. This illustrated that the cell was also sensitive to mechanical stimulation. One hypothesis for the control which the cell has over these processes is an electromagnetic regulation. This form of signalling has the advantage that it is exceedingly rapid relative to extracellular signalling. The cytoskeleton seems to play a key role in each of these mysterious examples of cellular signalling.

Consider the action potential. This electrical signal passes along the neural membrane driven by a cascade of sodium ions flowing into the cell and is switched off by a delayed flow of potassium ions out of the cell. While the action potential moves, there is no attenuation of the signal. Only toxins, which can disable the function of voltage-gated ion channels, are able to stop the progress of the action potential. Thus the behavior of the system appears soliton-like [253]. The cytoskeleton adopts a configuration in neurons where the orientation of MTs is parallel with uniform direction within the axon, along which the cell transmits signals to other cells, but the MTs adopt an aligned configuration with a non-uniform direction in the dendrites where the signals are received from other cells. Since there are molecular motors such as kinesin and dynein which move in opposite directions along MTs, one would suspect that there is another reason for the specific structure that is seen rather than simply the ability to transport goods using the MTs. MTs are known to be sensitive to both electric fields [332,333] and to magnetic fields [44] and align themselves such that they are parallel to the field lines [44]. The specific alignment in these two neuronal regions could be to make MTs insensitive to electric fields within dendrites but reinforce the susceptibility to electric fields within the axon. It has also been shown that in long cylindrical cells such as the geometry of an axon, electrical fields are able to penetrate most easily.

A key question in intracellular biological signalling is whether or not electron transport plays a role. The transfer of an electron between proteins results in a conformational change that has a physiological effect. In the case of ion channels, the donation or acceptance of electrons changes their internal electrostatistics and affects their function. This results in neuromodulation by changing in the response characteristics of the neuron and constitutes a reprogramming of neural networks. We will return to these issues in Chapter 4 of the book when we discuss in detail the various processes involved in the maintenance of living systems. We now turn our attention to the question of the origins of life on Earth and its evolution through the history of our planet.

1.5   Origin of Life

Initiation

The planet Earth was formed about 4.6 billion years ago as a result of accretion (inelastic collisions and agglomerations) of larger and larger rocky fragments gradually forming from the dust component of the gaseous dusty cloud which was the original matter of the Solar System. The 'Great Bombardment' ended only 3.9 billion years ago when the stream of meteorites falling onto the surface of the newly formed planet reached a more or less constant intensity. The first well-preserved petrified micro-stamps of relatively highly organized living organisms similar to cyanobacteria are about 3.5 billion years old, so life on Earth must have developed within the relatively short time of a few hundred million years.

Rejecting the hypothesis of an extraterrestrial origin of life, not so much for rational as for emotional reasons, we have to answer the question of origin of the simplest elements of living organisms: amino acids, simple sugars (monosaccharides) and nitrogenous bases. Three equally probable hypotheses have been put forward to explain their appearance [27]. According to the first and the oldest, these elements appeared as a result of electric discharges and ultraviolet irradiation of the primary Earth atmosphere containing mostly

CO_2 (as the atmospheres of Mars and Venus do today), and some amount of H_2O as well as strongly reducing gases CH_4, NH_3 and H_2S. According to the second hypothesis the basic elements of living organisms were formed in the space outside the orbits of large planets and transferred to Earth surface via collisions with comets and indirectly via carbon chondrites. The third hypothesis says that these elements appeared at the oceanic rifts that are the sites at which the new Earth crust is formed and where the water over-heated to 400 C containing strongly reducing FeS, H_2 and H_2S meets the cool water containing CO_2. Actually, the problem is still open as all the three hypotheses have been seriously criticized. Firstly, the primary Earth atmosphere might not have been reducing strongly enough. Secondly, the organic compounds from outer space could have deteriorated while passing through the Earth's atmosphere and, thirdly, the reduction of CO_2 in oceanic rifts requires non-trivial catalysts.

The three most important characteristics of life distinguishing it from among other natural phenomena have been aptly expressed by Charles Darwin, who created the theory of evolution that is so crucial for biology. Taking into account the achievements of the post-Darwinian genetics and biochemistry we can come up with the following definition:

- *Life is a process characterized by continuous: (1) reproduction, (2) variability and (3) selection (survival of the fittest).*

Using present-day language, to stay alive, individuals must have a replicable and modifiable *program*, a proper *metabolism* (a mechanism of matter and energy conversion) and a capability of *self-organization*.

The emergence of molecular biology in the 1950s brought answers to the question of the structure and functioning of the three most important classes of biological macromolecules: DNA (deoxyribonucleic acid), RNA (ribonucleic acid) and proteins. However, in the attempts at giving a possible scenario of evolution from small organic particles to large biomolecules, a classical chicken-and-egg problem was encountered. Here the question was what was first: the DNA that carried the coded information on enzymatic proteins controlling the physiological processes determining the fitness of an individual or the proteins which in particular enable the replication of DNA, its transcription into RNA and the translation of certain sequences of amino acids into new proteins (Fig. 1.4). This question was resolved in the 1970s as a result of the evolutionary experimentation in Manfred Eigen's laboratory [5]. The primary macromolecular system undergoing Darwin's evolution could be RNA. Single-stranded RNA is not only the information-carrier, a program or genotype, but thanks to a specific spatial structure it is also an object of selection or a phenotype. Equipped with the concept of a hypercycle [15] and inspired by Sol Spigelman, Manfred Eigen used the virial RNA replicase (Fig. 1.4b), a protein, to get new generations of RNA in vitro. It soon appeared that the complementary RNA could polymerize spontaneously, without any replicase, on the matrix of the already existing RNA as a template. Consequently, one

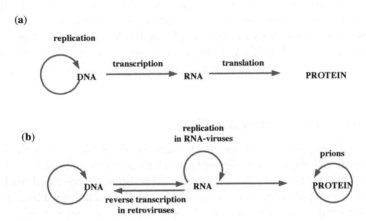

FIGURE 1.4: Processing genetic information. (a) The classical dogma: the information is carried by DNA which undergoes replication during the process of biological reproduction and transcription into RNA when it is to be expressed; gene expression consists in translation of the information written in RNA onto a particular protein primary structure. (b) Today's corrections to the classical dogma: also RNA can be replicated and transcribed, in the opposite direction, into DNA; the very proteins also can carry information as is assumed to take place in prion diseases.

can imagine the 'RNA World' composed only of nucleotides, their phosphates and their polymers - RNA's being the subject of Darwinian evolution, thus alive according to the definition adopted [19].

The concept of an RNA world is supported by many facts. Nucleotide triphosphates are highly effective sources of free energy. They fulfill this function as a relict in most chemical reactions of contemporary metabolism [262]. Dinucleotides play the role of co-factors in many protein enzymes. In fact, RNA molecules themselves can be enzymes [11]. It is becoming more common to talk about ribozymes. Contemporary ribosomes translating information from RNA onto a protein structure (see Fig. 1.4a) fulfill their catalytic function due to their ribosomal RNA rather than their protein components [29]. Finally, for a number of years now we have known the reverse transcriptase that transcribes information from RNA onto DNA (Fig. 1.4b). It appears, therefore, that RNA could possibly be a primary structure and DNA secondary. Modern organisms synthesize deoxyribonucleotides from ribonucleotides.

Machinery of Prokaryotic Cells

The smallest present-day system thought to be able to possess the key function of a living object, namely reproduction, is a cell. There is a sharp distinction between simple prokaryotic cells (that do not have a cell nucleus) and much more complex eukaryotic cells (with a well- defined nucleus). There is a body of evidence pointing to an earlier evolutionary emergence of prokaryotic cells. Eukaryotic cells are believed to have arisen as a result of mergers involving two or more specialized prokaryotic cells. Unfortunately almost nothing is known about the origin of prokaryotic cells themselves. The scenario that we present below is only an attempt to describe some key functional elements of the apparatus that each prokaryotic cell possesses. This is not a serious effort to reconstruct the history of life on Earth.

The world of the competing RNA molecules must have sooner or later reached a point at which a dearth of the only building material, nucleotide triphosphates, occurred. Those molecules that were able to provide themselves with adequate supplies of the building materials gained an evolutionary advantage. This, in turn, required a bag, a container in which RNA molecules with their supplies of material could protect themselves from the surroundings. In the liquid phase, such containers are formed spontaneously from phospholipids in which one of their parts is attracted and another repelled by water. In other words, these are amphipaticmolecules whose one part is hydrophilic and another hydrophobic. As a result of the hydrophobic part's movement away from water and the hydrophilic part's movement towards water, a lipid bilayer is formed without a boundary, i.e., a three-dimensional vesicle (Fig. 1.5). Since phospholipid vesicles can join in making bigger structures out of several small ones, they are important for the molecules of RNA that can divide and compete for food. Merging into bigger vesicles can be advantageous in foraging for food while dividing into small vesicles can be seen as giving rise to offspring. The phospholipid vesicle itself does not solve the problem since there must also be a way of selective infusion of nucleotides into its interior. Employing a new type of biomolecules, amino acids, some of which are hydrophilic and some other ones hydrophobic, however, solved this latter problem. Their linear polymers are called peptides and long peptides give rise to proteins. Proteins possess three-dimensional structures whose hydrophobicity properties depend on the order in which amino acid segments appear in the linear sequence. Such proteins may spontaneously embed themselves in a lipid bilayer and play the role of selective ion channels (see Fig. 1.5).

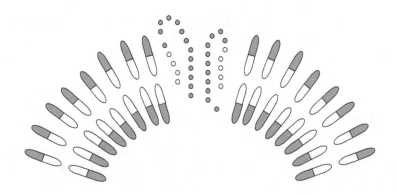

FIGURE 1.5: In a water environment, amphiphilic molecules composed of the hydrophilic part (shaded) and the hydrophobic part (white) organize spontaneously into bilayers closed into three-dimensional vesicles. Protein, a linear polymer of appropriately ordered hydrophilic (shaded circles) and hydrophobic(white circles) amino acids, forms a structure that spontaneously builds into the bilayer and allows selectively chosen molecules, e.g., nucleotide triphosphates, to pass into the lipid interior.

It appears therefore that the first stage in the development of a prokaryotic cell was probably the enclosure of some RNA molecules into phospholipid vesicles equipped with protein channels that enabled a selective transfer of triphosphate nucleotides into the interior region (Fig. 1.6a). The second stage must have been the perfection of these channels and a link between their structure and the information contained in the RNA molecules. This latter property would have gained selective significance. Selective successes could be scored by those RNA molecules which were able to translate a part of the information contained in the RNA base sequence into an amino acid sequence of an ion channel protein in order to synthesize it. This was the way to distinguish the so-called mRNA (messenger RNA) from tRNA (transfer

RNA) and from rRNA (ribosomal RNA). The mRNA carried information about the amino acid sequence in a protein while tRNA connected a given amino acid with the corresponding triple base set. The rRNA was a prototype of a ribosome, a catalytic RNA molecule that was able to synthesize into proteins the amino acids transported into its location by the molecules of tRNA. These amino acids had to be first recognized by triples of bases along the mRNA (Fig. 1.6b). The analysis of the nucleotide sequences in tRNA and rRNA of various origins indicates that they are very much alike and thus very archaic. The genetic code based on sequences of triples is equally universal and archaic. Contemporary investigations of both prokaryotic and eukaryotic ribosomes provide an ever-increasing body of evidence that the main catalytic role is played by rRNA and not the proteins contained within them.

However, proteins turn out to have much better catalytic properties than RNA. The key property is their high specificity vis a vis the substrate. Soon after, in the form of polymerases, they replaced RNA in the process of self-replication. On the template of RNA it was already possible to replicate sister RNA as well as DNA (deoxyribonucleic acid). DNA spontaneously forms a structure composed of two complementary strands (a double helix) and it turns out to be a much more stable carrier of information than RNA. This gives rise to the current method of transferring genetic information (see Fig. 1.6c) which goes as follows. Genetic information is stored on a double-stranded DNA, protein replicases duplicate this information in the process of cell division but protein transcriptases, if necessary, transcribe this information onto mRNA, which later, in the process of translation (partly ribozymatic and partly enzymatic), is used as a template to produce proteins. Transfer of information in the reverse direction from RNA to DNA via reverse transcriptases has been preserved as a living fossil in modern retroviruses.

Protein enzymes can carry out many useful tasks. For example, they can produce a much needed building material - triphosphate nucleotides, and in particular recycle them from the used up diphosphates and inorganic orthophosphate making use of saccharides as their source of free energy. Fig. 1.7 schematically illustrates the main metabolic pathways of energy and matter processing common to contemporary bacteria (prokaryotes) as well as animals and plants (eukaryotes). The central point at which many of these metabolic pathways converge is the pirogronian. It is easy to see the vertical path of glycolysis, the reduction of the most common monosaccharide, glucose, to pyruvate. It is equally easy to see the circular cycle of the citric acid that is connected with pirogronian through one or several reactions. The archaic origin of the main metabolic pathways is evident not only in their universality (from bacteria to man) but also in the presence in many of these reactions of nucleotide triphosphates, mainly ATP (adenosine triphosphate). If a given reaction is connected with the hydrolysis of ATP to ADP (adenosine diphosphate), it is marked in Fig. 1.7 by a letter P at the end of the reaction. If, on the other hand, a given reaction is linked to a synthesis of ADP and an orthophosphate group into ATP (a process called phosphorylation), it is

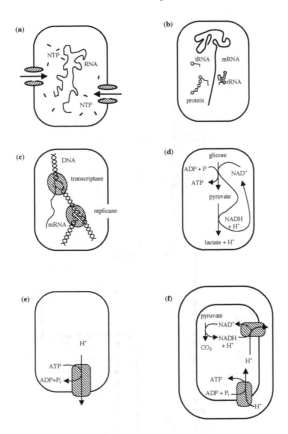

FIGURE 1.6: Development of the prokaryotic cell machinery. (a) The self-replicating RNA molecule with a supply of nucleotide triphosphates is enclosed in a vesicle bounded by a lipid bilayer with built-in protein channels that allow a selective passage of nucleotide triphosphates. (b) In an RNA chain a distinction is made between mRNA and various types of tRNA and rRNA, the latter being a prototype of a ribosome that was able to synthesize proteins according to the information encoded in an mRNA. Proteins produced this way turned out to be not only more selective membrane channels but also effective enzymes that could catalyze many useful biochemical processes. (c) A double-stranded DNA replaces RNA as a carrier of information. Protein replicates double this information in the process of division and protein transcriptases transfer this information onto mRNA. (d) Protein enzymes appear to be able to catalyze the process of lactose fermentation of sugars as a result of which the pool of high-energy triphosphate nucleotides can be replenished (mainly ATP) using low-energy di-phosphates (mainly ADP). The amount of oxidase (hydrogen acceptor) NAD^+ remains constant but the cell interior becomes acidic. (e) Proton pumps are created which are able to pump out H^+ ions into the cell exterior thanks to the ATP hydrolysis. (f) Other proton pumps use as fuel the hydrogen obtained from the decomposition of sugars through pirogronian to CO_2. Due to the presence of a wall or a second cell membrane the pumped out protons may now return to the cell interior through the first type of pumps which acting in reverse reconstruct ATP from ADP. A discovery is made of the membrane phosphorylation, which becomes the basic mechanism of bioenergetics in all living organisms today.

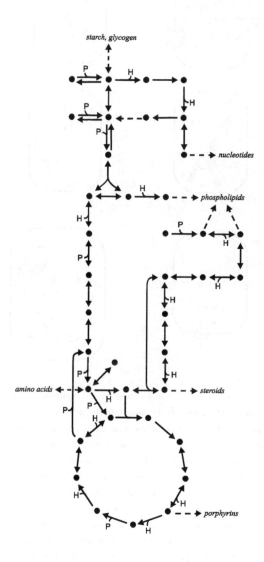

FIGURE 1.7: Outline of the main metabolic pathways. Substrates are represented by black dots and reactions, reversible or practically irreversible, each catalyzed by a specific enzyme, by arrows.

marked by a letter P at the start of a reaction.

From a chemical point of view, the process of going from glucose, $C_6H_{12}O_6$, to pyruvate, $CH_3-CO-COO^-$, is an oxidation reaction and it consists in taking hydrogen atoms from glucose. NAD (nicotinamide adenine dinucleotide) is a universal oxidant (an acceptor of hydrogen, i.e., simultaneously an electron and a proton). This is also a relict of the RNA world. The process of accepting by NAD^+ or $(NADP^+)$ two hydrogen atoms (one of which as a free proton H^+ that transfers the original charge of either NAD^+ or $NADP^+$) is marked in Fig. 1.9 by a letter H at the end of a given reaction.

An overall balance of the glycolysis reaction, i.e., an oxidation of glucose to a pyruvate, takes the form:

$$C_6H_{12}O_6 + 2NAD^+ + 2ADP + 2P_i \rightarrow$$
$$2CH_3 - CO - COO^- + 2NADH + 2H^+ + 2ATP + 2H_2O \quad (1.4)$$

During this process two molecules of NAD^+ are reduced by four atoms of hydrogen:

$$C_6H_{12}O_6 + 2NAD^+ \rightarrow 2CH_3 - CO - COO^- + 2H^+ + 2NADH + 2H^+ \quad (1.5)$$

Two protons are obtained from the dissociation of pyruvic acid into a pyruvate anion whereas another two protons transfer the original positive charge of NAD and the phosphorylation of two molecules of ADP to ATP according to the equation:

$$ADP + P_i + H^+ \rightarrow ATP + H_2O \quad (1.6)$$

In a neutral water environment ATP is present as an ion with four negative charges, ADP with three negative charges, and an orthophosphate P_i with two.

The primitive prokaryotic cells were properly equipped with the machinery of protein membrane channels able to select out of their environment only specific components. They also had protein enzymes to catalyze appropriate reactions. These cells became able to replenish their pool of nucleotide triphosphates at the expense of organic compounds of a fourth type - saccharides (see Fig. 1.6d). The oxidizer NAD^+ is recovered in the process of fermentation of a pyruvate into a lactate:

$$CH_3-CO-COO^- + NADH + H^+ \rightarrow CH_3-CHOH-COO^- + NAD^+ \quad (1.7)$$

This reaction is also utilized by present-day eukaryotic organisms whenever it is needed rapidly to obtain ATP under the conditions of limited oxygen supply.

The process of lactic fermentation that accompanies the process of phosphorylation of ADP to ATP with the use of sugars as a substrate has several drawbacks. In addition to its low efficiency (there is an unused lactate), it leads to an increased acidity of the cell. While sugars are neutral with a pH close to 7, lactate is a product of dissociation of lactic acid and in the process of breakdown of sugars, free protons H are released. This is a well-known experience during hard physical work and exercise when blood is unable to deliver a sufficient amount of oxygen to the muscles. The lowering of pH results in a significant slowdown or even a complete stoppage of the glycolysis reaction. For the decomposition of sugars to be effectively used in the production of ATP, a cell must find a different mechanism of fermentation whose product has a pH close to 7 or whose protons H^+ can be expelled outside the cell. This new type of fermentation was discovered by yeast, where it consists in the reduction of pirogronian to ethanol with a release of carbon dioxide in the process:

$$CH_3 - CO - COO^- + NADH + 2H^+ \rightarrow C_2H_5 - OH + CO_2 + NAD^+ \quad (1.8)$$

Before this took place, a proton pump was discovered utilizing the hydrolysis of ATP as a source of energy (see Fig. 1.6e). During the production of one molecule of ATP one hydrated proton H^+ is released inside the cell while the hydrolysis of one molecule of ATP results in the pumping outside the cell membrane of three hydrated protons H (see below 1.9a). The above process is hence still energetically favorable.

From the viewpoint of ATP production, a more efficient process appears to be a further oxidation of a pyruvate to an acetate and a carbon dioxide:

$$CH_3 - CO - COO^- + H_2O + NAD^+ \rightarrow CH_3 - COO^- + CO_2 + NADH + H^+$$
$$(1.9)$$

As a result of the above, a reduction of one molecule of NAD^+ is made by two atoms of hydrogen, and subsequently, in the citric acid cycle of Krebs (see Fig. 1.7). The end result is the release of carbon dioxide and water. The net balance in the Krebs cycle is given below:

$$[CH_3 - COO^- + H^+ + 2H_2O] + [3NAD^+ + FAD] + [GDP + P_i + H^+] \rightarrow$$
$$2CO_2 + [3NADH + 3H^+ + FADH_2] + [GTP + H_2O] (1.10)$$

Acetate enters into it bound to a so-called co-enzyme A (CoA) as acetyl CoA. During one turn of the Krebs cycle a further reduction takes place of three molecules of NAD^+ and one molecule of FAD (flavin adenine dinucleotide) involving eight atoms of hydrogen and a phosphorylation of molecule of GDP (guanosine diphosphate) to GTP (guanosine triphosphate). For better clarity of the overall reaction, we have used square half-brackets for the various subprocesses.

To discuss the economy of the Krebs cycle makes sense only when the cell is able to utilize the fuel in the form of hydrogen bound to the carriers NAD^+ and FAD for further phosphorylation of ADP to ATP. This became possible when a new generation of proton pumps was discovered that work not due to the ATP hydrolysis but as a result of the decomposition of hydrogen into a proton and an electron. These particles are further transported along a different pathway to the final acceptor, hydrogen, which in early stages of the biogenesis could be an anion of one of the inorganic acids.

The primitive bacterial cell was endowed with a cell membrane composed of peptidoglican, a complex protein-saccharide structure and later another, external cell membrane (see Fig.1.8). This facilitated an accumulation of protons in the space outside the original cell wall from which they can return to the cell interior using the proton pump of the first type (see Fig. 1.6f). This pump, working in reverse, synthesizes now ATP from ADP and an orthophosphate. This very efficient mechanism of membrane phosphorylation is universal being utilized by all present-day living organisms.

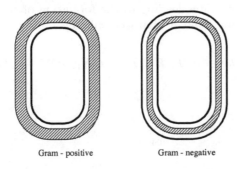

Gram - positive   Gram - negative

FIGURE 1.8: The structure of a bacterial cell is equipped with a cell wall composed of peptidoglycan, a complex protein-polysaccharide structure (shaded). It can also have a second, external membrane. The exposed thick peptidoglycan layer changes its color in the Gram dyeing procedure while the thin peptidoglycan layer which is covered by the external membrane does not change its color. Hence the bacteria are divided into Gram-positive and Gram-negative.

A more detailed explanation of the proton pump that utilizes the oxidation of hydrogen is shown in Fig.1.9b. In the original bacterial version this pump is composed of the two protein trans-membrane complexes: the dehydrogenase NADH and a reductase, for example the one taking the nitrate NO_3^- to nitrite NO_2^- . In the first complex, two hydrogen atoms present in the complex NADH - hydronic ion are transferred to the compound FMN (flavin mononucleotide). Later, after two electrons are detached, the hydrogens in the form of protons are transferred to the other side of the membrane. The two electrons are accepted in turn by one and then another iron-sulfur center (iron is reduced from the state Fe^{3+} to Fe^{2+}). Subsequently, at a molecule of quinone derivative Q, they are bound to another pair of protons that have reached the same site from the interior of the cell. An appropriate derivative of quinone Q is well soluble inside the membrane, and is an intermediary that ferries two hydrogen atoms between the two complexes. At the other complex, two hydrogen atoms are again split up into protons and electrons. The thus released protons are transferred to the exterior of the membrane and the electrons together with the protons from the interior of the membrane are relocated to the final acceptor site that can be a nitrate anion. The thus-created nitrite can oxidate another reaction that can be used by another reductase

$$NO_2^- \rightarrow N_2 \tag{1.11}$$

Alternatively, it can be involved with other inorganic anions: an acid carbide or sulfate in reactions leading to the formation of compounds with hydrogen: ammonia, methane or hydrogen sulfate:

$$NO_2^- \rightarrow NH_3, HCO_3^- \rightarrow CH_4, SO_4^{2-} \rightarrow H_2S \tag{1.12}$$

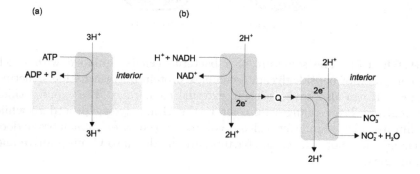

FIGURE 1.9:   Proton pumps transport free protons H^+ across the membrane from the cell interior to its exterior at the expense of the following chemical reactions: hydrolysis of ATP into ADP and an inorganic orthophosphate (a) or oxygenation of the hydrogen released in the decomposition of glucose to CO_2, and transported by NAD^+. (b) A derivative of quinone Q is an intermediary in the transport of hydrogen. This molecule is soluble inside the membrane and the oxidation is done for example through a nitrate NO_3^- that is reduced to nitrite NO_2^- . If pumps of both types are located in the same membrane, the first passing protons in the reverse direction can phosphorylate ADP to ATP.

The Photosynthetic Revolution

The Earth is energetically an open system and from the moment of its creation a substantial flux of solar radiation has been reaching it. Coupled with the rotational motion of the planet, this flux has been powering the machinery of oceanic and atmospheric motions. The primary energy sources for the newly emerged life on Earth were nucleotide triphosphates and small organic molecules such as monosaccharides that soon turned out to be exhaustible. Life became energetically independent only when it learned how to harness the practically inexhaustible solar energy source or more precisely this part of it that reaches the surface area of the oceans. The possibility of utilizing solar energy by living cells is linked to the use of chlorophyll as a photoreceptor [26]. Chlorophyll is a molecule that contains an unsaturated carbon-nitrogen porphyrin ring (see Fig. 1.10) with a built-in magnesium ion Mg^{2+} and phytol, a long saturated hydrophobic carbohydrate chain. The molecules of chlorophyll are easily excited in the optical range and equally easily transfer this excitation amongst them, creating in an appropriate protein matrix a light harvesting system. The last chlorophyll molecule in such a chain can become an electron donor replacing the fuel $NADH + H^+$ in a proton pump (see Fig. 1.9a).

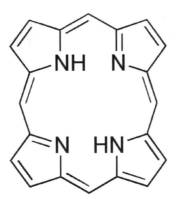

FIGURE 1.10: Porphyrin ring.

The first organisms that discovered this possibility were most likely purple bacteria. Their proton pumps are also two protein complexes built into the cell membrane (see Fig. 1.11). In the protein complex called type-II reaction center (RC) two electrons from the excited chlorophyll are transferred with two protons from the cell interior to a quinon derivative Q with a long carbohydrate tail, soluble in the membrane. Then, quinon Q reduced to quinol QH_2 carries the two hydrogen atoms inside the membrane to the next complex that contains a protein macromolecule called cytochrome bc1. This molecule catalyzes the electron transfer from each hydrogen atom onto another macromolecule called cytochrome c while the remaining proton moves to the extracellular medium.

Cytochromes are proteins that contain a heme that is a porphyrin ring with a built-in iron ion Fe^{2+}, that may also exist in a form oxidized to Fe^{3+}. Cytochrome c is a protein that dissolves in water and it removes electrons outside the cell membrane back to the reaction center. This completes the cyclical process during which two protons are carried outside the cell from the inside. An alternative source of electrons needed to restore the initial state of the reaction center used for example in sulfur purple bacteria can be the molecules of sulfurated hydrogen H_2S. We must stress here that contrary to the oxidation of NADH, oxidation of H_2S to pure sulfur is an endoenergetic reaction (consuming and not providing free energy) and it cannot be used in proton pumps. The proton concentration difference on each side of the cell membrane is further used by purple bacteria to produce ATP the same way as by non-photosynthetic bacteria.

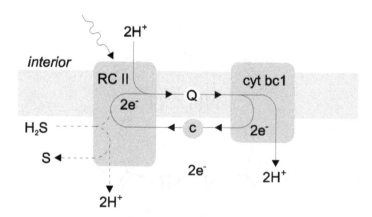

FIGURE 1.11: A proton pump in purple bacteria utilizing the energy of solar radiation. In the first protein complex called type-II reaction center (RC II) an electron from the excited chlorophyll molecule (the "primary donor") is transferred to a molecule of quinon derivative (Q) together with a proton taken out of the cell interior. The quinon molecule carries two hydrogen atoms formed this way to another protein complex containing cytochrome bc1. In this latter complex hydrogen atoms are again separated: a proton is moved outside the cell while an electron reduces a molecule of the water-soluble cytochrome c, which carries it back to the primary donor. An alternative source of electrons (broken line) for sulfur purple bacteria can be the molecule of sulfurated hydrogenH_2S.

An alternative way of using solar energy was discovered by green bacteria (see Fig. 1.12). Here, in the protein complex called type-I reaction center, an electron from photoexcited chlorophyll is transferred to the water-soluble protein ferredoxin. However, the lack of electrons in this chlorophyll is compensated right away uncyclically from sulfurated hydrogen decomposition. The electron carrier in ferredoxin is the iron-sulfur center composed of four atoms of Fe which are directly covalently bound with four atoms of S. After the reduction of iron, ferredoxin carries electrons to the next protein complex at which they bind to protons moving from the cell interior and reducing the molecules of $NADP^+$ (nicotinamide adenine dinucleotide phosphate) to $NADPH + H^+$. The entire system is not really a proton pump since there is no net proton transport across the cell membrane. It transforms light energy into fuel energy in the molecules of NADPH together with hydrated protons H^+ that carry the original charge of $NADPH^+$. This fuel is used in the synthesis of glucose from CO_2 and H_2O in the Calvin cycle whose overall balance equation takes the form:

$$[6CO_2 + 12NADPH + 12H^+] + [18ATP + 18H_2O] \rightarrow$$
$$[C_6H_{12}O_6 + 6H_2O + 12NADP^+] + [18ADP + 18P_i + 18H^+] \ (1.13)$$

This cycle is in a sense reverse to the Krebs cycle. Analogously to the Krebs cycle, we have used square brackets to denote summary component reactions in order to see more clearly the net reaction. In the Calvin cycle ATP is still being used. In the final balance, after the oxidation of glucose in the same way as for non-photosynthetic bacteria, an excess of ATP is produced.

Combining the two methods of using solar energy offers the optimal solution. This was discovered by cyanobacteria where cytochrome c has been replaced by plastocyanin (PC) and used as an electron carrier between a type-II reaction center and a type-I reaction center. The latter two are now being referred to as the photosystem II (PS II) and the photosystem I (PS I), respectively (Fig. 1.13). The electron carrier in plastocyanin is the copper ion Cu^{2+} reducible to Cu^+ and directly bound via four covalent bonds to four amino acids: cystein, metionin, and two histidines.

The greatest breakthrough came not as a result of the combination of the two photosystems but the utilization of water as the final electron donor (and a proton donor, hence a hydrogen donor). The dissociation of a hydrogen atom from a water molecule H_2O has turned it into molecular oxygen O_2 that is a highly reactive gas toxic to the early biological environment. Initially, molecular oxygen oxidized only iron ions Fe^{2+} which were soluble in great quantities in contemporary ocean water. As a result of this oxidation, ions of Fe^{3+} are formed which are poorly soluble. They sedimented, giving rise to the today's iron ore deposits. Simultaneously, an increased production of sugars from CO_2 and H_2O reduced the ocean's acidity causing a transformation of acidic anions of HCO_3^- into neutral ions of CO_3^{2-}, the latter reacted with the ions of Ca^{2+} initially present in high concentrations giving rise to

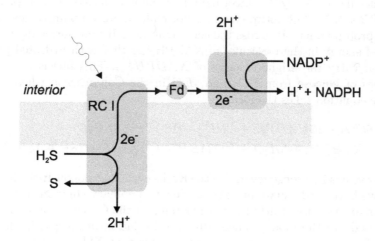

FIGURE 1.12:   The utilization of solar energy by green sulfur bacteria. In the first protein complex called type-I reaction center (RC I) an electron from the excited chlorophyll molecule is transferred to the water-soluble protein molecule of ferredoxin (Fd) which carries it to the complex of $NADP^+$ (nicotinamide adenine dinucleotide phosphate) reductase. The deficit electron in the initial chlorophyll is compensated in the process of oxidation of sulfurated hydrogen H_2S. The reduced hydrogen carrier $NADPH + H^+$ is utilized as fuel in the Calvin cycle that synthesizes sugar from water and carbon dioxide.

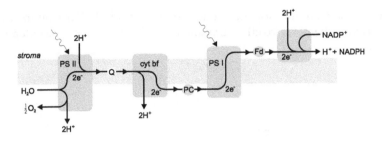

FIGURE 1.13: A proton pump using solar energy in the case of cyanobacteria can be thought of as a combination of the proton pump in purple bacteria (type-II reaction center called now photosystem II - PS II) and the photosynthetic system of green bacteria (type-I reaction center now called photosystem I - PS I). The coupling between the two systems is given by the water-soluble molecule of the protein plastocyanin (PC) with a copper ion being an electron carrier. The final electron donor is water H_2O which, after donating electrons and protons, becomes molecular oxygen O_2. The proton concentration difference between the two sides of the cell membrane is used to produce ATP via $H^+ATPase$ (see Fig. 1.9a) working in the reverse direction. In principle, an identical structure is that of the photosynthetic system in the thylakoid membrane of chloroplasts that are organelles of eukaryotic plant cells.

the sedimentation of insoluble calcium carbonate $CaCO_3$. Cyanobacteria's membranes captured the latter compound producing a paleobiological record of these processes in the form of fossils called stromatolites.

The formation of calcified stromatolites depleted the atmosphere from CO_2. When a deficit of compounds capable of further oxidation emerged, molecular oxygen O_2 started to be released into the atmosphere. Together with molecular nitrogen N_2 that was formed from the reduction of nitrates, they brought about the contemporary oxygen-nitrogen based atmosphere with only trace quantities of carbon dioxide. Life had to develop in a toxic oxygen environment from that point onwards and it managed solving this problem by discovering the mechanism of oxidative phosphorylation, which is used by modern aerobic bacteria and all higher organisms. In this mechanism, a proton pump using inorganic anions as final electron acceptors (see Fig. 1.9b) is replaced by a pump in which the final electron acceptor is molecular oxygen (see Fig. 1.14). Use has been made here of the protein complex with cytochrome bc1 transferring electrons from quinon Q to the water-soluble cytochrome c (see Fig. 1.11), a mechanism discovered earlier by purple bacteria. The source of electrons transferred to the quinon can be, as shown in Fig. 1.11, the fuel $NADPH + H^+$ generated in the process of glycolysis and in the Krebs cycle or, directly, $FADH_2$ (reduced flavin adenine dinucleotide) produced in one stage of the Krebs cycle (oxidation of succinate to fumarate). Electrons can also come from an inorganic source (chemotrophy). For example, to this end nitrifying bacteria oxidize ammonia to nitrate using molecular oxygen according to:

$$NH_3 \rightarrow NO_2^- \rightarrow NO_3^- \tag{1.14}$$

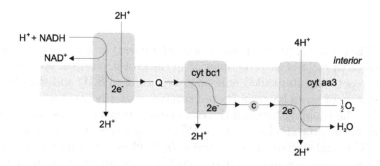

FIGURE 1.14:   Protein pump of heterotrophic aerobic bacteria. Electrons from the fuel in the form $NADH + H^+$ (produced in glycolysis and in the Krebs cycle) are transferred via quinon (Q) to the protein complex with cytochrome bc1 and then via cytochrome c to the complex with cytochrome aa3. The final electron acceptor is molecular oxygen O_2. During the transfer of two electrons along the membrane, six protons are pumped across it. The proton concentration difference between the two sides of the membrane is used to produce ATP by $H^+ATPase$ (see Fig.1.9a) working in reverse. This is in principle identical to the mechanism of oxidative phosphorylation in the mitochondrial membrane, which is an organelle present in all eukaryotic cells.

Nature has demonstrated here, as it has many times before and since, that it can use pollution of the environment to its advantage. It is interesting what use it will find for the countless tons of plastic bottles deposited in modern garbage dumps.

The Origin of Diploidal Eukaryotic Cells

In its 19th century interpretation, Darwin's theory of natural selection favoring the survival of the fittest could be readily associated with the contemporary struggle for survival in the early capitalist economy of that time. The latter could be described as ruled by the law of the jungle. This, in turn, became the foundation of the ideological doctrine of many totalitarian regimes on the 20th century political landscape. To the credit of the great biologist, Lynn Margulis [25], an emphasis has been given to the fact that survival can be accomplished not only through struggle but also through peaceful coexistence, called symbiosis in biology. Numerous indications have been found that support the significance of symbiosis in the formation of modern eukaryotic cells.

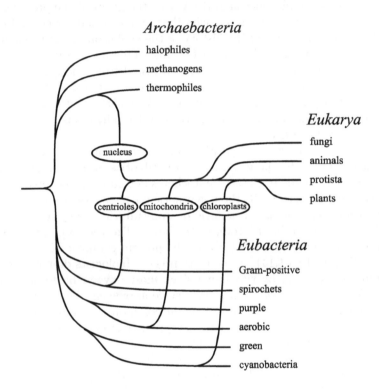

FIGURE 1.15: A simplified and somewhat hypothetical phylogenetic tree of living organisms on Earth. The first to be developed are the two groups of prokaryotic organisms *Archaebacteria* and *Eubacteria*. Then, as a result of the merger of prokaryotic cells with different properties, eukaryotic cells were formed (*Eucarya*). The latter ones further evolved either in an undifferentiated form as single cell organisms (kingdom of protista) or they differentiated into multi-cellular organisms (the kingdoms of heterotrophic fungi with multi-nuclear cells, heterotrophic animals and phototrophic plants).

Fig. 1.15 illustrates a simplified scheme of the phylogenetic tree of living organisms. There is a clear division between archaic bacteria (*Archaebacteria*) and true bacteria (*Eubacteria*) which must have emerged already in the earliest periods of life on Earth. The history of subsequent differentiation of prokaryotic organisms within these two groups, however, is not all that clear. Nowadays, the phylogenetic tree is obtained based on the differences in DNA sequences coding the same functional enzymes or ribozymes. The more differences are found in the DNA sequences, the earlier the two branches of the compared species must have split away. However, the results obtained by comparing, for example ribosomal RNA with the genes of the proteins in the photosynthetic chain, differ from each other by much and hence lead to very dissimilar reconstructions of the history of the evolution of the mechanism of photosynthesis [13, 42]. The reason for this ambiguity is due to the lateral gene transfer processes where genes are borrowed from one organism by another. Branches can not only split away but also merge with the passage of time. Therefore, the phylogenetic tree of *Eubacteria* shown in Fig.1.15 must be viewed with caution, especially since only the kingdoms essential to our discussion have been depicted.

As mentioned earlier, gram-positive bacteria have only one external membrane (hence are potentially sensitive to antibiotics and fortunately include most of the pathogenic bacteria). Spirochetes developed mechanisms of internal motion for the entire cell. Photosynthetic purple bacteria with the type-II reaction center can be either sulfuric or non-sulfuric. They must be evolutionarily close to aerobic bacteria since they utilize the same mechanism of reduction of the cytochrome c through the protein complex with cytochrome bc1 (see. Fig. 1.11 to 1.14). To be more precise, biologists do not distinguish a kingdom of bacteria by this name since most oxygen bacteria, including the best-known *Escherichia coli* , can survive in oxygen- deprived conditions. Green bacteria and cyanobacteria have in common the same mechanism of sugar photosynthesis with the use of the type-I reaction center.

The lateral gene transfer can be fully accomplished when several simple eukaryotic cells merge into one supercell. According to Lynn Margulis, this is the way in which eukaryotic cells were first formed. Most probably, first a thermophilic bacterium emerged with a stable genomic organization whose DNA was protected by proteinaceous histones that combined to form a prototype of chromatin. Such a bacterium then entered into a symbiotic arrangement with a spirochete that contained a motile apparatus formed from microtubules (see Fig. 1.15). This combination gave rise to a mitotic mechanism of cell division. In it, chromatin with a doubled amount of genetic material organizes itself after replication into chromosomes pulled in opposite directions by the karyokinetic spindles formed from the *centrioles* by the self-assembling microtubules that consume GTP as fuel. In the next stage, a cell containing a nucleus with chromatin has been able to assimilate several oxygen bacteria (see Fig. 1.15). The latter ones have been transformed into *mitochondria*, the cell's power plants that synthesize ATP via the mechanism of phosphorylative oxidation shown in Fig. 1.14. The thus formed cells *Eucarya* continued evolving (see Fig. 1.15) either in an undifferentiated form as single-cell organisms (the kingdom of protista) or in a differentiated form as multi-cellular organisms (the kingdoms of fungi and animals). All the above organisms were heterotrophs. The assimilation of prokaryotic cells of cyanobacteria as *chloroplasts* led to the formation of phototrophic single-cell organisms and multi-cellular plants.

So far, we have only talked about the symbiosis of different prokaryotic cells. An encounter of two organisms belonging to the same species could lead either to acts of cannibalism or to symbiosis. According to Lynn Margulis this type of encounter gave rise to the emergence of sex. A symbiotic cell becomes diploidal, i.e., contains two slightly different copies of the same genome. Obviously, reproductive cells nurtured by the parent organism before entering into new symbiotic arrangements are haploidal and contain only one copy of the genetic material. A reduction of the genetic information took place in the process of generating reproductive cells when meiotic division replaced mitotic division. The evolutionary advantage of sexual reproduction is due to the fact that in meiotic division a crossing-over takes place between maternal and paternal genes. Hence the genetic material undergoes a much faster variability compared to random point mutations. Moreover, such mutations are very seldom lethal.

Fig. 1.16 is a schematic illustration of an eukaryotic cell. It is composed of a system of lipid membranes spatially confining its various organelles: the nucleus, mitochondria, smooth and rough endoplasmic reticulum, the Golgi apparatus and lysosomes. The centrioles organize the motile system of the cell. The sketch shown here corresponds to an animal cell. Plant cells (see Fig. 1.17) contain an additional cell wall, vacuoles which are bubbles storing water and various additional substances in the form of grains, as well as chloroplasts which are large organelles made up of three layers of membranes that facilitate photosynthesis. The internal flattened chloroplast bubbles are called thylakoids and they can be viewed as removed mitochondrial crista (combs). Three types of organelles of eukaryotic organisms contain their own genetic material, which is different from that of the nucleus. As we mentioned earlier, eukaryotic systems emerged in the process of evolution due to the assimilation of previously developed prokaryotic cells. These are mitochondria that originated from oxygen bacteria, the centrioles originating from spirochetes and chloroplasts formed from blue bacteria.

The broad scheme of metabolism is illustrated in Fig. 1.7 where only the most important biochemical reactions have been shown. Today we know close to a hundred times as many reactions taking place in the cell [262]. For simplicity substrates are denoted by dots. A unique enzyme catalyzes each reaction, represented by either a uni- or a bi-directional arrow. Worth mentioning are repeated chains of sugar transformations (from the left side) and fatty acids (from the right side) and the characteristic closed Krebs cycle. It is also shown how connections are made with more complex reaction systems transforming other important bio-organic compounds such as: nucleotides, amino acids, sterols and porphyrins.

In each compartment of the cell specific reactions take place. These are:

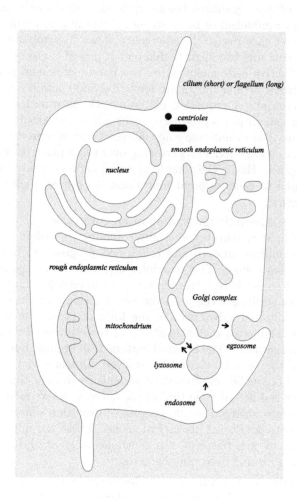

FIGURE 1.16:   Compartments of the eukaryotic cell. Solid lines represent membranes formed from two phospholipid layers and the neighboring compartments differ in the shading used.

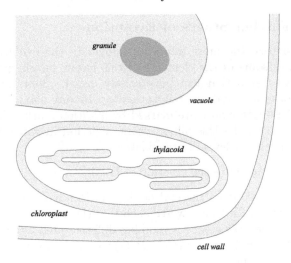

FIGURE 1.17: Additional organelles of the plant cell.

- in lysosymes biopolymer hydrolysis takes place,

- in the mitochondrial matrix Krebs reaction cycles and reactions of the Krebs cycle and fatty acid and amino acid degradation can be found,

- on interior surfaces of mitochondrial walls one finds oxidative phosphorylation reactions,

- on the membranes of the rough endoplasmic reticulum the processes of protein synthesis occur,

- on the membranes of the smooth endoplasmic reticulum the processes of lipid synthesis are observed,

- in the interior of the Golgi apparatus the synthesis of compound sugars takes place.

The sugars that have been synthesized in the Golgi apparatus, after a merger with membrane proteins, can be transported outside the cell in a process called exocytosis. The external cytoplasmic membrane, armed with such glycoproteins, plays the role of recognition and selective transport of substances from the external environment of the cell. In the cellular interior region, between various organelles, numerous transformations of simple sugars, amino acids and mononucleotides take place. This region is filled simultaneously with supramolecular structures of the protein cytoskeleton that provides the cell with a motile machinery.

Summary. Further Stages of Evolution

Let us summarize the most important stages in the evolution of life on Earth. We know some of the elements of this jigsaw puzzle quite well since they left visible traces that can be precisely dated. Some other elements are fairly hypothetical and are not backed up by any solid precisely dated discoveries. These latter ones are marked by question marks. We finish with a survey of events in the last billion years, a period in which life on Earth reached a supra-cellular level of organization.

- 4.6 billion years ago - the creation of Earth.

- 3.9 billion years ago - the end of the Big Bombardment. The surface temperature on Earth is lowered such that the early gaseous envelope of volcanic or comet/meteorite origin has differentiated into an atmosphere (mainly CO_2) and ocean (mainly H_2O) with the addition of some simple organic compounds.

- ? - the emergence of nucleotides and RNA enables the storage of information and self-replication subjected to Darwinian selection based on the survival of the fittest. Triphosphates of the nucleotides become the key source of free energy for the reactions of polymerization.

- ? - the emergence of the key elements of the cellular machinery for prokaryotic cells: the membrane, membrane channels, and the ribosome, DNA and RNA polymerases, proton pumps. Phospholipid bilayers with selective protein channels actively isolate the various types of RNA from their surroundings and protect a supply of the required nucleotides. Primitive ribosomes express the information contained in RNA in terms of the produced proteins. Proteins turn out to be more efficient as catalysts than RNA. The double-stranded DNA is found to be a more stable information carrier than RNA. Protein enzymes enable the use of sugars as a source of free energy in the reaction of phosphorylation in which di-phosphate nucleotides are recycled into their tri-phosphate forms. This is first carried out through the process of fermentation (substrate phosphorylation itself) and then in a more efficient process of membrane phosphorylation with inorganic anions playing the role of oxidants.

- 3.5 billion years ago - the discovery of phosphorylation is linked to a dissociation of H_2O. The resultant molecular O_2 oxidates Fe^{2+} to a poorly soluble Fe^{3+} while a sugar synthesis from CO_2 and H_2O lowers the acidity of the oceans leading to the elimination of an insoluble calcium carbide.

- 2 billion years ago - molecular oxygen starts to accumulate in large quantities in the atmosphere. The chemical machinery of oxidative phosphorylation is discovered.

- ? - a thermophilic bacterium enters into a symbiotic relationship with DNA, which is protected by histone proteins. A motile bacterium makes a symbiotic arrangement with a microtubular cytoskeleton. The cell nucleus is formed and the mechanism of mitotic cell division is carried out for the first time.

- 1.5 billion years ago - the symbiotic co-existence of a cell containing a nucleus with oxygen bacteria is established in which the latter play the role of mitochondria, or possibly with purple bacteria as chloroplasts. The modern eukaryotic cell is born.

- ? - sex is discovered and with it the amount of genetic information is doubled (a diploidal cell is formed) linked with meiosis and the possibility of recombination.

- 1.0 billion years ago - some cells that arose from cell division stop dividing thus leading to the emergence of an embryo that differentiates into a multi-cellular animal or plant organism.

- 550 million years ago - animals start developing skeletons out of calcium carbide or silicate thus enabling the formation of fossils that give us a lasting chronological record. The beginning of the Paleolithic era (the Cambrian explosion).

- 450 million years ago - the symbiosis of plants with fungi enables their emergence on land.

- 100 million years ago - the perfection of the most modern and effective ways of protecting the embryo (mammals with a placenta).

- 5 million years ago - first hominid forms are recorded.

- 100 thousand years ago - *Homo sapiens* walks for the first time on Earth.

1.6 Emergence, Intelligence and Consciousness

Gross Features of the Nervous System

The brain is probably the most complex structure in the known universe. It is complex enough to coordinate the fingers of a concert pianist or to create a three-dimensional landscape from light rays that impinge on a two-dimensional retina. While it is the product of millions of years of evolution, some of the structures unique to the human species have only appeared relatively recently. For example, only 100,000 years ago, the ancestors of modern man had a brain weighing only about 0.5 kg, roughly a third of the average weight at present. Most of this increased weight is associated with the most striking feature of the human brain, the cortex, the two roughly symmetrical, corrugated and folded hemispheres which sit astride the central core.

In evolutionary terms, all brains are extensions of the spinal cord. The distant ancestor of the human brain originated in the primordial soup some 500 million years ago. Life and survival in those early conditions was relatively simple and in consequence these precursor brains consisted of only a few hundred nerve cells. As the pre-historic sea-creatures evolved and became more complex, so too did the brain. A major change occurred when these early fish crawled out of the seas and onto the land. The increased difficulties of survival on land led to the development of the "reptilian brain". This brain design is still discernible in modern reptiles and mammals and hence offers a significant clue to our common evolutionary ancestry.

The next major transformation occurred with the emergence of the mammalian brain that came equipped with a new structure, the cerebrum (forebrain) along with its covering, the cortex. By that time, the brain consisted of hundreds of millions of nerve cells organized into separate regions of the brain associated with different tasks. Approximately 5 million years ago, another type of cortex appeared in the early man. In this brain, the surface of the cortex was organized into separate columnar regions less than 1 mm wide, each containing many millions of nerve cells or neurons. This new structure afforded a much more complex processing capability. Finally, about 100,000 years ago, this new cortex underwent rapid expansion with the advent of modern man. The cortex of the present day man contains approximately two-thirds of all neurons and weighs on average 1.5 kg, almost three times as much as one hundred thousand years ago.

In a nutshell, the human brain consists of three separate parts. The first one in the lower section, sometimes called the brain stem, consists of structures such as the medulla (which controls breathing, heart rate and digestion) and the cerebellum (which coordinates senses and muscle movement). Many of these features are inherited from the reptilian brain. The second segment appears as a slight swelling in lower vertebrates and enlarges in the higher primates and in man into the midbrain. The structures contained here link the lower brain stem to the thalamus (for information relay) and to the hypothalamus (which is instrumental in regulating drives and actions). The latter is part of the limbic system that, essentially in all mammals, lies above the brain stem and under the cortex and consists of a number of interconnected structures. These structures have been linked to hormones, drives, temperature control, emotion, and one part, the hippocampus to memory formation. Neurons affecting heart rate and respiration are concentrated in the hypothalamus and direct most of the physiological changes that accompany strong emotion. Aggressive behavior is linked to the action of the amygdala, which lies next to the hippocampus. The latter plays a crucial role in processing various forms of information as part of the long-term memory. Damage to the hippocampus produces the inability to lay down new stores of information (global retrograde amnesia). Much of the lower and mid-brain are rather simple systems that are capable of registering experiences and regulating behavior largely outside of any conscious awareness. The human brain is constructed according to the evolutionary development with the outer layer composed of the most recent brain structure, and the deeper layers consisting of structures inherited from lower creatures: reptiles and mammals. Finally, the third section, the forebrain, appears as a mere bump in the brain of the frog but expands into the cerebrum of higher life forms and covers the brain stem in a shape resembling the head of a mushroom. It has further evolved in humans into the walnut-like configuration of left and right hemispheres. The highly convoluted surface of the hemispheres, the cortex, is about 2 mm-thick and covers a total surface area of about $1.5m^2$. The structure of the cortex is extremely complicated. It is here that most of the "high-level" functions associated with the mind are carried out. Some of its regions are highly specialized. For example, the occipital lobes located near the rear of the brain are associated with the visual system. The motor cortex helps coordinate all voluntary muscle movements.

More neurons may be dedicated by the brain to connect to certain regions of the body than others. For example, the fingers have many more nerve endings than the toes. However, there is an approximate symmetry between the left and right hemisphere. Indeed, there are two occipital lobes, two parietal lobes and there are two frontal lobes. This symmetry is not entirely exact. For instance, the area associated with language appears only on the left hemisphere. The frontal lobes occupy the front part of the brain behind the forehead and comprise the portion of the brain most closely associated with "control" of responses to inputs from the outside. They are closely correlated with making decisions and judgments. In most people, the left hemisphere dominates over the right in deciding which response to make. Since the frontal lobes occupy 29% of the cortex in our species (as opposed to 3.5% in rats and 17% in chimpanzees, for example), they are often regarded as an indication of the evolutionary development. In individuals with normal hemispheric dominance, the left hemisphere, which manages the right side of the body, controls language and general cognitive functions. The right hemisphere, controlling the left half of the body, manages non-verbal processes, such as attention or pattern recognition. Although the two hemispheres are in continual communication with each other, each acting as independent parallel processors with complementary functions, the dominant left-hemisphere appears most closely associated with consciousness.

While the structural organization of the brain and the nervous system has been known for many years, the nature of the processes involved in the functioning of the nervous system is still enigmatic today, especially in regard to consciousness. What we do know today is that the basis of the functioning of the nerves is electrical signal propagation that has been thoroughly investigated by generations of neurophysiologists. Numerous aspects of the functioning of the nervous system have been modeled mathematically [23]. The history of the development of conceptual understanding of neurophysiology is briefly outlined below [32].

The idea that nerves functioned by conducting "electrical powers" was first advanced by the Englishman Stephen Hales. Experimental support for this concept came from Swammerdam in Holland and Glisson at Cambridge who showed that the muscles did not increase in volume when they contracted concluding that the "humor" must be "etherial" in nature. Hence, Hales' electricity seemed to be a good candidate. In 1786, Luigi Galvani, while dissecting the muscles of a frog, happened to touch the nerve to the muscles with his scalpel while a static electrical machine was operating close by causing the electricity to go down the nerve and producing muscle contraction. Based on this observation and further experiments, he concluded that the electricity was generated within the animal's body and he called this electricity "animal electricity" identifying it with the long sought for "vital force". His fellow Italian Alessandro Volta repeated and confirmed Galvani's observations, at first agreeing with his conclusions of "animal electricity," but later opposing it claiming that the current was due to the presence of a bimetal. This set off a scientific controversy that occupied the life sciences for the next century and a half. Eventually, Humboldt's publication in 1797, just before Galvani's death, clearly established that both Volta and Galvani were simultaneously right and wrong. Bimetallic electricity existed but so did animal electricity.

In the 1830's another Italian, Carlo Matteucci, demonstrated that an electrical current was generated by injured tissues and that serial stacking of such tissue could multiply the current similarly to adding more bimetallic elements to a Voltaic pile. The current was continuously flowing and the existence of at least this type of "animal electricity" was unequivocally proven. However, it was not located within the central nervous system. Later, Du Bois-Reymond discovered that when a nerve was stimulated, an electrically-measurable impulse was produced at the site of stimulation and then traveled rapidly down the nerve producing muscular contraction. He thus unlocked the mystery of the nerve impulse, the basic mechanism of information transfer in the nervous system. In addition to an impulse, a resting potential was measured as a steady voltage observed on unstimulated nerve or muscle. Following Du Bois-Reymond's discovery, von Helmholtz succeeded in measuring the velocity of the nerve impulse as 30 m/s. Julius Bernstein confirmed von Helmoltz's velocity measurement. Moreover, in 1868 he presented his theory of nerve action and bioelectricity, which became the cornerstone of all modern concepts in neurophysiology. The "Bernstein hypothesis" stated that the membrane of the nerve cell is able to selectively pass certain kinds of ions. Situated within the membrane was a mechanism that separated negative from positive ions, permitting the positive ones to enter the cell and leaving the negative ions in the fluid outside the cell. When equilibrium had been reached an electrical potential would then exist across the membrane, the "trans-membrane potential." The nerve impulse was simply a localized region of "depolarization," or loss of this trans-membrane potential, that traveled down the nerve fiber with the membrane potential being immediately restored behind it. Bernstein also postulated that all cells possessed such a trans-membrane potential, similarly derived from separation of ions, and he explained Matteucci's current of injury as being due to damaged cell membranes "leaking" their trans-membrane potentials.

In 1929, Hans Berger discovered the electroencephalogram which has become one of the standard testing and diagnostic procedures in neurology. In the following decade Burr conducted experiments on the DC potentials measurable on the surface of a variety of organisms relating changes in these potentials to physiological functions including growth, development and sleep. He formulated the concept of a "bioelectric field" generated by the sum total of electrical activity of all the cells of the organism, and postulated that the field itself directed and controlled these activities. Later, Leao demonstrated that depression of activity in the brain was always accompanied by the appearance of specific type DC potentials, regardless of the cause. Gerard and Libet expanded this concept in a series of experiments in which they concluded that the basic functions of the brain, excitation, depression and integration, were directed and controlled by these DC potentials. However, the sources for the DC potentials could not be the trans-membrane potentials. At the junction between the nerve and its end, the microscope had revealed a gap, henceforth called the synapse. In the 1920's, Otto Loewi experimentally demonstrated that the current transmission across the synaptic gap was due to a chemical acetylcholine that was released into the gap and that it subsequently stimulated the receptor site on the end organ. Finally, the broad outlines of the Bernstein hypothesis were proven by Hodgkin, Huxley and Eccles in the 1940's. Using microelectrodes that could penetrate the nerve cell membrane, in a protocol called the space-clamp technique, they demonstrated that the normal trans-membrane potential is produced by sodium ions being excluded from the nerve cell interior, and when stimulated to produce an action potential the membrane permits these ions to enter.

However, it still was not clear how neurons were integrated to work together in order to produce a coherent brain function. In the 1940's neurophysiologists Gerard and Libet performed a series of experiments on the DC electrical potentials in the brain. In some areas of the brain steady or slowly varying potentials oriented along the axono-dendritic axis were measured. These potentials changed their magnitude as the excitability level of the neurons was altered by chemical treatment. In other experiments, Gerard and Libet found slowly oscillating potentials and "traveling waves" of potential change moving across the cortical layers of the brain at speeds of approximately 6 cm/s. These wave phenomena were interpreted as actual electrical currents that were flowing outside the nerve cells through the brain and that these extra-cellular electrical currents exert the primary controlling action on the neurons.

It soon became apparent that the "circuitry" of the brain was not a simple "one-on-one" arrangement. Single neurons were found to have tree-like arrangements of dendrites with input synapses from numerous other neurons.

Dendritic electrical potentials were observed that did not propagate like action potentials, but appeared to be additive; when a sufficient number were generated the membrane depolarization reached the critical level and an action potential would be generated. Other neurons were found whose action potentials were inhibitory to their receiving neurons. Graded responses were discovered in which ion fluxes occurred across the neuronal membranes and while insufficient in magnitude to produce an action potential, still produced functional changes in the neuron. The complexity of function in the brain was found to be enormous. Furthermore, it was discovered that only about 10% of the brain was composed of neurons with the remainder made up of a variety of "perineural" cells, mostly glia cells that did not possess the ability to generate action potentials. Hence, they were assumed to have the function of protecting and nourishing the nerve cells. Later on, the role played by the glia as "supporting" cells that constituted 90% of the total mass of the brain began to be questioned. Electron microscopy revealed close and involved associations between the glia and the neurons as well as between the glia cells themselves. The analog of the glia cell, the Schwann cell, was found to invest all peripheral nerve fibers outside of the brain and spinal cord. It was deduced that they are in continuous cytoplasmic contact along the entire length of each nerve. Biochemical changes were found to occur in the glia concurrent with the brain activity and strong evidence was found that glia cells were involved in the process of memory storage. While the action potential of the neuron did not seem to influence the glia cells, the reverse appeared possible. It emerged that the extra-neuronal currents originally described by Libet and Gerard could be associated with some electrical activity in these non-neuronal cells themselves.

It is feasible that the electromagnetic energy present in the nervous system is a major factor exerting controlling influence over growth and development processes. For example, the nerves, acting in concert with some electrical factors of the epidermis, produce a specific sequence of electrical potential changes that cause limb regenerative growth in some animals. In animals not capable of limb regeneration, this specific sequence of electrical changes is absent. This leads us to consider the topic of bioelectricity and biomagnetism from a general viewpoint.

Biocomputing

The current explosion of interest in the future of computing is strongly motivated by an imminent approach of the limit of classical computing as extrapolated from Moore's Law stating that the number of transistors that can be fabricated on commercially available silicon integrated circuit doubles every 18-24 months. This amazingly fast trend towards miniaturization has been valid in the microelectronics sector for close to four decades. Today, the smallest available silicon chips contain up to 100 million transistors on a few cm^2 of a wafer translating into linear dimensions on the order of 200 nm or less. To reach the dimensions of small clusters of several atoms, approximately 2 nm in length requires a 10,000-fold miniaturization of micro-circuitry, and, according to Moore's Law, is expected to occur some time between 2019 and 2028. However, there is a growing concern heretofore unexplored technologies will have to be found even before this limit is reached. There is already substantial effort underway in an area referred to as quantum computation since nanometer-size objects reach into the realm of quantum mechanics and entirely different physical laws apply. Unfortunately, practical considerations such as the so-called entanglement of the system's wave function with the environment pose a serious challenge to any practical applications of quantum computing.

Simultaneously with an effort to build the first quantum computer, a quest has been pursued to use biological materials provided to us by nature itself, or perhaps in combination with silicon-based technology, to come up with not only smaller electronic devices but also with devices that are more flexible structurally and functionally. It is hoped that one day soon a biological computer will be built that is fast, small and evolvable. Of course, we are already intimately familiar with its prototype, our brain, which can be held as proof of concept. The human brain is composed of 10 billion nerve cells interconnected with as many as 1000 neighboring neurons communicating via signals in the form of electric potential differences that travel along axons, with speeds in the m/s range. We know that waves of electric activity are correlated with the brain's cognitive functions but do not know which structures and/or process are responsible for consciousness or even what constitutes memory. This is sometimes referred to as the mind-body problem. Without getting too far into this hotly debated topic, we might just add that there is a camp of researchers galvanized by the Oxford mathematician Sir Roger Penrose who believes that the fundamental nature of the mental processes lies in quantum mechanics. If true this would, in a way, provide a neat conceptual link between the two routes towards a new type of computing: nano-scale in-silico computing and also nano-scale biological or in vivo computing. The process of nerve excitation involves a passage of the electrical signal from one nerve cell to another in the form of synaptic transmission. Synapses are connections between nerve cells, their axons or dendrites and form nanometer size gaps which are crossed by neurotransmitter molecules stored in vesicles that open when stimulated. Here again is a point of contact with quantum mechanics. In addition there is a very intricate structure of protein filaments filling both the nerve cell's body and its axons. Inside the axons one finds a parallel architecture of microtubular bundles interconnected with other proteins, a structure that resembles parallel computer wiring [21] leading to the hypothesis that this microtubular structure may be involved in subcellular (nano-scale, possibly quantum) computation. Brown and Tuszyński [46] demonstrated theoretically their feasibility as information storing and processing devices. There is a great promise that protein networks are strongly involved in both information processing and storage.

This then suggests that living cells perform significant computational tasks. If so, can we reconcile laws of thermodynamics and information theory with our knowledge of cell biology? Furthermore, can we gain insights into the inner workings of the cell such that in the future hybrid computers can be designed that harness the power of biological computational elements being integrated with silicon? Although computer scientists and biochemists have not found a clear path from the test tube to the desktop, what they have found amazes and inspires them. Viewed through the eyes of computer scientists, evolution has produced the smallest, most efficient computers in the world, namely the living cell.

The various points of reference regarding the nature of the living state undoubtedly reflect the prevailing Zeitgeist of the period in which a given theory has been created. The viewpoint of representing the cell as a machine, or even a factory, closely mirrors the worldview of the industrial revolution of the 19th century. Likewise, the currently popular opinion that living cells are intensely engaged in some type of computation is closely linked with the technological revolution ushered into in the late 20th century as a result of the proliferation of computer technology. Both points of view have merits, i.e., the cell obeys the laws of physics such as the first law of thermodynamics and hence can be viewed as a thermodynamic machine and simultaneously reduces local entropy (creates information) by creating structural and functional order. In other words, it creates and maintains information. Furthermore, it most certainly processes information and engages in signaling thereby actively performing computation. It is safe to say that living cells can be viewed as both micro-factories (with nano-machines performing individual tasks) and biological computers whose nano-chips are the various proteins and peptides in addition to DNA and RNA. Most of the cell is what we might call hardware while a small fraction is software (for example the genetic code in the DNA that instructs for the synthesis of proteins). Probably only a small fraction of the cell can be seen as pure information content. Is there something else in living systems that neither machines nor computers possess? Probably yes. At least two properties distinguish animate matter from inanimate objects: procreation and autonomy expressed by free will (to move against the whims of thermal noise, in the very least).

On a more practical note, can biomimetics be used to enhance our computational capabilities? The answer is yes, although progress in this area has been slow. In general terms, a chemical computer is one that processes information by making and breaking chemical bonds, and it stores logic states or information in the resulting chemical (i.e., molecular) structures. A chemical nano-computer would perform such operations selectively among molecules taken just a few at a time in volumes only a few nanometers on a side. An alternative direction has been to adapt naturally occurring biochemicals for use in computing processes that do not occur in nature. Important examples of this are: Adleman's DNA-based computer, Birge's bacteriorhodopsin-based computer memories as briefly discussed below.

Adleman first used DNA, to solve a simple version of the "traveling salesman" problem where the task is to find the most efficient path through several cities. Adleman [1] demonstrated that the billions of molecules in a drop of DNA contained significant computational power. Digital memory can be seen in the form of DNA and proteins. Exquisitely efficient editing machines navigate through the cell, cutting and pasting molecular data into the stuff of life. Has evolution produced the smallest, most efficient computers in the world? Even if this is an exaggeration, the innate intelligence built into DNA molecules could help fabricate tiny, complex structures using computer logic not to crunch numbers but build things. Furthermore, DNA computers may use a billion times less energy than electronic computers, while storing data in a trillion times less space. Moreover, computing with DNA is highly parallel: in principle there could be billions upon trillions of DNA or RNA molecules undergoing chemical reactions, performing computations, simultaneously. Molecular biologists have already established a toolbox of DNA manipulations, including enzyme cutting, ligation, sequencing, amplification and fluorescent labeling. The idea behind DNA computing springs from a simple analogy between the following two processes: (a) the complex structure of a living organism ultimately derives from applying sets of simple instructed operations (e.g., copying, marking ,joining, inserting, deleting, etc.) to information in a DNA sequence, (b) any computation is the result of combining very simple basic arithmetic and logical operations. Eric Winfree intends to create nanoscopic building blocks out of DNA that are designed to carry out mathematical operations by fitting together in specific ways. DNA is not the only candidate for a biological computer chip. Birge [6] proposed the use of the light-sensitive protein dye bacteriorhodopsin, that is produced by some bacteria. He and his collaborators have shown that it could provide a very high density optical memory that could be integrated into an electronic computer to yield a hybrid device of much greater power than a conventional, purely electronic computer. In conclusion, electronic computers assembled using DNA and run on organic nutrients instead of electricity are another science-fiction idea that may soon become reality.

Chapter 1 Questions and Problems

QUESTION 1.1 (a) What is the role of genes in the evolutionary process? (b) Is evolution a random process? (c) Comment on the values in the table below, which shows the percent difference in amino acid sequences between horse and human proteins. Do these values support your answer to Part (b)? How might this type of data have led to the definition of a "molecular clock"?

TABLE 1.2: Amino Acid Sequences in Horse and Human Proteins

Protein	Number of Amino Acids	% Difference in Amino Acid Sequence
Histone	103	0%
Cytochrome C	104	12%
Hemoglobin	146	18%
Fibrinopeptide (clotting agent)	20	86%

QUESTION 1.2 How would you monitor metabolic reactions in a cell? What are the advantages/disadvantages of your approach?

QUESTION 1.3 Find 2 high quality websites that illustrate some modern advances in biology, and describe their utility.

QUESTION 1.4 In "The Origin of Eukaryotic Cells", Lynn Margulis provides a plausible explanation of organelles via the endosymbiotic theory. The endosymbiotic hypothesis states that mitochondria, chloroplasts and other organelles were formed from free living prokaryotic ancestors, such as purple bacteria, that combined with a host cell. Purple bacteria are considered to be evolutionarily close to aerobic bacteria, such as *Escheria coli*, since they use the same mechanisms in biological processes. Compare an *E.coli* cell with a human mitochondrion and in point form note the similarities between the two.

PROBLEM 1.5 Large bacteria normally have the shape of cylinders of (fixed) radius a and variable length L.

(a) Prove that the surface-to-volume ratio of these bacteria does not depend on the length of the bacterium.

(b) Assume that the bacteria feed by absorbing glucose through their surface. Explain why a large cylindrical bacterium can do this more efficiently than a large spherical bacterium. Do this by comparing the surface areas of cylindrical and spherical bacteria with the same volume (so they have the same food requirement). Assume that the spherical bacterium has a radius of 10^{-5} m and the cylindrical bacterium has a radius of 10^{-6} m.

PROBLEM 1.6 Cell culture A has a population of 1.00×10^3 ; culture B has a population of 1.50×10^3. Culture A has a growth constant of 0.0500 day^{-1}. Culture B doubles in size every 3.00 days. How long will it be until culture B has π times as many cells as culture A?

PROBLEM 1.7 You are working with two cell cultures, X and Y. Culture X has a growth constant of 0.0124 day^{-1}, and a present population of 1286. Culture Y has a decay constant of magnitude 0.186 day^{-1}, and a population of 1573.

(a) Plot a graph of the populations of both cultures over the next 5 days, and also over the 5 previous days.

(b) From your graph, estimate when the populations will be equal.

(c) Check your estimate in (b) by calculating when the populations will be equal.

(d) From your graph, estimate when there were twice as many cells in Y as in X.

(e) Calculate when there were twice as many cells in Y as in X.

PROBLEM 1.8 Yet another culture, C12, has a growth constant which is 0.200 day^{-1} when supplied with proper nutrients. Without nutrients, the population decays by a factor of 3.50 every 6.00 days. Exactly 1 day ago, culture C12 was provided with enough nutrients for 4.00 days. Its present population is 852.

(a) Plot a graph of the population, starting with time $t = 0$ at 1 day ago, and ending 11.0 days from now.

(b) From your graph, estimate when the population will have declined by 40% from its present value.

(c) Calculate when the population will have declined by 40% from its present value.

Chapter 1 References

1. Adleman, L. *Science* **266**, pp. 1021-1023, 1994.

2. Alberts, B., Bray, D., Lewis, J., Raff, M., Roberts, K., and Watson, J.D., *Molecular Biology of the Cell.* Garland Publishing, London, 1994.

3. Amos, L.A. and Amos, W.B. *Molecules of the Cytoskeleton.* Macmillan Press, London, 1991.

4. Bennett, C. H. *Dissipation, Information, Computational Complexity and the Definition of Organization.* *Emerging Syntheses in Science,* Pines, D. (ed.), Addison-Wesley, Redwood City, CA, pp. 215-233, 1985.

5. Biebricher, Ch. K. and Gardiner, W. C. *Biophys. Chem.* **66**, 179-192, 1997.

6. Birge, R. *American Scientist,* **July-August**, 348-355, 1994.

7. Bolterauer, H., Limbach, H.J. and Tuszyński, J.A. *Bioelectrochem Bioenerg.* **48(2)**:285-95, 1999.

8. Brandhorst, B.P. and McConkey, E.H. *J. Mol Biol.* **85**, 451-563, 1974.

9. Bras, W. *An X-ray Fibre Diffraction Study of Magnetically-Aligned Microtubules in Solution,* PhD Thesis, Liverpool John Moores University, October 1995.

10. Brown, J.A. and Tuszyński, J.A. *Phys. Rev. E* **56**, 5834-5840, 1997.

11. Cech, C. R. *Sci. Amer.* **255**, 76-84, 1986.

12. De Robertis, E.D.P. and De Robertis, E.M.F. *Cell and Molecular Biology,* Saunders College, Philadelphia, 1980.

13. Doolittle, W. F. *Science* **284**, 2124-2128, 1999.

14. Dreyfuss, G., Swanson, M.S., Pinol-Roma, S. and Burd, C.G. *Annu. Rev. Biochem.* **62**, 289-321, 1993.

15. Eigen, M. and Schuster, P. *The hypercycle: A principle of natural self organization,* Springer, Berlin, 1979.

16. Elsasser, W.M. *Atom and Organism: A New Approach to Theoretical Biology.* Princeton University Press, Princeton, 1966.

17. Fausett, L. *Fundamentals of Neural Networks*, Prentice-Hall, 1994.

18. Fröhlich, H. *Int. J. Quantum Chem.* **2**, 641, 1968.

19. Gesteland, R. E., Cech, T. R., and Atkins, J. F., eds. *The RNA World.* Cold Spring Harbor Laboratory Press, New York, 1999.

20. Gurney K. *An Introduction to Neural Networks*, UCL Press, 1997.

21. Hameroff, S. *The Ultimate Computing.* Elsevier, Amsterdam, 1987.

22. Haykin, S. *Neural Networks*, 2nd Edition, Prentice Hall, 1999.

23. Keener, J.P. and Sneyd, J. *Mathematical Physiology*, Springer-Verlag, Berlin, 1998.

24. Maniotis, A.J., Chen, C.S., and Ingber, D.E. *Proc. Natl. Acad. Sci. USA*, **94**, 849-854 (1997).

25. Margulis, L. *Symbiotic Planet. A New Look at Evolution.* Sciencewriters, Amherst, 1998.

26. Nitschke, W. and Rutherford, A.W. *TIBS* **16**, 241-245, 1991.

27. Orgel, L. E. *TIBS* **23**, 491-495, 1998.

28. Peyrard, M. ed. *Nonlinear Excitations in Biomolecules.* Springer-Verlag, Berlin, 1995.

29. Ramakrishnan, V. and White, S. W. *TIBS* **23**, 208-212, 1998.

30. Ray, T. S. *An Approach to the Synthesis of Life. Artificial Life II*, C. G. Langton et al. (Eds.), Addison-Wesley, Redwood City, CA, 371-408, 1992.

31. Scott, A.C. *Stairway to the Mind: The Controversial Science of Consciousness.* Springer-Verlag, New York, 1995.

32. Selden, G., Becker, R.O., and Guarnaschelli, M.D. (Editor), *The Body Electric: Electromagnetism and the Foundation of Life*, William Morrow and Co. Publishers, 1985.

33. Shannon, C.E. and W. Weaver. *The Mathematical Theory of Communication* (5th ed.). University of Illinois Press, Chicago, 1963.

34. Stryer, L. *Biochemistry*, 4th Ed., Ch. 8, Freeman, New York, 1995.

35. Szilard, L. *Z. Phys.* **53**, 840, 1929.

36. Vassilev, P.M., Dronzine, R.T., Vassileva, M.P., and Georgiev, G.A. *Biosci. Rep.* **2**, 1025-1029, 1982.

37. Vater, W., Stracke, R., Böhm, K.J., Speicher, C., Weber, P., and Unger, E. "Behaviour of individual microtubules and microtubule bundles in electric fields." *The Sixth Foresight Conference on Molecular Nanotechnology.* Santa Clara, CA (USA) 1998.

38. von Neumann, J. *Theory of Self-Reproducing Automata.* (Ed. by A. W. Burks), Univ. of Illinois Press, Champaign, 1966.

39. Waldrop, M. M. *Complexity: The Emerging Science at the Edge of Order and Chaos,* Simon & Schuster, New York, 1992.

40. Wiener, N. *Cybernetics: or Control and Communication in the Animal and Machine,* M.I.T. Press, New York, 1961.

41. Wolfram, S. *Cellular Automata and Complexity: Collected Papers,* Addison-Wesley, Reading MA, 1994.

42. Xiong, J. et al. *Science* **289**, 1727-1730, 2000.

43. Zeleny, M. (Ed.) *Autopoiesis: A Theory of Living Organization.* North-Holland, New York, 1981.

Chapter 2

What Are the Molecules of Life?

2.1 Nucleic Acids, DNA, RNA

Animate matter is almost exclusively built up from six elements:

- *hydrogen* H (60.5%), *oxygen* O (25.7%), *carbon* C (10.7%), *nitrogen* N (2.4%), *phosphorous* P (0.17%) and *sulfur* S (0.13%).

In brackets, the percentage atomic abundance is given for particular elements in soft tissues of the mature human body (33). The remaining 0.4% are elements playing the role of *electrolytes*:

- calcium Ca^{2+} (0.23%), sodium Na^+ (0.07%), potassium K^+ (0.04%), magnesium Mg^{2+} (0.01%) and chloride Cl^- (0.03%).

Two transition metal ions that carry electrons are:

- iron Fe^{3+} Fe^{2+} and copper Cu^{2+} Cu^+,

and some *trace elements*:

- Mn, Zn, Co, Mo, Se, J, F, ...

are much less abundant. At the lowest level of chemical organization, animate matter is composed of a limited number of standard building blocks. One can divide them into six classes:

- *carboxylic acids*, with a general formula R-COOH, which in neutral water dissociate to anions $R - COO^-$,

- *alcohols*, with the general formula R-OH,

- *monosaccharides (hydrocarbons)* with a general formula $(CH_2O)_n$, generally with n = 5 (pentoses) and n = 6 (hexoses),

- *amines*, with a general formula $NH_2 - R$, which in the neutral water environment are protonated and become cations $NH_3^+ - R$,

- *nitrogenous heterocycles*,

- *phosphates*, with a general formula $R - PO_3^{2-}$, which are doubly dissociated in the neutral water environment,

- *hydrosulfides* with a general formula $R - SH$.

Here, R denotes additional chemical entities. These can be shorter or longer hydrocarbon chains or ring structures as shown in Fig. 2.1 in the case of carboxylic acids and in Fig. 2.2 in the case of alcohols.

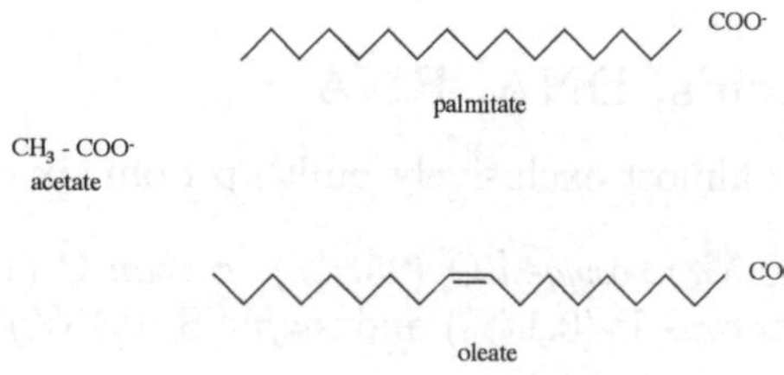

FIGURE 2.1: Examples of the three most important kinds of carboxylic acids with purely carbohydrate substituents.

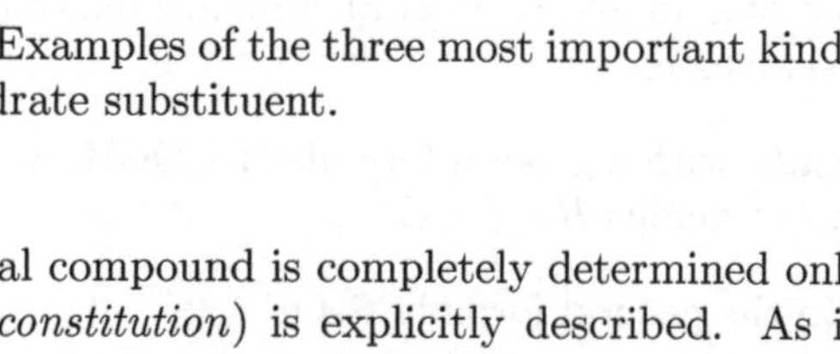

FIGURE 2.2: Examples of the three most important kinds of alcohols with a purely carbohydrate substituent.

A given chemical compound is completely determined only if its system of chemical bonds (*constitution*) is explicitly described. As it often happens, two or more different compounds (referred to as *isomers*) can have the same atomic composition, but different arrangements in space. To simplify notation of structural formulas we omit the symbols of carbon and hydrogen atoms (C and H, respectively) in larger molecules and leave only a lattice of covalent bonds between carbon atoms. It is assumed that at all vortices of such a lattice the carbon atoms are complemented by an appropriate number of hydrogen atoms to satisfy the fourfold carbon valency.

In Fig. 2.3 the three most important examples of monosaccharides are shown. In the neutral water environment they form a ring structure closed by the oxygen atom coming either from the *aldehyde* $-CHO$ or the *ketone* $-CO$ group. Nitrogenous heterocycles are modifications of aromatic hydrocarbons in which a few hydrogenated carbons CH are replaced by nitrogen N.

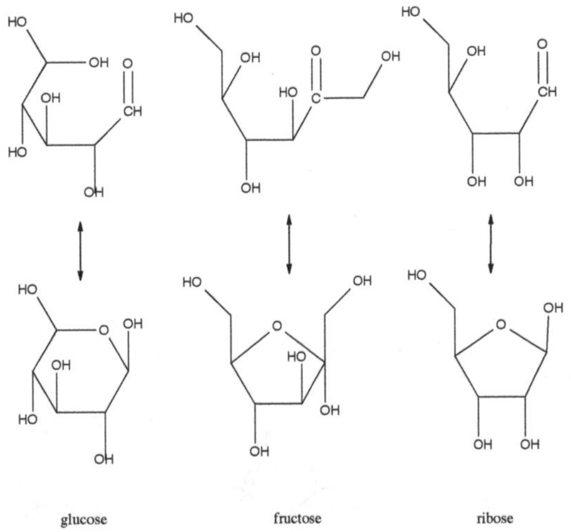

glucose   fructose   ribose

FIGURE 2.3: Three examples of monosaccharides: glucose and fructose (hexoses), and ribose (a pentose). The rings of glucose and ribose are closed by oxygen coming from the aldehyde group whereas the ring of fructose is closed by oxygen coming from the ketone group.

Some more important examples are given in Fig. 2.4. The simplest example of a phosphate is the *orthophosphate* $H - O - PO_3^{2-}$, a doubly dissociated anion of the orthophosphoric acid that is often referred to in biochemistry as an *inorganic phosphate* (P_i). Phosphates will be considered further on when we examine phosphodiester bonds. Hydrosulfides $R - SH$, which are structurally similar to hydroxides (alcohols) $R - OH$, maintain important chemical differences such as the ability to form thioesters, $R - CO - S - R'$, but also (after reduction) disulfide, $R - S - S - R'$, bonds.

Very often elementary organic building blocks belong simultaneously to two or even three classes listed above, which means that the substituent R, besides being a hydrocarbon constituent, also consists of some additional functional group or it is itself such a group. Thus, we can have simple carboxylic acids with ketonic -CO , hydroxylic -OH or protonated aminic $-NH_3^+$ groups (*keto*, *hydroxy* and *amino acids*, respectively) or, e.g., amines of heterocycles or

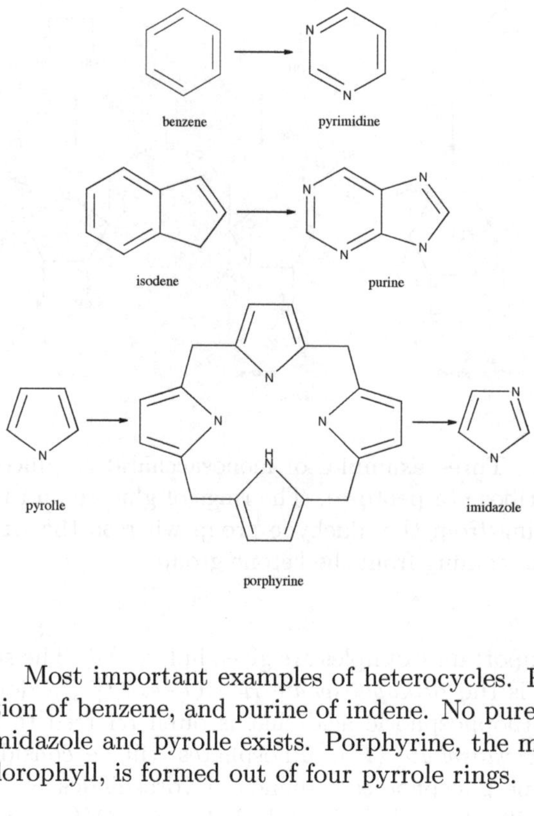

FIGURE 2.4:   Most important examples of heterocycles. Pyrimidyne is a direct modification of benzene, and purine of indene. No purely hydrocarbon homologue of imidazole and pyrolle exists. Porphyrine, the main component of heme and chlorophyll, is formed out of four pyrrole rings.

phosphates of saccharides and alcohols (Fig. 2.5).

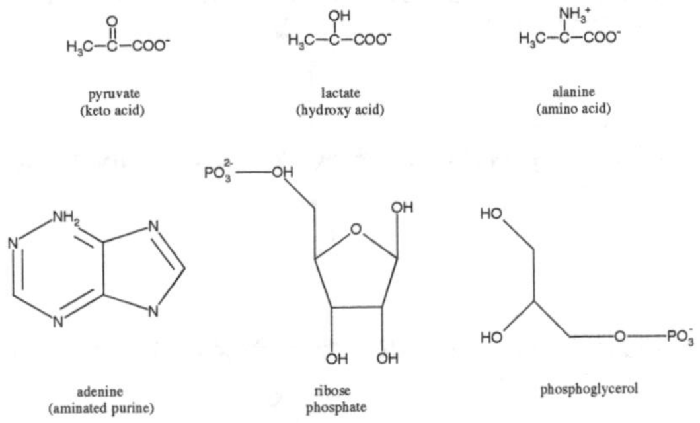

FIGURE 2.5: Some organic compounds with two functional groups.

2.1.1 Chemical Bonds and Bond Energies

Chemical bonds are the major force holding atoms together within a molecule. They may be classified according to their energy levels or how the bonds are made. Chemical bonds usually have a specific inter-atomic distance (bond length) and maintain specific angles relative to each other. However, for bonds with smaller energies, these restrictions become less strict. This is especially true in macromolecules, which are the molecules commonly found in biological systems. In macromolecules the collective contributions from a large number of bonds with small energies become important.

2.2 Generalized Ester Bonds

Of special importance in organic chemistry is the reaction between a carboxylic acid and an alcohol, the product of which is referred to as an *ester*, with a characteristic *ester bond* $-COO-$ (Fig. 2.6a). In particular, *lipids* are esters of fatty acids and glycerol (Fig. 2.7a). The alcohol in the esterfication reaction can be replaced by amine to form an *amide*, with an *amide bond* $-CONH-$ (Fig. 2.6b). Meanwhile, the carboxylic acid can itself be replaced by a phosphate, to form what is called a *phosphodiester bond* (Fig. 2.6c). For example, *phospholipids* are lipids in which one fatty acid chain is replaced by some phosphate with such a phosphodiester bond (Fig. 2.7b).

Monosaccharides are molecules that behave both as alcohols and as car-

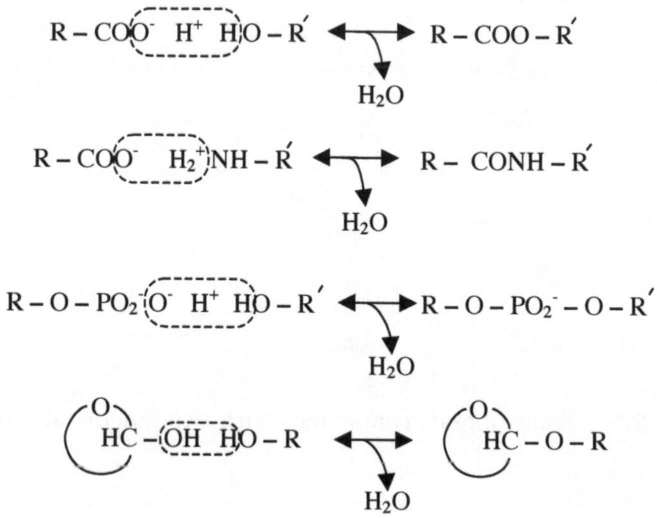

R – COO⁻ H⁺ HO – R′
R – COO – R′
H2O
R – COO⁻ H2⁺NH – R′
R – CONH – R′
H2O
R – O – PO2⁻O⁻ H⁺ HO – R′
R – O – PO2⁻ – O – R′
H2O
O
HC – OH HO – R
O
HC – O – R
H2O

FIGURE 2.6: Formation of the ester bond (a) and its generalization to amide (b), phospodiester (c) and glycosidic (d) bonds. The circular ring shown represents any monosaccharide ring closed by the oxygen atom neighboring the carbon atom taking part in the glycosidic bond. The carboxylic and phosphate groups are assumed to be dissociated.

R1 – COO⁻ H⁺ HO
R2 – COO⁻ H⁺ HO
R3 – COO⁻ H⁺ HO
3 H2O
R1 – COO
R2 – COO
R3 – COO
R1 – COO
R2 – COO
O – PO2⁻O⁻ H⁺ HO – R3
H2O
R1 – COO
R2 – COO
O – PO2⁻ – O – R3

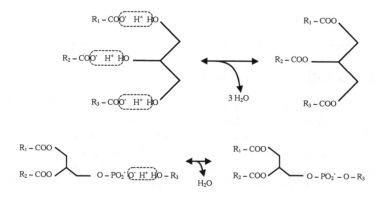

FIGURE 2.7: Formation of lipid (a) and phospholipid (b).

boxylic acids. From their open structure (Fig. 2.3) it follows that the carboxylic properties have $-OH$ groups bound to the carbon atom neighboring the oxygen atom closing the saccharide ring. The generalized ester bond formed by such a group with another hydroxylic group $-OH$ of the alcoholic properties is referred to as *glycosidic bond* (Fig. 2.6d). The alcohol hydroxylic group $-OH$ can be replaced by the imino group $-NH-$ of some nitrogen heterocycles being derivatives of pyrymidine and purine (Fig. 2.4). Compounds with ribose-forming glycosidic bonds and such nitrogenous bases (imino groups are good proton acceptors) are of great biological importance and are referred to as nucleosides. Compounds consisting of ribose phosphate and nitrogenous bases are referred to as nucleotides. Fig. 2.8 illustrates the most common nucleotide: adenosine monophosphate, or AMP. All canonical nitrogenous bases entering the canonical nucleotides shall be presented later in Section 2.7.

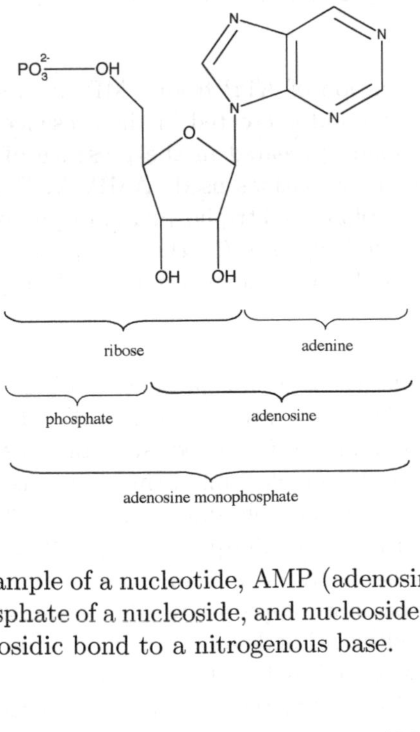

FIGURE 2.8: Example of a nucleotide, AMP (adenosine monophosphate). A nucleotide is a phosphate of a nucleoside, and nucleoside is a monosaccharide ribose linked by glycosidic bond to a nitrogenous base.

The binding of a nucleoside *monophosphate*, e.g., AMP, to *pyrophosphate* $HO - P_2O_5^{3-}$ (triply dissociated anion of pyrophosphoric acid, referred to as *inorganic diphosphate*, PP_i) results in a phosphodiester bond and the formation of a nucleotide *triphosphate*, in our case *ATP* (adenosine triphosphate, Fig. 2.9a). ATP is a universal donor of energy in biochemical processes. The hydrolysis of ATP to ADP (adenosine *diphosphate*, Fig. 2.9b) results in the liberation of an orthophosphate $HO - PO_3^{2-}$ (inorganic phosphate P_i) and is often coupled to endoergic reactions which could not proceed otherwise.

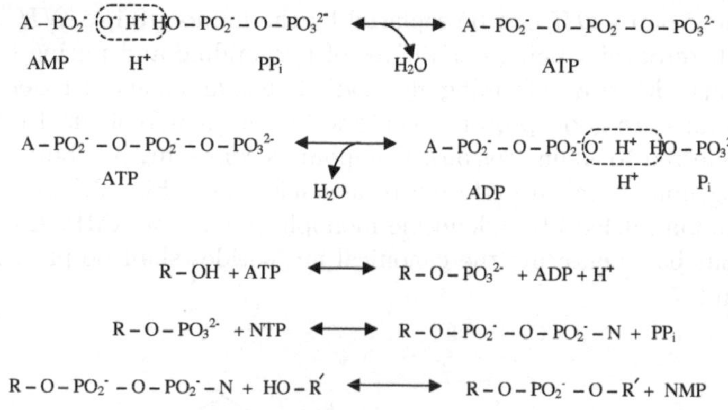

$$A - PO_2^- \; \{O^- \; H^+ \; H\}O - PO_2^- - O - PO_3^{2-} \quad \longleftrightarrow \quad A - PO_2^- - O - PO_2^- - O - PO_3^{2-}$$

AMP　　　　H$^+$　　　　PP$_i$ 　　　　　　H$_2$O 　　　　　　ATP

$$A - PO_2^- - O - PO_2^- - O - PO_3^{2-} \quad \longleftrightarrow \quad A - PO_2^- - O - PO_2^- \{O^- \; H^+ \; H\}O - PO_3^{2-}$$

ATP 　　　　　　　　　 H$_2$O 　　　　 ADP 　　　 H$^+$ 　　 P$_i$

$$R - OH + ATP \quad \longleftrightarrow \quad R - O - PO_3^{2-} + ADP + H^+$$

$$R - O - PO_3^{2-} + NTP \quad \longleftrightarrow \quad R - O - PO_2^- - O - PO_2^- - N + PP_i$$

$$R - O - PO_2^- - O - PO_2^- - N + HO - R' \quad \longleftrightarrow \quad R - O - PO_2^- - O - R' + NMP$$

FIGURE 2.9: Formation of ATP from AMP (a) and of ADP from ATP (b). Any phosphoester bond is created in the presence of ATP (c) and also any phosphodiester bond is created in the presence of ATP (or, more generally, NTP, d). The abbreviations used: AMP, ADP and ATP - adenosine monophosphate, diphosphate and triphosphate, respectively, NMP and NTP - nocleoside mono- and triphosphate, P_i - the orthophosphate anion ("inorganic phosphate") and PP_i - the pyrophosphate anion ("inorganic diphosphate").

The equilibrium of all the reactions presented in Fig. 2.6 is strongly shifted towards hydrolysis rather than synthesis of the bond. Hence, the formation of the generalized ester bond usually proceeds simultaneously with the hydrolysis of ATP along a quite different reaction pathway that is shown in Fig. 2.6. In particular, the formation of any phosphate $R - O - PO_3^{2-}$ takes part in the presence of ATP, and not orthophosphate (Fig. 2.9c). It should be stressed that phosphates are not salts of orthophosphoric acid and a term such as *phosphoester* would be more appropriate, if not unconventional. Also the formation of phosphodiester bonds does not necessarily follow the scheme in Fig. 2.6c. In the presence of nucleotide triphosphates, it follows the two-step scheme presented in Fig. 2.9d, the liberation of nucleotide monophosphate and pyrophosphate.

Three kinds of biological macromolecules: polysaccharides, proteins and nucleic acids are built out of the elementary entities: monosaccharides, amino acids and nucleotides, linked by generalized ester bonds. *Polysaccharides* are linear or branched polymers of monosaccharides linked by glycosidic bonds (Fig. 2.10a). Small polysaccharides are called *oligosaccharides* and very small ones, *disaccharides, trisaccharides,* etc. *Proteins* are polymers of amino acids $NH_3^+ - R_iCH - COO^-$ (cf. Fig. 2.5) linked to the linear chain by amide bonds (Fig. 2.10b). All canonical side chains R_i of amino acids shall be listed in Section 2.6. Small proteins are called *(poly)peptides,* thus an amide bond is often referred also to as a peptide bond.

Nucleic acids are linear polymers of nucleotides linked by a phosphodiester (Fig. 2.10c). Note that *dinucleotides,* often occurring as cofactors of protein enzymes that carry hydrogen (electrons jointly with protons, e.g., NAD^+ - nicotinamide adenine dinucleotide and FAD - flavin adenine dinucleotide or as small molecular groups (e.g., coenzyme A)) are two nucleotides linked by a phosphotriester bond rather than diester bond (see Fig. 2.10). A more detailed introduction to the foundations of organic chemistry can be found in many standard textbooks. We recommend, e.g., Applequist, DePuy and Rinehart [17].

2.3   Directionality of Chemical Bonds

The spatial structure of biomolecules, so important for their function, is related to directional properties of chemical bonds and those are, in turn, determined by the spatial distribution of electronic states of the constituent atoms. The hydrogen atom contributes only one s-orbital to chemical bonding while carbon, nitrogen and oxygen atoms contribute one s-orbital and three p-orbitals; the phosphorous and sulfur atoms can contribute as many as five d-orbitals. An outline of the spatial distribution of the electron probability density in the s- and p-orbitals is given in Fig. 2.11a. One s-orbital and three p-orbitals can hybridize into four orbitals of tetrahedral symmetry (sp^3 hybrydization), one s-orbital and two p-orbitals can hybridize into three orbitals of trigonal symmetry (sp^2 hybridization) and one s-orbital and one p-orbital can hybridize into two orbitals of linear symmetry (sp hybridization). The three types of hybridized orbitals are presented in Fig. 2.11b.

The carbon atom has four electrons in the outer shell that are organized in four sp^3 hybridized orbitals and it needs four additional electrons for those orbitals to be completely filled up. As a consequence, it can bind four hydrogen atoms, each giving one electron and admixing its own s-orbital to the common bonding orbital, each of axial symmetry (see Fig. 2.12). The nitrogen atom has five electrons in four sp^3 hybridized orbitals and it needs only three additional electrons; thus it binds three hydrogen atoms. The ammonia molecule that is formed like the methane molecule also has tetrahedral angles between the bonds. However, one bond is now replaced by the lone electron pair (see Fig. 2.12). The oxygen atom has six electrons in four sp^3 hybridized orbitals; thus it binds two hydrogen atoms and has two lone electron pairs (Fig. 2.12).

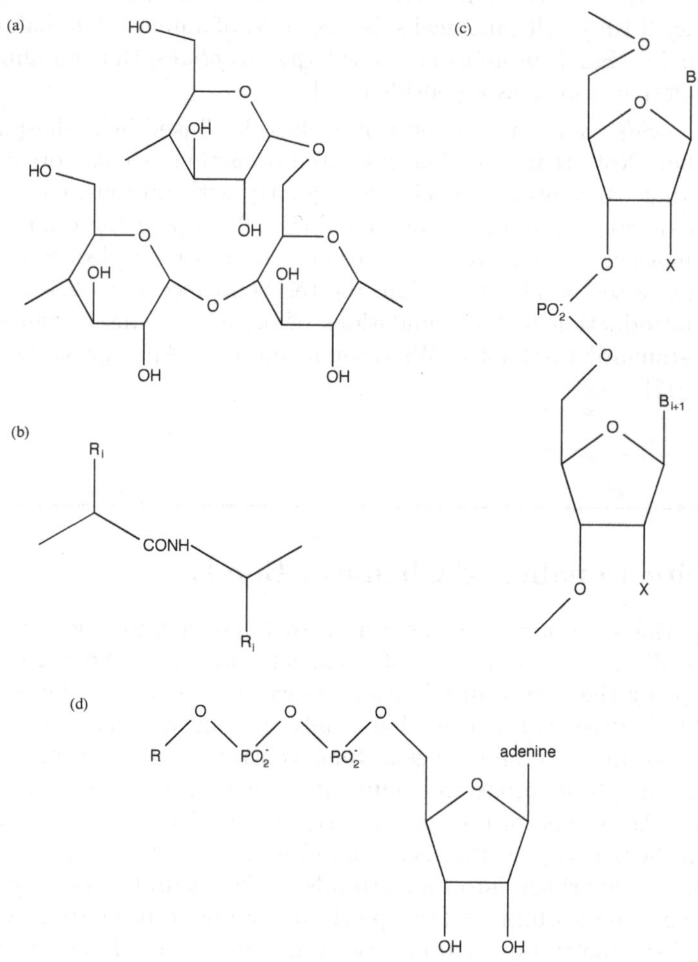

FIGURE 2.10: (a) Polysaccharides (hydrocarbons) are polymers of monosaccharides linked by glycosidic bonds, non-branched, as in the case of amylase, a type of starch - the main storage polysaccharide of plants, or branched, as in the case of amylopectin, other type of starch, or glycogen - the main storage polysaccharide of animals. (b) Proteins are polymers of amino acids linked by amide (peptide) bonds. (c) Nucleic acids are linear polymers of nucleotides linked by phosphodiester bond. In RNA (ribonucleic acid), X denotes the hydoxyl group OH whereas in DNA (deoxyribonucleic acid), hydrogen H. (d) Phosphotriester bond in dinucleotides usually formed out of adenosine and other, usually highly modified nucleotide R.

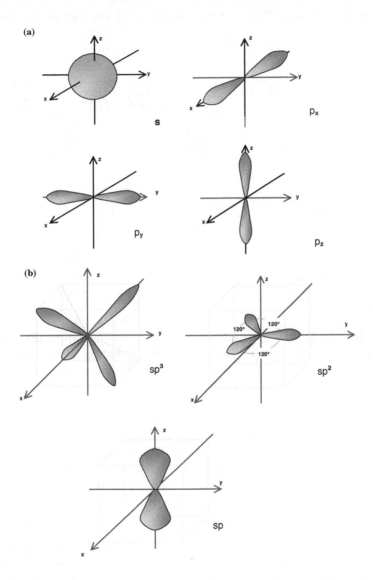

FIGURE 2.11: (a) An outline of the spatial distribution of the electron probability density in the s-orbital and three p-orbitals p. (b) The four orbitals of the tetrahedral sp^3 hybridization, three of the trigonal sp^2 hybridization and two of the linear sp hybridization.

Molecular and Cellular Biophysics

As a consequence of sp^3 hybridization, the nitrogen and oxygen atom orbitals have highly directed negative charge distributions leading, as we shall see, to hydrogen bonding which is very important for the spatial structure of all biomolecules.

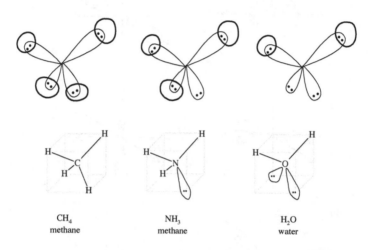

CH₄
methane

NH₃
methane

H₂O
water

FIGURE 2.12: The spatial structure of methane, ammonia and water molecules is determined by the tetrahedral sp^3 hybridization of the central atom orbitals. A 'cloud' with two dots denotes the lone electron pair.

By replacing one hydrogen atom in methane, ammonia or water by the methyl group CH_3 one obtains the spatial structures of ethane, methylamine and methanol, respectively (Fig. 2.13). The methyl group CH_3 can rotate freely around the $C - C$, $C - N$ or $C - O$ bonds. The lowest value of the potential energy of the molecule is found when looking along the rotational axis. The electron densities of one triple of atoms or free electron pairs are located between the electron densities of the other triple of atoms.

In the case of ethane, methylamine and methanol, a 120° rotation does not lead to a structural change of the molecule. However, if in the other triple there are free electron pairs or atoms other than hydrogen, a rotation around the central covalent bond may lead to a new *conformational state* of the molecule. Such a state cannot be reconstructed from the original state by either a translation or a rigid body rotation. Different conformational states become geometrically significant for long molecular chains, e.g., carbohydrates, which can exist both in maximally stretched linear forms and as folded clusters. A rotation of each covalent bond in the chain allows one conformational *trans* state and two *gauche* states (Fig. 2.14a). If we ignore

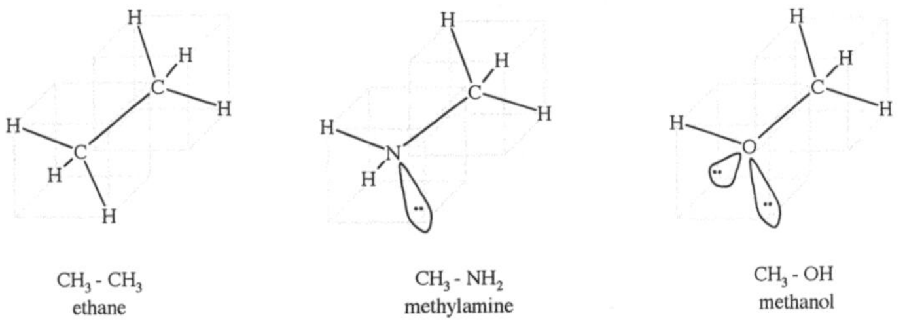

CH$_3$ - CH$_3$
ethane

CH$_3$ - NH$_2$
methylamine

CH$_3$ - OH
methanol

FIGURE 2.13: The spatial structures of ethane, methylamine and methanol.

steric constrains (i.e., exclude "volume effects"), that very rapidly emerge in the case of longer chains and constitute a separate significant problem, the differences between equivalent conformational states of a single bond amount to several kJ/mol while the potential energy barrier height is on the order of 10 do 20 kJ/mol (see Fig. 2.14b). This corresponds to four to eight times the value of the average thermal energy at physiological temperatures $k_B T$ = 2.5 kJ/mol. The probability of a random accumulation of such an amount of energy in one degree of freedom, defined by the Boltzmann factor, i.e., the exponential of its ratio to $k_B T$, $exp(-\Delta/k_B T)$, equals between 10^{-2} and 10^{-4}. The latter value multiplied by the average frequency of thermal oscillations, $10^{13}s^{-1}$, gives between 10^{11} and 10^9 random local conformational transitions per second at physiological temperatures. A conformational state of a molecule is, therefore, not very stable.

In the case of closed chains (unsaturated cyclic carbohydrates or monosaccharides - see Fig. 2.3), 120 degree rotation around individual bonds are not possible without breaking them. Hence, conformational transitions involve much smaller rotations that are simultaneously applied to many bonds. These processes are referred to as ring puckering since an entirely flat conformation becomes energetically unstable. In the case of a five-atom ring we distinguish the conformations of an *envelope* and a *half-chair* while in the case of a six-atom ring we see the conformations of a *chair*, a *twist* and a *boat* - Fig. 2.15.

Not all transitions between different energetically stable geometrical structures can be achieved by simple rotations around covalent bonds involving small energy barrier crossings. A mutual exchange of hydrogens with hydroxyl groups of monosaccharide rings (see Fig. 2.3) is not possible without covalent bond breaking and its subsequent restoration and hence requires an energy of about 300 kJ/mol (see Table 2.1). This is one and a half orders of magnitude greater than the energies mentioned earlier. Various geometrical forms of chemical molecules with the same chemical formulas but that require bond breaking and bond restoration to convert from one to the other are referred to as *isomers*. It is easy to imagine a large diversity of

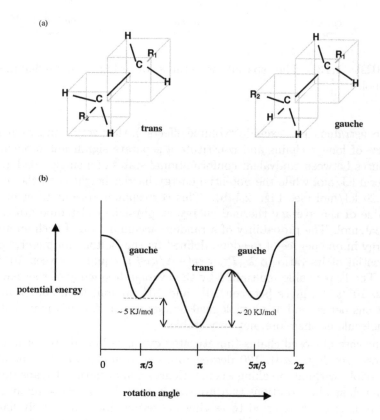

FIGURE 2.14: (a) Conformational states related to the rotation around one covalent bond. The symbols R_1 and R_2 denote atoms or molecular groups (in the case of longer chains) other than atomic hydrogen. (b) The potential energy related to the rotation of a single covalent bond. The three minima corresponding to three conformational states: one *trans* and two *gauche* (see the neighboring diagram) are separated by energy barriers with heights between 10 and 20 kJ/mol.

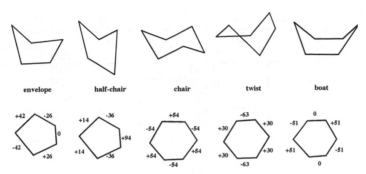

FIGURE 2.15: The *envelope* and *half-chair* conformational states of a five-atom ring and *chair*, *twist* and *boat* conformations of a six-atom ring. The angles between neighboring covalent bonds only slightly differ (within 5°) from the undistorted tetrahedral angle. Each stable conformation can be envisaged as the puckering of a planar but unstable conformation resulting from rotations around subsequent bonds by angles whose values are given in the second line in this figure.

such isomers even for simple monosaccharides. They differ not only in the location of hydrogen atoms and hydroxyl groups but also in the position of the oxygen bridge. For example, hexases can form both five- and six-atom rings (see the structure of glucose and fructose in Fig. 2.3). Note that $kJ/mol = 1.04 \times 10^{-2} eV/bond = 0.40kT/bond$ at 298 K.

Half of the monosaccharides are a simple mirror image of their counterparts. Such isomers are called *enantiomers* and the phenomenon itself is referred to as chirality. This is a symmetry breaking of the handedness (left/right invariance) analogous to the human hands and hence the origin of the name from the Greek word *cheir* (hand). Ignoring specific relations to double bonds, stero-isomerity is usually linked to the existence of at least one carbon atom in the molecule whose all four covalent bonds lead to different atoms or groups of atoms. Chiral molecules are optically active and they twist the light's polarization plane either to the left or to the right.

Enantioners are divided into D- and L-type. A definition for these types is based on the simplest monosaccharide, glyceraldehyde (Fig. 2.16). D-type monosaccharides are derivatives of the D-glyceraldehyde; similarly, L-type monosaccharides are derivatives of the L-glyceraldehyde. By definition the D-glyceraldehyde twists the polarization plane to the right but this is not the general case for all D-type monosaccharides. Also, all amino acids except for glycine are chiral which is linked to the fact that the central carbon C^α is a chiral carbon. Therefore, we have both D-amino acids and L- amino acids. Chemists adopted an unambivalent way of transferring the determination of D and L- type enantiomers from the defining glyceraldehydes onto other com-

TABLE 2.1:   Properties of Biologically Important Bonds

Kind of bond	Bond distance (nm)	Bond energy (kJ mol^{-1})
C-C	0.154	350
C=C	0.133	610
N-N	0.145	161?
N=N	0.124	161?
C-N	0.147	290
C=N	0.126	610
C-O	0.143	350
C=O	0.114	720
C-S	0.182	260
C=S		480
O-H	0.096	463
N-H	0.101	390
S-H	0.135	340
C-H	0.109	410
H-H	0.074	430
H-O	0.096	460

pounds different from monosaccharides. This is a little too complicated for our purposes, so we will not dwell on it here. We only wish to mention that all biologically active monosaccharides are D-type while all biologically active amino acids are L-type.

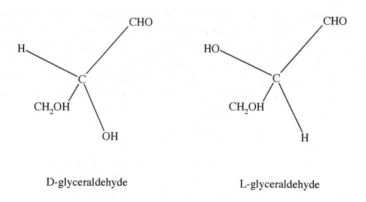

D-glyceraldehyde   L-glyceraldehyde

FIGURE 2.16:   Two enantiomers of glyceraldehyde.

Let us now consider the bonds formed by electron orbitals in the process of trigonal hybridization sp^2. Using these orbitals two carbon atoms can form σ bonds (with axial symmetry) both amongst themselves and with the other four atoms that lie in the same plane. The remaining p-orbitals that are perpendicular to this plane, one for each carbon atom, then form a second, slightly weaker bond between the two carbon atoms. This is called a π bond (see Fig. 2.17a). In this way a *planar* molecule of ethylene is formed in which two carbon atoms are bound together via a *double* bond. If one carbon atom is replaced by a nitrogen atom, with an extra electron, and simultaneously one hydrogen atom is replaced by a lone electron pair, we obtain an imine molecule. If, on the other hand, one carbon atom is replaced by an oxygen atom with two excess electrons and simultaneously two hydrogen atoms by two free electron pairs, we obtain a formaldehyde molecule (Fig. 2.17b).

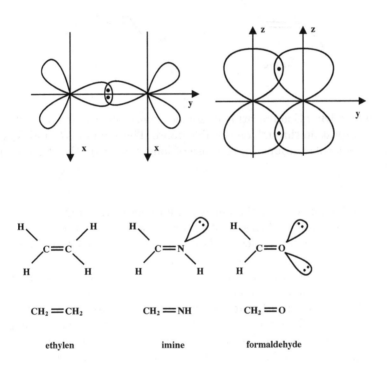

FIGURE 2.17:   (a) A double bond consists of a σ-bond formed by orbitals in the process of trigonal hybridization sp^2 and a π-bond formed from orbitals perpendicular to the previous ones. (b) Planar particles of ethylene, imine and formaldehyde.

Double bonds can exist in longer chain molecules, for example unsaturated carbohydrates. For each double bond there can be two energetically stable spatial structures each of which is obtained from the other by a 180° rotation. A rotation around a double bond requires a temporary breakage of the π-

bond, and hence an energy of approximately 100 kJ/mol (see Table 2.1). The two structures are referred to as *configurational isomers*: *trans* and *cis* (see Fig. 2.18). Commonly, π-bonds can be found in a delocalized form. In the

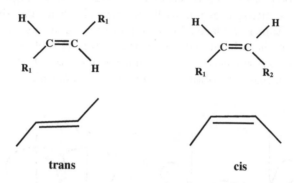

trans   **cis**

FIGURE 2.18: Configurational isomers *trans* and *cis* are related to each other via a rotation through one double bond. The symbols R_1 and R_2 denote atoms or molecular groups (in the case of longer chains) other than hydrogen atoms.

historically famous case of the benzene molecule C_6H_6 (Fig. 2.19a) one can consider two symmetrical Kekule structures whose every other carbon-carbon bond is a double bond. Which of these two structures is actually adopted by the molecule? Are both of them represented statistically in the population of identical molecules with the same probability? Answers to these questions are given by quantum mechanics. The actual state of each benzene molecule is a quantum-mechanical linear combination of the two states defined by the Kekule structures. The π-bond is not localized on every other carbon pair but rather delocalized over the entire ring. An electric current circulates around the ring and can be induced by magnetic fields and other electric currents. Due to such planar interactions in benzene rings and also in other aromatic heterocycles (e.g., in purin and pyramidine bases), their compounds have a tendency for parallel stacking. Effectively, each carbon pair is allocated one half of a π-bond, which justifies the commonly accepted notation using a broken line (Fig. 2.19a). The π-bond between carbon and oxygen in a disso-ciated carboxyl group $-COO^-$ is also delocalized (Fig. 2.19b). This group can be envisaged as a planar, negatively charged plate that has a rotational degree of freedom around the axis that connects one carbon atom to the next atom. The phosphodiester bond (Fig. 2.19c) exhibits similar behavior ex-cept the phosphorus atom is connected to the rest of the molecule via two single bonds, each of which provides a rotational axis. The delocalization of

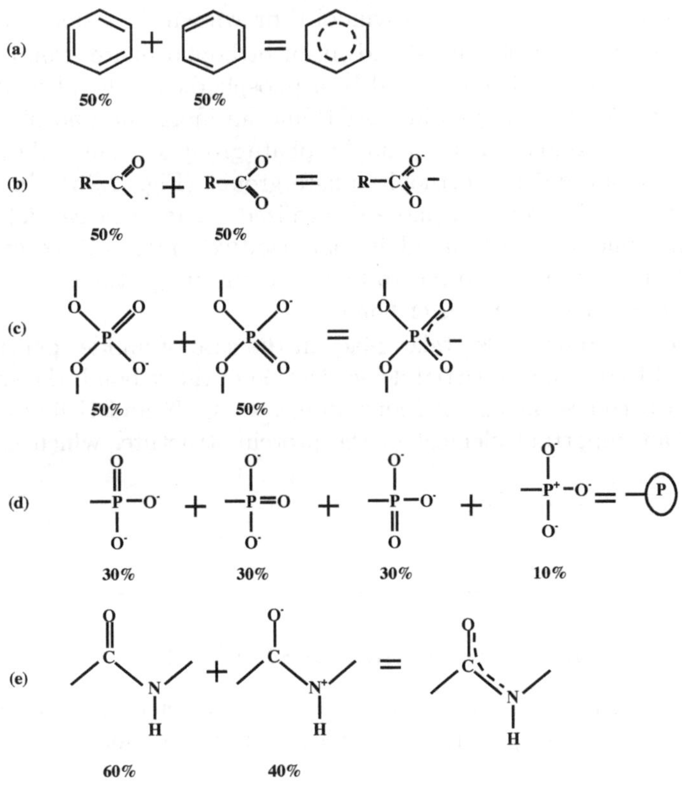

FIGURE 2.19: Delocalization of double bonds in the case of benzene (a), carboxyllic anion (b), phosphodiester bond (c), orthophosphate anion (d) and amide (peptide) bond (e).

the π-bond in the case of the orthophosphate group $-PO_3^{2-}$ is a little more complicated. A double bond can be established here between phosphorous and each of the three oxygen atoms. In addition, there is also the possibility of an electron transfer state involving an electron from the phosphorus atom onto a hitherto neutral oxygen atom. The actual state of the group is a linear combination of all four possibilities (see Fig. 2.19d). Instead of drawing broken lines for partial bonds between phosphorus and the three equivalent, partially negatively charged oxygen atoms, biochemists represent the entire orthophosphate group using the symbol P drawn within a small circle.

The delocalization of the π-bond in an orthophosphate group requires a high-energy expense that is stored in a phosphodiester bond in ATP. As a result of the ATP hydrolysis into ADP and an inorganic phosphate Pi, two phosphodiester bonds and one orthophosphate group are replaced by one phosphodiester bond and two orthophosphate groups (Fig. 2.9b). In this latter arrangement π-electrons are more delocalized. The negative delocalization energy in connection with an additional negative energy of hydration causes the products of ATP hydrolysis to have a lower energy than the products of hydrolysis of other general ester bonds.

An electron transfer also takes place in the case of amide (peptide) bonds (see Fig. 2.19e). Due to partial delocalization of the π-bond, the structure of this bond becomes planar. All four atoms: O, C, N and H lie in one plane which is an important element of the protein structure, which is discussed later.

2.4　Types of Inter-Atomic Interactions

In this section we discuss in more detail the various types of inter-atomic interactions that take part in the formation of biomolecular structures.

2.4.1　Ionic Interactions

Ionic interactions are probably one of the most basic interactions between matter. Ionic bonding involves ions of opposite charges, which form a bound state as a result of the Coulomb interaction between them. The respective energies are in the range of a few eV up to 10 eV. This is strong enough to resist dissociation at room temperature; in fact only temperatures into the thousands of kelvins can break up ionic bonds, well above physiological thresholds. If two charges of q_1 and q_2 are separated by distance r in an area of space with dielectric constant ε, the force F between them is

$$F = \frac{q_1 q_2}{4\pi\varepsilon\varepsilon_0 r^2} \qquad (2.1)$$

If the charges are opposite, the force will be attractive and vice versa. The potential energy between the two charges may be considered as the amount of work necessary to bring the charges together from the infinite distance to a finite distance r (see Fig. 2.20).

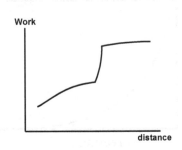

FIGURE 2.20: The amount of work necessary to bring two charges together from infinity to a finite distance.

For example, the binding energy of two charged particles such as Na^+ and Cl^- can be calculated through

$$\Delta W = -\frac{q_1 q_2}{4\pi\varepsilon\varepsilon_0 r} \tag{2.2}$$

where r in this case is $2.4\mathring{A}$ giving $\Delta W = 5.5eV$.

The polarity of a bond depends on the differences in the electronegativities of the two elements forming the molecule. When the two charges have opposite signs, the potential energy will be negative at a finite distance and 0 at infinity.

When two ions are very close to each other, repulsive forces from the overlapping electron clouds and the opposing nuclear charges result. The potential energy function of these conservative forces can be expressed by the formula,

$$PE = be^{-\frac{r}{a}} \tag{2.3}$$

where a and b are constants. A graphical example of this function is given in Fig. 2.21. The net potential energy, U, is, then,

$$U = (\frac{q_1 q_2}{4\pi\varepsilon\varepsilon_0 r}) + be^{-\frac{a}{r}} \tag{2.4}$$

and this is shown graphically for NaCl, in Fig. 2.22.

The equilibrium distance and the corresponding potential energy vary according to the state of NaCl. The potential energy, U, of NaCl in its crystalline form is greater than in NaCl vapor, as is the bond distance between sodium and chlorine ions. This is due to the increased repulsive energy that results when the ions are confined to the crystal lattice.

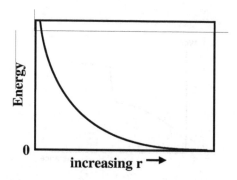

FIGURE 2.21:   Potential energy curve for two equally charged ions close to each other.

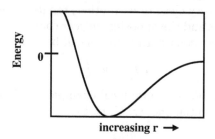

FIGURE 2.22:   Net potential energy for NaCl.

TABLE 2.2: Potential Energies and Bond Distance

	in vapor	in crystal
r	2.51 Å	2.81 Å
U	-118 Kcal/mole	-181 Kcal/mole

2.4.2 Covalent Bonds

Covalent bonds arise in pairs of neutral atoms such as two hydrogens or two carbon atoms. This interaction is electromagnetic in nature and results from a charge separation between the nucleus and its electronic cloud. These two atoms share some electrons such that the energy is reduced and so is the average separation between their nuclei. All the elements which are linked by covalent bonds are either electropositive (Li, Na, K, Rb, Be, Mg and Ca) or electronegative (O,S, Se, F, Cl, Br, I). When a covalent bond is formed between two atoms of the same element, the electron pair is shared equally by both atoms. On the other hand, when a covalent bond is formed between two atoms of unequal electro-negativity, the electron pair is attracted towards the atom possessing higher electronegativity, e.g., in the C-Cl bond, the pair of electrons in the covalent bond is closer to Cl than to C. This leads to a partial positive charge on C and a partial negative charge on Cl giving a partial ionic character to such covalent bonds. The energy of covalent bonds is still quite high. For example, the H-H bond has a binding energy of 4.75 eV while the C-C binding energy is 3.6 eV, which is on the same order as the binding energy of NaCl. The binding energy of most other covalent bonds is also in this energy range, which at room temperature is large compared to thermal energy k_BT. The covalent bond is responsible for maintaining the structural integrity of DNA, proteins and polysaccharides (chain-like sugar molecules). In a covalent bond, electrons shuttle back and forth constantly between positively charged nuclei and hence the covalent bond is electrostatic in origin with a characteristic distance d = 1Å.

In 1916 W. Kossel proposed the theory of the electrovalent bond, i.e., bonds formed through the loss or gain of electrons. G. N. Lewis proposed the idea of non-polar bond formation through the sharing of electrons between atoms in 1918. Subsequent development of quantum mechanics explained the quantization of electron energy levels in atoms and the structure of atomic orbitals and consequently provided a basis for the theory of molecular orbitals by W. Heitler and F. London in 1927. The distribution of electrons in an atom was determined through the Pauli exclusion principle formulated in 1924. The theory states that no two electrons can have the same quantum state in an atom. Since the spin of the electron is quantized, and it can only adopt values of $+\frac{1}{2}$ or $-\frac{1}{2}$, each electron orbital can be occupied by no more than two electrons. For an atom in its ground state, the electronic energy levels will be filled starting with the lowest energy. For example, a carbon atom has 6 electrons which may be distributed as follows:

$$C^{12} : (1s)^2(2s)^2(2p)^2 --- (1s)^2(2s)^2(2p_x^1, 2p_y^1) \qquad (2.5)$$

Alternatively, for an excited electron, the distribution of electron density can overlap that of other energy levels, and the constructive and destructive interference between them gives rise to a new electron density distribution:

$$C^{12} : (1s)^2(2s)^2(2p)^2 --- (1s)^2(2s)^2(1p_x^1, 2p_y^1, 2p_z^1) \qquad (2.6)$$

Therefore the outermost electrons may form different types electronic orbitals. When two atoms are bound together in a covalent bond(s) by sharing electrons, the outermost electrons of each atom may be found in bonding orbitals, anti-bonding orbitals and non-bonding orbitals. For example, when two hydrogen atoms form a hydrogen molecule, fusion of two s-orbitals forms a sigma orbital. The newly created wave functions for the bonding and anti-bonding may be expressed, using the method of *linear combination of atomic orbitals* (LCAO), as, for bonding: $\Psi(A) + \Psi(B)$, while for anti-bonding: $\Psi(A) - \Psi(B)$. When the covalent bond is formed with two $2p_x$ orbitals, they will form $s2p$ and s^*2p orbitals. By pairing $2p_z$ or $2p_y$ orbitals, they may form p_y2p and p_y^*2p orbitals.

As an example, consider bond formation between carbon and hydrogen. The outermost electrons of the carbon atom may have different configurations. When carbon is interacting with four hydrogen atoms, it may be assumed that all the molecular bonding orbitals must be alike. Thus, there have to be 4 electrons of similar energies.

$$C^{12} : (1s)^2(2s)^1(2p_x)^1(2p_y)^1(2p_z)^1 \qquad (2.7)$$

Since it is assumed that 2s and 2p electrons are at about the same energy level, the wave functions of all these four electrons may considerably mix and four new hybrid electronic orbitals are formed in sp hybridization. The wave functions of these orbitals are,

$$\Psi(t_1) = a\Psi(2s) + b\Psi(2p_x) \qquad (2.8)$$

Here, a and b are the electron density distribution constants for each wave function. There will be four of these wave functions representing four electrons and they are directed as far away as possible from each other in a tetrahedral arrangement. When a carbon atom interacts with non-equivalent atoms, it may experience another form of hybridization, trigonal hybridization, in which only $(2s)$, $(2p_x)$ and $(2p_y)$ electrons are involved. Consider,

$$C^{12} : (1s)^2(2s)^1(2p_x)^1(2p_y)^1(2p_z)^1 \qquad (2.9)$$

The molecular orbital functions for these electrons may be represented by

$$\Psi(h_1) = a_i\Psi(2s) + b_i\Psi(2p_x) \qquad (2.10)$$

Note that the electrons in $2p_z$ orbital are free to form resonating orbitals. It is also noteworthy that the angle between planar orbitals is 120 degrees.

In covalent bonds the electron density may be uniformly distributed about the bonding orbitals, but if any of the atoms is electro-negative or positive, there will be an uneven distribution of electron density. The result will be a bond possessing contribution of both covalent and ionic bond characteristics. Consider the bonding of H and Cl atoms. Expressing electrons with the usual Lewis diagram dots, there will be resonance between the two structures,

$$H : Cl \Leftrightarrow H^+ + Cl^- \tag{2.11}$$

The bonding wave function may be expressed as

$$\Psi = \Psi(covalent) + a\Psi(ionic) \tag{2.12}$$

The relative amount of ionization depends on the type of atom to which hydrogen is bound. Some examples are given in Table 2.3.

TABLE 2.3:   Relative Ionization

Molecule	HF	HCl	HBr	HI
% ionic character	60	17	11	0.5

Obviously the degree of ionic character depends upon the electro-negativity of the atoms. This trend may be understood by considering the distribution of electrons in these atoms. In the following table the electronic distribution is described for some atoms in the order of decreasing electro-negativity.

TABLE 2.4:   Electronic Distribution

F	$(1s)^2(2s)^2(2p)^5$
Cl	$(1s)^2(2s)^2(2p)^6(3s)^2(3p)^5$
Br	$(1s)^2(2s)^2(2p)^6(3s)^2(3p)^6(3d)^{10}(4s)^2(4p)^5$
I	$K - L - M(4s)^2(4p)^6(4d)^{10}(5s)^2(5p)^5$

The tendency of these atoms is to attract additional electrons to satisfy the sixth electron position in the p orbitals. The strength of this attraction, i.e., electro-negativity, will be greater in the atoms with smaller number of electrons, which will shield the effect of the nuclear charge. A similar table of electron distribution may be made for atoms common in biological systems.

Both covalent and ionic bonds discussed here have bond energies as large as hundreds of kcal/mol in vacuum. While the strength of covalent bonds remains relatively invariant over different environmental conditions, the strength

TABLE 2.5: Electron
Distribution for Biological Atoms

H	$(1s)^1$
C	$(1s)^2(2s)^2(2p)^2$
N	$(1s)^2(2s)^2(2p)^3$
O	$(1s)^2(2s)^2(2p)^4$
P	$K - (3s)^2(3p)^3$

of ionic bonds is not, and is sensitive to the environment's dielectric constant in particular. In the covalent bonds, bond distances and bond angles are well defined. Depending upon the number of electrons shared by the atoms, there will be single, double and triple bonds. The rotation of atoms around the latter two bonds is restricted.

2.4.3 Free Radicals

During chemical reactions a covalent bond may be cleaved in various ways. When the bond breaks, the shared electron pair may remain with any one of the atoms or it may become distributed evenly between the two. The former situation is referred to as the heterolytic cleavage while the latter is called a homolytic cleavage. In the homolytic cleavage case, the atoms possessing unpaired electrons are called free radicals and are highly reactive. The unpaired electron in a free radical possesses a magnetic moment and the radiation of microwave frequency induces transitions between magnetic energy levels of these particles. The study of these magnetic energy levels and transitions between them through Electron Spin Resonance (ESR) is a valuable source of information about the system.

When the nature of the bond becomes closer to that of the covalent bond, the influence of other atoms of the environment diminishes. But replacement of an electron acceptor with other molecules forces the donor molecule to form hydrogen bonds with the new acceptor molecules. While ionic and covalent bonds are the major bonds involved in constructing the architectural framework of macromolecules, there are other types of bonds with significantly smaller energies that contribute simply because there are so many of them. Below we discuss some of the other bonds that are commonplace in biological systems.

2.4.4 Van der Waals Forces

Van der Waals Forces refer to the interactions between neutral molecules. Energies involved in Van der Waals interactions are in the range of 0.5 to 4.5 kJ/mol, i.e., 100 times weaker than a covalent bond and 5-10 times weaker than a hydrogen bond. They are, however, responsible for producing inter-

TABLE 2.6: Data for Different Types of Bonds

Donor-Acceptor	Molecules	Distance (Å)	Energy(Kcal/mole)
- O H ... O <	H_2O	2.76	4.5
- O H ...O = C <	CH_3COOH	2.8	8.2
> N H ... O = C<	Peptides	2.9	
> N H ... N <	Nucleotides	3.1	1.3
> N H ... F -	NH_4F	2.63	5.0
> N H ... S <		3.7	

molecular cohesion in liquids and some solids. They fall into the following categories:(a) **Keesom forces** are due to two permanent dipoles that are attractive and favor parallel arrangement along the axis joining the two molecules. (b) **Debye forces** are forces between a dipolar molecule that induces a dipole moment of a non-polar molecule leading to an attraction between the two. (c) Finally, an important category of such forces is called the **London or dispersive forces** and they arise due to the induced dipole moment formation in non-polar molecules, e.g., group IV halides such as CF_4 and CCl_4. Interestingly, these London interactions may have bond energies comparable to the dipole-dipole energy. These forces are the main source of attraction between non-polar or weakly polar chemical species. The London dispersion interaction appears to arise from non-statistical time dependent distribution of charges on the molecules, thus creating temporary induced dipoles. The exact nature of this interaction is not well understood, but the magnitude of its potential energy follows $1/r^6$.

In order to describe dipolar interactions physically, we introduce some background information (see also [25]). As we have seen in ionic interactions, a dipole interacts with ionic charges and other dipoles in its vicinity. Consider a dipole due to two opposite charges q and $-q$ separated by a distance d.

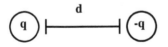

FIGURE 2.23: Dipole due to two opposite charges.

The dipole moment, p, for this dipole is,

$$p = qd \tag{2.13}$$

If the magnitude of the elementary charges is $e = 1.6 \times 10^{-19}C$ and the distance separating them is: $d = 1\text{Å} = 10^{-10}m$, then the dipole moment is: $p = 1.6 \times 10^{-29}Cm$. Since $3.33 \times 10^{-30}Cm = 1$ *debye*, $p = 4.8$ *debye*.

Molecular dipole moments are typically on the order of one elementary charge e times one angstrom, i.e., about 10^{-29} Cm. The dipole moment of ethanol is, for instance, $5.6 \times 10^{-30} Cm$. Water is polar but it has a more complex charge distribution than ethanol, namely a negatively charged oxygen with two positively charged hydrogen atoms sticking out making an angle of $104°$ with each other. The dipole moment of a water molecule amounts to 1.85 *debye* resulting from a 1.6 *debye* per each O-H bond, the two being at 104 degrees to each other. In addition to permanent dipole moments present in numerous bonds, induced dipole moments arise in others when they are subjected to external electric fields. Larger symmetrical molecules such as methane may contain bonds with dipole moments that cancel out due to their symmetrical arrangement.

A dipole may interact with ions and other dipoles with some definable potential energy as shown below. Consider first a dipole interacting with a charge q_1,

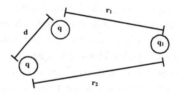

FIGURE 2.24:   Dipole interacting with a charge q_1.

The potential V between the dipole and the charge results from the interactions between charges $-q_2, -q_1$, and q_2 and q_1. Thus,

$$V = \frac{q_1(-q_2)}{4\pi\varepsilon\varepsilon_0 r_1} + \frac{q_1 q_2}{4\pi\varepsilon\varepsilon_0 r_2} \qquad (2.14)$$

Since,

$$r_1^2 = (\frac{d}{2})^2 + r^2 - 2(\frac{d}{2})r \cos q \qquad (2.15)$$

and assuming that d is significantly smaller than r, we find that

$$r_1 = r - (\frac{d}{2}) \cos q \qquad (2.16)$$

and

$$r_2 = r + (\frac{d}{2}) \cos q \qquad (2.17)$$

Substituting these relations to the previous equation and making further approximations, we obtain

$$V = -\left(\frac{q_1 q_2}{4\pi\varepsilon\varepsilon_0 r_2}\right) d \cos q \qquad (2.18)$$

The potential energy between two dipole moments, m_1 and m_2, may be calculated in a similar manner as

$$V = \frac{-2m_1 m_2}{4\pi\varepsilon\varepsilon_0 r^3} \qquad (2.19)$$

when $d << r$. It is noteworthy that the potential energy is inversely proportional to r^3. For two permanent dipoles at an arbitrary angle to each other, their interaction is given by the familiar dipole-dipole formula and is illustrated schematically in Fig. 2.25, which shows the force between two permanent electric dipoles p_1 and p_2 separated by a distance r in (a) general orientation, (b) oriented along the lines joining them in the same sense, and c) oriented perpendicular to the line joining them in the same sense.

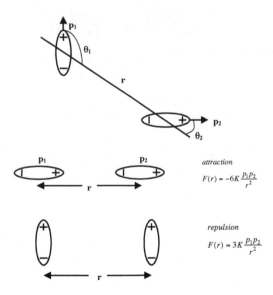

FIGURE 2.25: Interaction of two permanent dipoles.

Molecules and bulk matter, which have diffusive charges, can have a dipole moment induced in them if they are exposed to an external electric field, such as that due to a point charge. The induced dipole moment, m_{ind}, is given by

$$m_{ind} = a\left(\frac{q_1}{4\pi\varepsilon\varepsilon_0 r^2}\right) \tag{2.20}$$

where a is the polarizability of the medium.

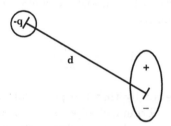

FIGURE 2.26: Induced dipole moment.

The potential energy between the induced dipole moment and the charge acting on it is

$$V = -\left(\frac{m_{ind}q_1}{4\pi\varepsilon\varepsilon_0 r^2}\right) = -\left(\frac{aq_1^2}{4\pi\varepsilon\varepsilon_0^2 r^4}\right) \tag{2.21}$$

It should be noted that the influence of distance is now $1/r^4$. Namely, the magnitude of potential energy diminishes more rapidly as the distance between the two increases.

Important biomolecules such as the carbohydrates are neither charged nor polar. They interact through a net induced dipole- induced dipole attraction, resulting from the distortion of electron distributions. This is called a Van der Waals interaction and it is always attractive. Two molecules separated by a distance R attract each other by an attractive potential energy:

$$U_{VW}(R) = -\frac{A}{R^6} \tag{2.22}$$

The parameter A, whose units are Nm^7, depends on the chemical structure of the two molecules but its value is generally on the order of $10^{-77} Nm^7$. The following table gives values of A between identical molecules in water:

Note that the table goes from small to big molecules and that A increases with molecule size. When two atoms or molecules get so close that the electron clouds overlap, then the electrostatic interaction changes character and becomes highly repulsive, a force which is known as steric repulsion.

The short range repulsive force can be included by adding a repulsive term to the Van der Waals potential giving

$$U_{VW}(R) = -\frac{A}{R^6} + \frac{B}{R^{12}} \tag{2.23}$$

We call this combination formula the Lennard-Jones potential. The empirical values of the coefficients A and B for some very important atomic pairs are listed in Table 2.8:

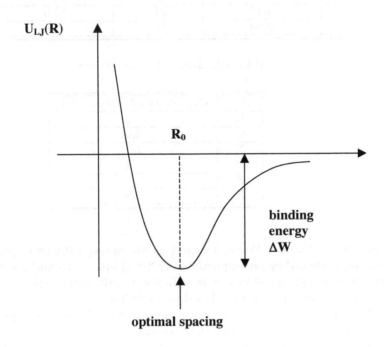

FIGURE 2.27: The Lennard-Jones potential.

The depth of the potential energy curve $U_{LJ}(R)$ at its minimum point, which is about 1 kcal/mol, is the strength of the Van der Waals bond. Van der Waals bonds are somewhat weaker than polar and hydrogen bonds but they play a similar role in molecular biology, namely as a source of weak,

TABLE 2.7: The Coefficient A
for Different Molecules

Molecule	A in 10^{-77} Nm^7
Water	2.1
Benzol	4.3
Phenol	6.5
Diphenylanilin	14.4

TABLE 2.8: Empirical Values of Coefficients A and B

Bond	R_{min} (angstroms)	A (kcal/mol)	B(kcal/mol) 10^{-4}
C-C	3.40	370	28.6
C-N	3.25	366	21.6
C-O	3.22	667	20.0
N-N	3.10	363	16.0
C-H	2.90	128	3.3

TABLE 2.9: Van der Waals Radii

Atom	Van der Waals Radius ($\breve{A}$)
H	1.2
C	2.0
N	1.5
O	1.4
S	1.85
P	1.9

temporary links. Van der Waals interactions are non-specific and apply to any two atoms when they are approximately 3-4 Å apart. To find a contact distance between any two atoms, it is sufficient to add their respective Van der Waals radii, some of which are listed in Table 2.9.

The Van der Waals bond energy for a pair of atoms is on the order of 1 kcal/mol which is only slightly more than the average thermal energy at room temperature (0.6 kcal/mol); consequently they are important only when a sufficient number of these bonds are made simultaneously.

2.5 The Hydrogen Bonds and Hydrophobic Interactions

The fact that oxygen and nitrogen atoms possess two or one lone electron pairs, respectively, bears an enormous impact on the spatial organization of the four most important classes of biomolecules: lipids, hydrocarbons, proteins and nucleic acids. The negative lone electron pairs distributed in tetrahedrally oriented σ-orbitals attract positively polarized hydrogen atoms of other molecules, and result in the formation of *hydrogen bonds*. The energy of hydrogen bonds is comparable to the potential barrier heights for rotations around single covalent bonds, i.e., it ranges between 10 and 20 kJ/mol. The processes of reorganization of the system of hydrogen bonds, their breakage and restoration possibly in new locations, take place at a rate that is comparable to conformational transitions and for all intents and purposes are indistinguishable from them.

Of all non-covalent bonds, the hydrogen bond is one of the most common. It can be either inter-molecular or intra-molecular. Dimerization of formic acid is a good example of the inter-molecular type of hydrogen bond.

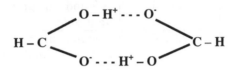

FIGURE 2.28: Dimerization of formic acid.

Hydrogen is normally placed between relatively weakly electronegative atoms (N, C, F). Note that despite its large electronegativity, Cl does not form a hydrogen bond. The hydrogen bond originates from the interaction between a dipole and a charge. However, their closeness results in quantum mechanical interactions in which electrons are shared on both electronegative atoms. Thus the bond distances and angles among the atoms are well defined. In small molecules, the atoms are oriented linearly, due primarily to the repulsive force of the electronegative atoms, but in larger molecules, there may be sufficient strain to bend them.

Hydrogen bonds are formed between a hydrogen nucleus of one molecule and the unbounded electrons of the electronegative atom of another molecule. Generally, the hydrogen bond is characterized as a proton shared by two electron pairs. The formation of H bonds is an exothermic process while the opposite is true of the dissociation of H bonds. The strengths of these bonds are on the order of a fraction of an eV, i.e., hydrogen bonds are some 20 times weaker than ionic or covalent bonds. For example, the energy of the hydrogen bond in the C-H...N complex is approximately 0.2 eV, while that in the C-H...O complex is 0.25eV, H-F...H (10 kcal/mol, H-O...H 7.2 kcal/mol, H-N...H 2kcal/mol). Typically the energy of an H-bond is in the range of 2-8 kcal/mol, i.e., it is intermediate between van der Waals and covalent bond strengths. The bond lengths vary between 2.3 and 3.4 Å as shown below.

All molecules can be classified into four groups as follows:

- Molecules with one or more donor groups and no acceptor groups (e.g., acetylenes)

- Molecules with one or more acceptor groups but no donor groups (e.g., ketones, ethers, esters, aromatics)

- Molecules with both donor and acceptor groups (alcohols, water, phenols)

TABLE 2.10: Bond Lengths for Various Bond Types

Bond Type	Bond Length	Location
O-H...N	2.78 Å	
O-H...O	2.72 Å	- often found in proteins as either hydroxyl-hydroxyl or hydroxyl-carbonyl bonds
N-H...N	3.07 Å	- found in proteins as amide-imidazole
N-H...O	2.93 Å	- found in proteins as amide-hydroxyl and amide-carbonyl bonds
N-H...S	3.40 Å	- found in proteins as amide-sulphur
C-H...O	3.23 Å	

- Molecules with neither donor nor acceptor groups (e.g., saturated hydrocarbons)

Only molecules in the last group are unable to make H-bonds. Molecules in the third group can self-associate by H-bonding with themselves. There are two types of H-bonds: intramolecular and intermolecular. The latter results in association of molecules and is affected by steric factors. It leads to a change in the number, mass, shape and electronic structure of the molecular systems involved.

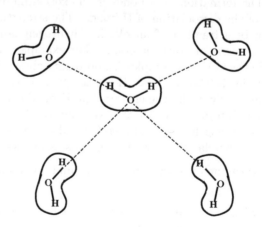

FIGURE 2.29: Hydration of H_2O by H_2O: hydrogen bond.

Interactions between biomolecules take place often via H-bonds that are very important in biological systems due to the abundance of water in the cell. One key role of H-bonds is in the protein folding phenomenon that will be discussed later. Water constitutes large percentage of the surface of the Earth and the bodies of living organisms. In fact, the presence of water in liquid and ice forms makes the Earth unique. Water has a high heat of fusion (80 cal/g) and a large dielectric constant (80 at $10°C$). Having a high number of hydrogen bonds also causes water's high surface tension. Meanwhile, the density of water reaches a maximum of about 1.0 g/ml at $4°C$. All these properties make water an ideal solvent for both organic and inorganic substances, in addition to making it an excellent supporting medium for living organisms. Water molecules carry a dipole moment and interact through the dipolar electrical fields they produce. As a consequence, the electric field of a charged protein polarizes the water molecules in its neighborhood, which in turn affects the electric field. In addition, both blood and the fluid inside living cells (cytoplasm) are electrolytes. The free ions of electrolytes further modify the electrical field. Since the dielectric constant of water is close to 80, the electrostatic interactions between ions in a water environment are reduced accordingly. This has important consequences, one of which is the fact that the ionic binding energy is reduced by a factor of 80. For instance, instead of $8.8 \times 10^{-19} J$ for the binding energy of NaCl in a vacuum, in water we find only about $10^{-20} J$ or 5 kcal/mol, using d = 2.4Å and $q_1 = -q_2 = -e$. The ionic binding energy in water is only about ten times the thermal energy $k_B T$. In many cases, biochemical reactions involve ATP to ADP conversion as a source of energy that releases about 7.3 kcal/mole. That's enough to break an ionic bond in water but not sufficient to break an ionic bond in vacuum.

Water is unique among the family of $R - H_n$ compounds. Relative to this family of compounds, water has high or a large: melting point, boiling point, heat of vaporization, specific heat capacity, heat of vaporization, and surface tension, etc. All these properties may be attributed to the effects of hydrogen bonding.

TABLE 2.11:   Properties of Various $R - H_n$ Compounds

Compounds	Molec. weight	T_{melt} ($°C$)	T_{boil} ($°C$)	H_{vap} (cal/g)	Heat Cap. (cal/g)
CH_4	16.04	-182	-162	149.0	
NH_3	17.03	-78	-33	326.4	1.23
H_2O	18.02	0	100	661.9	1.00
H_2S	34.08	-86	-61	160.4	
Organic Liq.				100	0.5

An important distance that relates the magnitude of the electrostatic energy in water to the thermal energy $k_B T$ is the so-called Bjerrum length [32]. The Bjerrum length l_B is the distance at which the electrostatic energy of two monovalent ions equals $k_B T$. It follows that

$$\frac{1}{4\pi\varepsilon\varepsilon_0}\frac{e^2}{l_B} = k_B T \qquad (2.24)$$

For water at room temperature, the Bjerrum length is about 7.2Å. The electrostatic interaction energy between two monovalent ions is thus negligible compared to the thermal energy if their separation is large compared to the Bjerrum length.

While the ionic (or polar) bond is greatly reduced by surrounding water molecules, this is not so true for the covalent bond. The strength of the covalent bond in water is about the same as in vacuum because the covalent bond involves the motion of an electron between two closely adjacent nuclei such that it is unaffected by the surrounding water molecules. As a result there are two very different energy scales available to biomolecules: (a) covalent bonds with energies in the 50-100 kcal/mole range and (b) ionic bonds in the 5 kcal/mole range. Covalent bonds cannot be broken by thermal fluctuations and the covalent bonding energy is large compared to the chemical energy stored in one ATP to ADP conversion. Hence, covalent bonds are used as permanent links keeping macromolecules together. The much weaker polar bonds can be broken by an ATP to ADP conversion, by thermal fluctuations and also by light adsorption. They are useful for temporary links, e.g., to control the folding of proteins or the mutual binding of proteins, or the binding of proteins to DNA.

From the study on heat of sublimation it has been discovered that the energy required to remove water molecules from ice is 12.2 kcal/mol. Similarly, from studies with methane, which has no hydrogen bonding, 2 kcal/mol is required. This energy may be from the contribution of London dispersion forces or translational enthalpy. Thus the energy contributed by hydrogen bonds should equal to 10 kcal/mol for water or 5 kcal/mol for each hydrogen bond. The accepted value now is closer to 4.5 kcal/mol.

Water is the simplest system in which the structure and dynamics of hydrogen bonds plays a fundamental role. Each oxygen atom, in addition to covalent bonds within the same water molecule, can form two hydrogen bonds with other water molecules (Fig. 2.30). The structure of crystallized ice with completely saturated hydrogen bonds is highly ordered and has low density. To determine which structure is more stable at a given temperature T, one uses the condition of free energy F minimum

$$F = E - TS \qquad (2.25)$$

in which the internal energy E favoring order competes with the entropy S that favors disorder. The higher the density of the system, the lower the

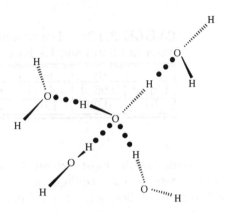

FIGURE 2.30:   System of hydrogen bonds in water.

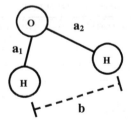

FIGURE 2.31:   The structure of water in liquid and ice form.

internal energy E, and the more hydrogen bonds formed the lower the entropy S, hence the greater the value of the term $-TS$.

The structure of water in liquid and ice form has been summarized in Table 2.12.

TABLE 2.12:   The Structure of Water in Liquid and Ice Form

	a_1	a_2	b
Liquid	0.96 Å	1.41 Å	104° 27
Ice	0.99	1.61	109°

When each water molecule has the most extended and straight hydrogen bond arrangement around tetrahedrally configured oxygen, the structure of ice has been obtained using crystallography, and a bulk density of $0.9 g/cm^3$ is predicted. As the temperature or pressure is changed this conformation collapses and the density increases. When the temperature is increased, the collapse of hydrogen bonds is compensated by random thermal motion of water molecules and the bulk volume tends to increase. Thus, the highest density of water happens to take place at $4°C$. It is also possible that this density reflects the formation of a clathrate with one water molecule in its center. Some structural orderliness is maintained even at temperatures as high as at $83°C$. The source of this orderliness, after long-standing disagreements in this field, is now assumed to come from quickly exchanging formation of water between the cluster and surrounding free forms. The size of cluster at different temperatures has also been estimated (see Table 2.13).

TABLE 2.13:   Size of Cluster at Different Temperatures

Temp (°C)	Number of molecules per cluster		Mole fraction	
	H_2O	D_2O	H_2O	D_2O
0	91		0.24	
4	117		0.23	
10	72	97	0.27	0.25
20	75	72	0.29	0.27
30	47	56	0.32	0.30
40	38	44	0.34	0.32
50	32	35	0.36	0.34
60	28	29	0.38	0.36

It is evident that formation of ice distorts both the bond distances and the bond angle, b. The molecular orbital configuration is tetrahedral around each

O atom.

$$O : (1s)^2(2s)^2(2p)^4 H : (1s)^1 \tag{2.26}$$

Eight electrons, six total in (2s) and (2p) orbitals of oxygen and two in the (1s) orbital of each of the two hydrogen atoms, circulate around the oxygen atom. There are two pairs of lone pair electrons near the oxygen making the region possess a $2 \times (-0.17e)$ charge. On the other hand, two pairs of bonding electrons lose electrons to the electronegative oxygen, and consequently near each hydrogen, there will be a charge of $+0.17e$. Thus, water molecules can become both hydrogen donors and acceptors for the hydrogen bond formation.

Thermodynamically, the interactions between non-polar groups and polar groups are unfavored. Thus, non-polar groups in water tend to stick together to reduce the energy of mixing. This unfavorable mixing, however, introduces interesting thermodynamic changes in water, e.g. a decrease in entropy in response to the more ordered form and change in its molar heat capacity arising from the increased hydrogen bonds. For example, Table 2.14 illustrates the change of thermodynamic parameters when an organic molecule is introduced into water:

TABLE 2.14: Change in Thermodynamic Parameters with Introduction of an Organic Molecule into Water

	Methane into water	Benzene into water
G	3.0 kcal/mol	4.3 kcal/mol
S	-18 cal/mol deg	-14 cal/mol deg
H	-2.6 kcal/mol	0 - 0.6 kcal/mol

It is the entropic term, rather than the enthalpic term, that makes the mixing of a non-polar molecule in water non-spontaneous.

As early as 1945, Fran and Evance postulated formation of icebergs around hydrophobes leading to the suggestion by Klotz and Luborsky (see [175]) that the iceberg around a molecule may contribute to its stabilization. Nemethy and Scheraga [239] proposed that a water molecule can have a co-ordination of 5, i.e., four hydrogen bonds and one hydrophobic pocket.

The potential to form four hydrogen bonds may be interrupted when the water molecule is mixed in with other water molecules or a non-polar solvent. The number of hydrogen bonds for a water molecule may vary, as shown below in Table 2.15.

The system of hydrogen bonds in water determines its key biologically significant properties: high specific heat capacity, high latent heat of melting, high electric permittivity and its specific dynamical properties which will be mentioned later. This system also determines the water solubility properties of various molecules. Molecules capable of forming hydrogen bonds with wa-

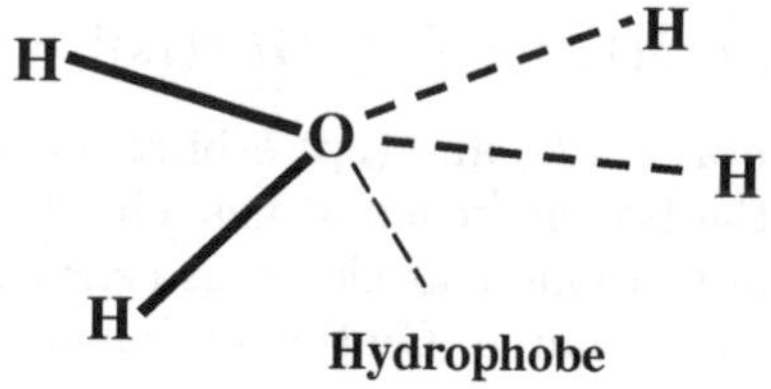

Hydrophobe

FIGURE 2.32:   Water molecule with four H bonds and one hydrophobic pocket.

TABLE 2.15:   Number of Hydrogen Bonds for a Water Molecule

Number of H-bonds	in pure water	in non-polar solution
4	23%	43%
3	20	6
2	4	18
1	23	12
0	29	21

ter, e.g., sugars or alcohols, increase the disorder in the system of hydrogen bonds and hence increase entropy leading to a free energy reduction. Thus the process of solvation is thermodynamically favorable in this case. Molecules which do not form hydrogen bonds but are electrically charged or at least have a high dipole moment reduce the electrostatic energy of the system and their solvation is also thermodynamically favorable in spite of introducing order into the hydrogen bond distribution. Molecules which do not form hydrogen bonds and are uncharged and non-polar, e.g., long carbohydrate chains or aromatic rings, only order the water environment (Fig. 2.22) but do not contribute to the system's energy and hence are not water soluble. We refer to them as *hydrophobic* (fearing water), in contrast to *hydrophilic* soluble particles.

2.5.1   Polysaccharides

In carbohydrates, the ring-oxygen atom is an H-bond acceptor and each hydroxyl group is associated with two H-bonds, one a donor and the other an acceptor. A single sugar molecule, referred to as a monosaccharide, has the chemical formula $(CH_2O)_n$ where n=5 or 6. Sugars are the primary source of energy in animal biological systems. If not used up as an energy source, they are linked up into long, branched chains called polysaccharides. H bonds with distances between 2.7 and 3.0 $\mathring{A}$ are common in disaccharides and polysaccharides. The order of individual sugars in a long chain is not precisely determined or encoded. Likewise, their sequence is not passed on genetically. Specialized enzymes repeatedly add sugars to the end of a growing chain. Each type of sugar is linked to a different type of enzyme. Polysaccharides are neutral and unreactive hence ideally suited as energy storage molecules. An additional role sugars play is as a structural element, for example as the cellulose fibres in wood and, of course, as a component of the DNA molecule. Examples of disaccharides include sucrose and lactose. Glucose is a common food molecule in cells and can be stored in a polysaccharide form glycogen in animal cells or as starch in plant cells. Cellulose fibrils aggregate into microfibrils held to one another by extensive hydrogen bonds. The core of a microfibril possesses a perfect 3D crystalline lattice.

2.6 Amphipatic Molecules in Water Environments

Interesting physical phenomena take place when *amphipathic* molecules, containing both a hydrophilic and a hydrophobic moiety, are placed in an aqueous environment [310]. Among biological systems such amphipatathic molecules are phospholipids (see Fig. 2.7b) and glycolipids. Their polar *head* is hydrophilic and two hydrocarbon *tails* are hydrophobic. In water environment, to minimize free energy, amphipathic molecules spontaneously organize into spherical micelles (at low concentrations) or *bilayers* (at higher concentrations). These structures allow the molecules to have their hydrophilic head groups facing outside and hydrophobic tails inside (Fig. 2.33). Bilayers close up to form three-dimensional *vesicles* which, when sufficiently large, contain a hierarchy of internal vesicles and are referred to as *liposomes*. When the amount of solvent becomes too small, liposomes unfold to form *lamelles*, in which subsequent bilayers are placed parallel to each other (Fig. 2.33).

Vesicles and liposomes are *lyophilic* (they like solvents) *colloidal* 100 - 500 nm diameter particles and when dispersed in water form a spatially inhomogeneous *dispersive structure* called *sol*. A decrease of water content results in an unfolding process of liposomes into lamelles and a transition of sol into a spatially homogeneous *lamellar phase* (Fig. 2.34). In the lamellar structure, molecular orientation is ordered but their spatial arrangement is not. The lamellar phase is a special example of a liquid crystal. Removing more water molecules from the systems causes excessive proximity of the polar or similarly charged hydrophilic groups which destabilizes the lamellar phase. Smaller surfaces of the head groups in contact form a *cubic phase* or *hexagonal phase*, in which bilayers are again replaced by spherical (in the first case) or cylindrical (in the second case) micelles (see Figs. 2.34 and 2.35). In both the sol phase and in the three liquid crystalline phases individual molecules

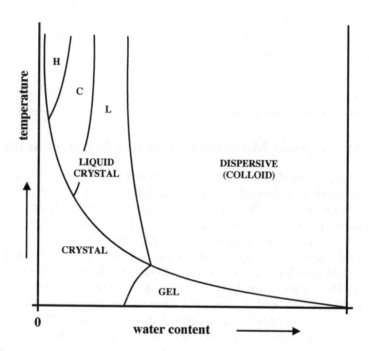

FIGURE 2.33:   Structures created by amphipathic molecules in water environment: from micelles through vesicles to liposomes and lamelles. White circles denote hydrophilic head groups while line segments denote one or two hydrophobic tails.

retain their translational degrees of freedom. In the sol phase, additional motion of entire micelles, vesicles or liposomes is allowed. A lowering of the temperature causes a freezing of translational degrees of freedom, and sol undergoes a phase transition into gel and a liquid crystal into a normal crystal (see Fig. 2.34).

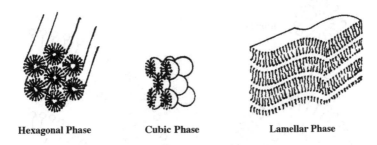

Hexagonal Phase   **Cubic Phase**   **Lamellar Phase**

FIGURE 2.34: Sketch of the phase diagram of an amphipathic molecule-water system as a funciton of temperature and water content. L, C and H stand for lamellar, cubic and hexagonal phase, respectively (see Fig. 2.36).

For biological systems, the sol phase is optimal since it contains vesicles or liposomes of appropriate sizes. The stability and mechanical properties of a bilayer at physiological temperatures are controlled by an appropriate chemical composition of phospholipids and glycolipids and, additionally, by cholesterol (see Fig. 2.2) which stiffens the bilayer. Such a lipid bilayer integrated with the built-in protein molecules forms a *biological membrane*. Proteins of the membrane can perform various functions such as those of immobilized enzymes, channels, pumps, receptors, signal generators and constituents of a membrane skeleton. An important immunological role is played by carbohydrates of glycolipids and glycoproteins on the external face of the cytoplasmic membrane.

2.6.1 Fatty Acids

One of the principal constituents of the plasma and other membranes is the family of double-chain lipids in the form of carboxylic acids with a chemical formula RCOOH, where R stands for a hydrocarbon chain. Most of the fatty acids found in cell membranes have only a few double bonds and these are restricted to the cis and trans forms, usually the former type. The presence of cis double bonds creates a kink in the chain.

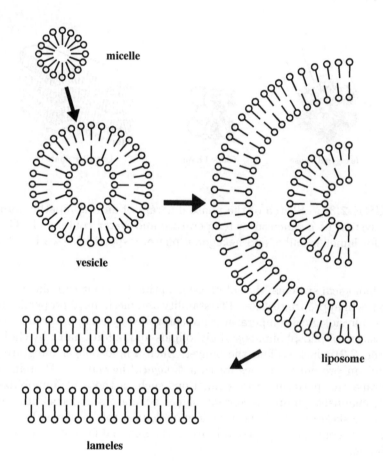

FIGURE 2.35: Three possible liquid crystalline phases of the amphipathic molecule-water system.

2.6.2   Lipids

Lipids, commonly known as fats and oils, are composed of a small hydrophilic head carrying two or three long hydrophobic tails. When placed in water, lipids spontaneously aggregate into a droplet to hide their tails pointing them inwards with their heads facing outwardly. In a cell, lipids form bilayers where tails pack side by side and heads are located on the two faces making contact with the surrounding water. Cells use lipid bilayers as their primary boundaries that seal in their interior macromolecules such as nucleic acids and proteins. Lipid bilayers are dynamic and fluid since their individual component molecules have freedom to rotate about their long axes while the tails can fluctuate in their positions and twist. Lipid molecules may also migrate in two dimensions within the layer. In cell, membranes contain protein molecules embedded into a lipid bilayer. The role of membrane proteins is to facilitate transport, signalling and molecular recognition.

The key components of cell membranes are phospholipids obtained as a result of a reaction between a fatty acid and glycerol according to the formula:

$$R_1 O'H + R_2 CO''OH - > R_2 CO''O'R_1 + H_2 O \qquad (2.27)$$

where the primes on O indicate the origin of the various oxygen atoms.

While three fatty acids are linked to the glycerol backbone in a triglyceride, only two are linked in a phospholipid. The OH group in glycerol is replaced with a phosphate group that connects to the polar head group. The polar head groups come in several varieties, the most common include choline, glycerol, ethanolamine and serine. The generic phospholipid has a hydrophilic polar head group and two non-polar hydrocarbon chains which are hydrophobic. Thus when placed in an aqueous environment, the head groups are attracted to water while the chains avoid water exhibiting so-called amphiphilic behaviour. Depending on the solutions' composition, various mesoscopic structures can form in water/amphiphile mixtures. These include a lipid bilayer and micelles.

2.7   Structure of Proteins

There are 20 'canonical' amino acids; some others occur but very rarely [304]. Their side chains are listed in Fig. 2.36. Three amino acids play a special role in the spatial structure of proteins. We refer to them as *structural* amino acids. Let us begin with *glycine* (abbreviation: Gly or G) whose side chain reduced to a single hydrogen atom. The small side chain results in little steric hindrance, and enables almost free rotation of the main chain about the neighboring $C^\alpha - N$ and $C^\alpha - C$ bonds. The site at which glycine is situated behaves like a ball-joint in the polypeptide chain. *Proline* (Pro or P), on the contrary, being in fact an imino acid, attaches the main chain through the two bonds (Fig. 2.27b) which makes it locally rigid and looped in a defined manner. The third structural amino acid, *cysteine* (Cys or C), is able to form, after oxidation, relatively strong covalent bonds with each other (disulfide bonds, Fig. 2.27c) and pieces together distant sites of a single polypeptide chain as well as separate chains composing the protein macromolecule.

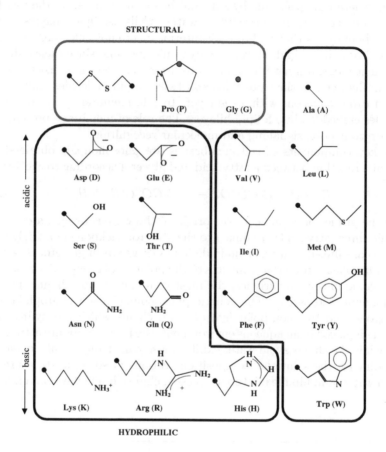

FIGURE 2.36:   Side chains of 20 canonical amino acids and their partition into three main groups. The dot denotes central carbon atom C^α .

As Table 2.16 illustrates, some amino acids are charged and hydrophilic. Others are rich in carbon and strongly hydrophobic. Some are rigid and others are flexible. Some are chemically reactive and others are relatively inert. Each amino acid has the general chemical formula: $RCH(NH_2)COOH$ where R stands for a residue. By giving up a water molecule at each connection, amino acids can form a valence bonded chain in the form of a polypeptide or a protein. For example, myoglobin has the chemical formula $C_{738}H_{1166}FeN_{203}O_{208}S_2$.

TABLE 2.16: The twenty naturally occurring amino acids are listed along with their one and three letter codes as well as whether they have a polar character and whether they are charged. *Histidine has a pKa of 6.5 and consequently will be protonated and positively charged should the pH of the cytoplasm dip below this value.

1 Letter	3 Letter	Amino Acid	Polar	Charge	R-group
A	Ala	Alanine	no	no	methyl
C	Cys	Cysteine	yes	no	Thiol
D	Asp	Aspartic Acid	no	-e	
E	Glu	Glutamic Acid	no	-e	
F	Phe	Phenylalanine	no	no	aromatic, Toluene
G	Gly	Glycine	no	no	proton
H	His	Histidine	no	+e*	
I	Ile	Isoleucine	no	no	sec-Butyl
K	Lys	Lysine	no	+e	
L	Leu	Leucine	no	no	Isobutyl
M	Met	Methionine	no	no	
N	Asn	Asparagine	yes	no	amide connected at alpha-C
P	Pro	Proline	no	no	Cyclopentyl Amine
Q	Gln	Glutamine	yes	no	amide connected at beta-C
R	Arg	Arginine	no	+e	
S	Ser	Serine	yes	no	1 Alcohol
T	Thr	Threonine	yes	no	2 Alcohol
V	Val	Valine	no	no	Isopropyl
W	Trp	Tryptophan	no	no	
Y	Tyr	Tyrosine	yes	no	aromatic, para-methylphenol

All amino acids are built from a central alpha-carbon bonded to four different groups listed below.

The alpha on the carbon indicates the priority position from which the numbering follows for all subordinate groups. The four substituents connected to C^α are:

- the alpha proton or hydrogen (-H),
- a side chain -R that gives rise to the chemical variety of the amino acids,
- the carboxylic acid functional group (-COOH),
- and the amino functional group ($-NH_2$).

The α carbon is the asymmetric center of the molecule for all 20 amino acids except glycine, which has only a proton as its side chain. The configuration about the α carbon center must be the L-isomer for proteins synthesized on the ribosome.

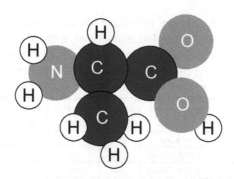

FIGURE 2.37: The four substituents connected to C^α .

All but one of the amino acids can be written in the form $NH_2 - RCH - COOH$. In the 20 amino acids designed by nature, what varies is the side chain. Some side chains are hydrophilic while others are hydrophobic. Since these side chains stick out from the backbone of the molecule, they help determine the properties of the protein made from them. Most naturally-occurring amino acids are the L- form, whereas synthetically-produced amino acids give a 50:50 mixture. Since these molecules are mirror images of each other, it is impossible to rotate one molecule to make it look like the other.

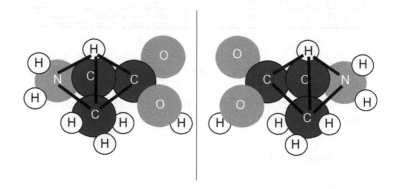

FIGURE 2.38: Naturally-occurring and synthetically-produced amino acids.

The side chains exhibit a wide chemical variety and can be grouped into three categories: non-polar, uncharged polar, and charged polar. The simplest amino acid is glycine. Alanine, valine, leucine, isoleucine, and proline are amino acids whose side chains are entirely aliphatic. Alanine has a methyl group as its side chain. Valine has two methyl groups connected to the β-carbon, and this residue is said to be β-branched. Leucine has one more carbon atom in the side chain than valine, so that two methyl groups are attached to the C_γ. Leucine and isoleucine are isomers whose only difference in structure is the position of the methyl groups. Isoleucine is a β-branched amino acid and has a second asymmetric center at the β-carbon. Proline contains an aliphatic side chain that is covalently bonded to the nitrogen atom of the α-amino group, forming an imide bond and leading to a constrained 5-membered ring.

Depending on the nature of the side chain groups, three classes of amino acids can be distinguished:

- Amino acids with both H-bond donor and acceptor groups: histidine, lysine, aspartic acid, glutamic acid, serine, threonine, tyroine and hydroxyproline.

- Amino acids with donor groups only: arginine, tryptophan and cysteine.

- Amino acids with acceptor groups only: asparagine and glutamine.

Side chains that are generally non-polar have low solubility in water because they can engage only in van der Waals interactions with water molecules. The remaining amino acids contain hetero-atoms in their side chains, including threonine, asparagine, glutamine, tyrosine and tryptophan. Serine, threonine and tyrosine contain hydroxyl groups so they can function as both hydrogen bond donors and acceptors, and threonine also has a methyl group, making it β- branched. The benzene ring in tyrosine permits stabilization of the anionic phenolate form upon loss of the hydroxyl proton, which has a pKa near 10. Serine and threonine cannot be deprotonated at ordinary pH values. Asparagine and glutamine side chains are relatively polar in that they can both donate and accept hydrogen bonds. The nitrogen and proton of the tryptophan indole side chain can also take part in hydrogen bond interactions.

Another group of polar residues can have a full, formal charge depending on pH. These include lysine, arginine, histidine, aspartic acid, glutamic acid and cysteine. Lysine and arginine are two basic amino acids that can have a positive charge at the end of their side chain. The lysine α-amino group has a pKa value near 10. Histidine is another basic residue with its side chain organized into a closed ring structure that contains two nitrogen atoms. One of these nitrogens already has a proton on it, but the other one has an available position that can take up an extra proton and form a positively charged histidine group with a pKa of about 6. Aspartic acid and glutamine acid differ only in the number of methylene $(-CH_2-)$ groups in the side chain, with one and two methylene groups, respectively. Their carboxylate groups are extremely polar, can both donate and accept hydrogen bonds, and have pKa values near 4.5.

A group of three amino acids that all have aromatic side chains are phenylalanine, tryptophan and tyrosine. The aromatic ring of phenylalanine is like that of benzene or tolylene. It is very hydrophobic, and chemically reactive under extreme conditions although its ring electrons are readily polarized. The side chain of histidine is arguably considered aromatic.

Finally, the α-amino and α-carboxylate groups of amino acids can also ionize, with pKa's of 6.8-7.9 and 3.5-4.3, respectively, for the aliphatic amino acids. Note that in an aqueous environment, both amino and carboxylate groups are charged. Nearby charged side chains can alter the pKa's of these groups. Each amino acid incorporated into a polypeptide chain is referred to as a residue. Thus, only the amino- and carboxyl- terminal residues possess available α-amino and α-carboxylate groups, respectively.

In proteins, the alpha-carboxyl group of one amino acid is joined to the alpha-carboxyl group of another amino acid by a peptide bond (also called an amide bond). To form a protein, its amino acids engage peptide bonds involving the group $O = C - N - H$ through dehydration synthesis. Some proteins contain only one polypeptide chain while others, such as hemoglobin, contain several polypeptide chains twisted together. The sequence of amino acids in each polypeptide or protein is unique giving it its own, unique 3-D shape or native conformation.

2.7.1   The Polypeptide Chains

In order to form a polymeric chain, amino acids are condensed with one another through dehydration synthesis. This reaction occurs when water is lost between the carboxylic functional group of one amino acid and the amino functional group of the next to form a C-N bond. These polymerization reactions are not spontaneous. However, they can be arranged to occur through the energy-driven action of the ribosome. Ribosomes are complexes of proteins and RNA that translate a gene sequence in the form of mRNA into a protein sequence. The 20 amino acids listed above are encoded by the genes and incorporated by the ribosomal machinery during protein synthesis.

A polypeptide chain (see Fig. 2.39) consists of a regularly repeating part called the main chain, and a variable part, consisting of distinctive side chains. The main chain is called the backbone. In some cases even a change in one amino acid in the sequence can alter the protein's ability to function. For example, sickle cell anemia is caused by a change in only one nucleotide in the DNA sequence that causes just one amino acid in one subunit of the hemoglobin polypeptide to be changed. Due to this alteration, the whole red blood cell ends up being deformed and unable to carry oxygen properly. Note that this mutation also makes the carrier less susceptible to malaria.

The peptide unit is rigid and planar. The hydrogen of the amino group is

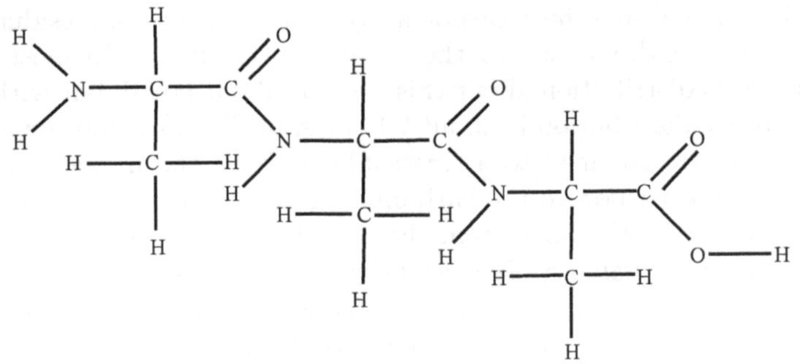

FIGURE 2.39: A polypeptide chain.

almost always trans to the oxygen of the carbonyl group with no freedom of rotation about the bond between the carbonyl carbon atom and the nitrogen atom of the peptide unit due to the double bond present whose length is 1.32 Å. However, there is a lot of rotational freedom about the single bonds involving the carbonyl carbon group and an alpha- carbon and a nitrogen. As a polypeptide chain forms, it twists and bends into its native conformation.

The peptide bond between two amino acids is a special case of an amide bond flanked on both side by α-carbon atoms. Peptide bond angles and lengths are well known from many direct observations of protein and peptide structures. These bond lengths and angles reflect the distribution of electrons between atoms due to differences in polarity of the atoms, and the hybridization of their bonding orbitals. The two more electronegative atoms, O and N, can bear partial negative charges, and the two less electronegative atoms, C and H, can bear partial positive charges. The peptide group consisting of these four atoms can be thought of as a resonance structure. Thus, the peptide bond has partial double bond character, accounting for its intermediate bond length. The formation of a peptide bond is an endoergic reaction and needs free energy which is usually released in GTP (guanine triphosphate) hydrolysis.

Like any double bond, rotation about the peptide bond angle ω is restricted, with an energy barrier of 3 kcal/mole between cis and trans forms. These two isomers are defined by the path of the polypeptide chain across the bond. While there is restricted rotation about the peptide bond, there is free rotation about the four bonds to the α-carbon of each residue. Two of these rotations are of particular relevance for the structure of the polypeptide backbone. Successive α-carbons in the chain (i, i+1) are on the same side of the bond in the cis isomer as opposed to the staggered conformation of the trans isomer. For all amino acids but proline, the cis configuration is greatly unfavoured because of steric hindrance between adjacent side chains. Ring closure in the proline side chain draws the β-carbon away from the preceding residue, leading to lower steric hindrance across the X-pro peptide bond. In most residues, the trans to cis distribution about this bond is about 90-10, but with proline, the trans to cis distribution is about 70-30. Also like any other double bond, certain atoms are confined to a single plane about the peptide bond. The bond from the α-carbon to the carbonyl carbon of that residue is given the name ψ . Similarly, the bond from the α-carbon to the amino group of that residue is given the name ϕ. Because C^α is one of the six planar atoms of the peptide group, rotation about ψ or ϕ flanking C^α rotates the entire plane of the peptide group. Since the entire plane rotates on either side of C^α, certain values of the angles cannot be achieved due to steric occlusion. The allowed regions of ϕ, ψ space differ for each amino acid because of the restriction due to C^α and its substituents.

2.7.2   Proteins

Proteins are the most abundant macromolecules in the living cells, present in their membranes, cytosol, cell organelles and chromosomes. They constitute about 50 percent of the dry mass of a living cell. Like nucleic acids, proteins are long, unbranched molecular chains. Proteins are polymers made up of combinations involving any of the 20 amino acids, each of which is a monomer. Proteins are the most structurally complex macromolecules known. Each type of protein has its own unique structure and function. Human hair, skin and fingernails consist largely of a protein called keratin. Fibroin is the protein that makes up the silk of cocoons and spider webs. Sclerotin forms the external skeletons of insects. Structural proteins such as collagen and elastin and adhesive proteins such as fibronectin and laminin are abundant in cells. The collagen protein is really a family of 15 or more different types, most of which bundle together to form fibrils whose diameters range between 10 and 300 nm. Collagen is the most abundant protein in mammals. Elastin preferably forms sheets and cross-linked networks giving the connective tissue its elasticity. Adhesive proteins are found in the matrix and their purpose is to bind the matrix to cells (fibronectin) and to bind connective tissue to the lamina (laminin). These are but a few examples of 'structural proteins'. Functional proteins, on the other hand, play crucial roles in dynamic processes within living cells. For example, insulin regulates the metabolism of glucose; rhodopsin converts incoming light in the retina of the eye to ionic signals in the optic nerve. Actin and myosin generate forces in muscle cells. Dynein is an energy-producing component of the cytoskeleton. Kinesin carries intracellular transport. Functional proteins catalyze nearly all the metabolic processes in the cell. Each functional protein has a specific conformation that is best suited for its function.

The three-dimensional shape adopted by a protein in water is crucial to its function. For example, enzymes are proteins which, by positioning a target molecule next to the proper reactive group, regulate chemical reactions usually increasing their rates by large factors. Structural proteins, on the other hand, have binding properties that promote the formation of large aggregates for strengthening and support roles in the cell. Proteins also act as carriers of cargo for directed transport inside the cell. Many hormones are proteins sending molecular messages between specific locations in the body. Proteins, due to their diversity and functionality, perform most of the typical tasks of the cell. While a typical bacterium may have a few thousand kinds of proteins, the human body has tens of thousands of protein kinds. All structural and functional properties of proteins derive from the chemical properties of the polypeptide chain.

Successive amino acids are numbered starting from the N terminus. Most proteins are heteropolymeric (i.e., they contain most or all the different amino acids). Only rarely do regions of proteins consist of sequences composed of just a few amino acids. Any region of a typical protein will therefore have a chemically heterogeneous environment. Proteins that assume a quaternary structure often have multiple active sites, and therefore identical chemical environments (i.e., hemoglobin or myoglobin).

The levels of structural organization in globular proteins can be listed as follows:

Primary (amino acid) sequence → secondary structure → supersecondary

structure → domain → globular protein → aggregate

In a nutshell, the *primary structure* is defined as the linear sequence of amino acids in a polypeptide chain. It represents the covalent backbone and the linear sequence of the amino acid residues in the peptide chain. One of the peptide chains is called the N-terminal and the other the C-terminal. The *secondary structure* is characterized by spatial organization in terms of alpha helices, beta sheets, random coils, triple helices and an assortment of other motifs usually in localized regions of the macromolecule. The secondary structure refers to certain regular geometric figures of the chain. It is entirely due to the H-bonding between the $C = O$ and $N - H$ groups of the peptide bonds. The beta pleated sheets are more common in structural proteins while in globular proteins they are interconnected by alpha-helix segments. Such combinations of secondary structures are called supersecondary structures. Further folding of polypeptide chains results in the formation of *tertiary structures*. Tertiary structure results from long-range contacts within the chain. The latter ones are stabilized by a combination of van der Waals, hydrogen, electrostatic and disulfide bonds. The *quaternary structure* is the organization of protein subunits, or two or more independent polypeptide chains. Quaternary structure represents aggregates in which a number of globular proteins are bound by non-covalent interactions or disulphide bonds spontaneously forming oligomers of varying sizes ranging from dimmers to dozens of monomers. Below we provide a detailed description of protein's structural elements.

The sequence in which the individual amino acids of definite side chains

occur along the main chain or chains of protein is strictly fixed and genetically determined. The concept of a primary structure is identical to that of
a chemical structure (constitution). This determines completely the system
of covalent bonds in the protein macromolecule and also includes information about disulfide bridges which form spontaneously (or can also be formed
enzymatically) during protein synthesis on the ribosome. However, after the
completion of this process, the majority of proteins are subjected to additional
enzymatic chemical modifications. These modifications may involve cutting
off some fragments of the main chain, methylation of some charged side chains
(leading to charge neutralization), phosphorylation of side chains ending with
hydroxyl groups (which endows the originally neutral chains with negative
charge). Enzymatic proteins often form permanent bonds with different prosthetic groups (co-enzymes), while the external proteins of cell membranes
undergo extensive glycolization. The process of biosynthesis of protein ends
with a spontaneous formation of supermolecular structures, e.g., multienzy-
matic complexes.

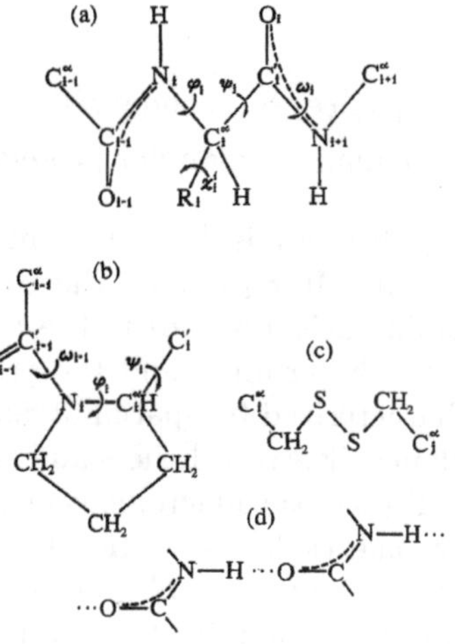

FIGURE 2.40:   Chemical structure of a protein.

The spatial structure of the protein macromolecule with a given primary
structure is determined by local conformations of the main and side chains
as well as a system of non-covalent (secondary) bonds [241]. The commonly
accepted notation for dihedral angles in the polypeptide chain is given in
Fig. 2.40a. In the absence of any steric hindrances, rotation about the single
bonds $C-C$, $C-N$, $C-O$ and $C-S$ allows three stable local conformations
(one *trans* and two *gauche*, cf. Fig. 2.14a) unless it is a rotation of planar

carboxylic, amidic or aromatic rings which allows two stable conformations. Individual conformations can differ in energy by a few kJ/mol and are separated by barriers whose heights range from 10 to 20 kJ/mol originating from van der Waals and electrostatic multipolar interactions (Fig. 2.14b).

Only rotations of covalent bonds distant from C^α atoms, being the branching sites of the polypeptide chain, are usually unhindered, obviously if one does not take into account the long-range excluded volume effects. This applies to angles χ_1, χ_2, χ_3, ..., describing internal conformations of longer side chains if not branched, as well as to the angle ω describing the local conformation of peptide bonds. The peptide bond C-N, as discussed in Section 2.3, is in part a double bond; thus it may exist as two configurational isomers trans and cis (Fig. 2.18). The potential energy barrier height is determined by delocalization energy and approaches 80 kJ/mol. This is the reason why peptide bonds occur as a single local configurational isomer, almost exclusively as the trans form. The only exception to the rule is the peptide bond neighboring a proline residue (Fig. 2.19b) in which the part of the polar structure is smaller thereby lowering the *trans-* to *cis* -transformation barrier to 50 kJ/mol.

Note that there is an energy barrier between the α-helical region of the ϕ , ψ space and the β-strand region. Thus, direct conversion between α- and β- structures is restricted even though most residues are allowed in both regions. Since the restrictions on ϕ,ψ space arise in part from steric hindrance between side chains and the backbone, this same steric hindrance is the origin of α and β secondary structures. There is no sequence dependence on the steric restrictions of the α and β space because ϕ , ψ restrictions arise within each residue rather than between residues. However, a sequence of residues that all have similar allowed ϕ, ψ space can give rise to a chain segment that forms α or β structures. Thus, these secondary structures owe their formation to both backbone and side chain steric restrictions. The allowed regions of ϕ ,ψ space for each amino acid are displayed on Ramachandran plots [284]. The allowed regions can be defined in terms of the energetic cost that must be paid to enter a disallowed region, or in terms of the limiting so-called hard-sphere boundary when atoms clash. For β-branched residues the restrictions are severe, and only a small fraction of ϕ, ψ space is allowed. Valine and isoleucine have access to only about 5% of all ϕ, ψ space. However, all residues have access to at least part of the most favorable regions of ϕ, ψ space in the upper and lower left of the plot. It turns out that these two regions correspond to combinations of ϕ and ψ angles that characterize the two common regular secondary structures that can be adopted by the polypeptide backbone, the α-helix and the β-strand. The presence of bulky side chains of all amino acids always affects steric hindrance for angles ϕ and ψ making the corresponding local conformations mutually dependent. In Figs. 2.41a and 2.41b the region sterically allowed for angles ϕ and ψ (the Ramachandran map) is shown, separately, for glycine and an arbitrary side chain different from glycine. Glycine behaves really like a ball-joint whereas for the remaining side chains three distinct, much smaller sterically allowed regions are seen,

corresponding to three distinct, already cooperative conformations of the pair
(ϕ, ψ). Strictly speaking, because also rotation about the angle χ_1 is hindered
for some values of angles ϕ and ψ, the whole triple (ϕ, ψ, χ_1) should be treated
as a single unit. A number of sterically allowed local conformations for the
angle χ_1 is shown on a background of the Ramachandran map in Fig. 2.41c.

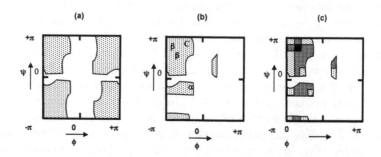

FIGURE 2.41: Sterically allowed regions for angles ϕ and ψ (the Ramachandran map) for glycine (a) and an arbitrary side chain different from
glycine (b). Configurations of α-helix, parallel and antiparallel β-sheets, and
triple helix of collagen (C) are indicated. (c) Number of sterically allowed conformations for the angle χ_1 according to the value of ϕ and ψ. For residues
with one side carbon atom third from C_α, blackened, heavy shaded and light
shaded regions correspond to three, two and one conformation, respectively.
For residues with two side carbon atoms third from C^α (threonine, valine and
isoleucine), they correspond to three, two and no conformations, respectively.
For alanine three rotational conformations are allowed everywhere.

The secondary bonds in proteins are mainly hydrogen bonds. Their energy,
10 to 20 kJ/mol, is comparable to potential barriers for transitions between
the local conformations of protein chains. It is reasonable to distinguish hydrogen bonds within the polypeptide backbone from those formed by the side
chains. The hydrogen bonds within the main chain (chains) link nitrogen and
oxygen atoms of distinct peptide groups (Fig. 2.19d). A regular pattern in
which hydrogen bonds are organized in the polypeptide backbone without reference to the side chain types is traditionally known as a secondary structure
of protein. Two main secondary structures are distinguished. The *α-helix*
is a helical arrangement of a single polypeptide chain with hydrogen bonds
between each carbonyl group at position i and the peptide amine at position
$i+4$ (Fig. 2.42a). The *helix* structure looks like a spring. The most common
shape is a right handed α-helix defined by the repeat length of 3.6 amino acid
residues and a rise of 5.4 A per turn. Thus residues $(i+3)$ and $(i+4)$ are
closest to residue (i) in the helix. The pitch and dimensions of the helix also

bring the peptide dipole moments of successive residues into proximity such that their opposite charges neutralize each other substantially in the middle of a helix. The pitch and dimensions of the helix also bring the amide proton of residue $(i+3)$ or $(i+4)$ into proximity to the carbonyl oxygen of residue (i) such that a hydrogen bond can form. All peptide group hydrogen bond donors and acceptors are satisfied in the central part of the helical segment, but not at the ends. While structural evidence indicates that these hydrogen bonds are highly populated in helical segments of proteins, their contribution to helix stability is less clear since donors and acceptors would be satisfied by hydrogen bonding to water in non-helical structures. The structure of an alpha helix was first deduced by Pauling and Corey before it was observed in x-ray data. The collagen helix is a third periodic structure and it is specialized to provide a high tensile strength of collagen.

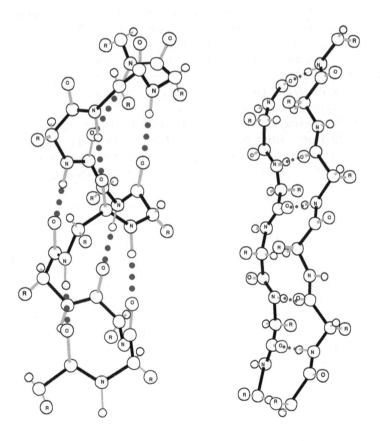

FIGURE 2.42: The structure of an α-helix (a) and an antiparallel β-pleated sheet.

β strands are the other regular secondary structure that proteins form. These are extended structures in which successive peptide dipole moments alternate direction along the chain. Because it is an extended structure, ϕ, ψ steric hindrance is reduced in the β strand, and the β region of ϕ, ψ space is larger than the α region. Two or more strand segments can pair by hydrogen bonding and dipolar interactions to form a β-sheet. Unlike helical segments, all peptide group hydrogen bond donors and acceptors are satisfied not within but between β-strand segments. Thus individual β-strands do not have an independent existence. Also unlike a helical segment, adjacent strands of a sheet can come from sequentially distant segments of the chain; rarely, this can occur even within one strand of a sheet.

In β-pleated sheets, hydrogen bonds are also evident but they link different strands of a polypeptide chain placed in a parallel or anti-parallel manner (Fig. 2.42b for anti-parallel β-pleated sheet). β sheets can consist of either parallel or anti-parallel strands, or a mixture of the two. In purely anti-parallel sheets, segments that are sequentially next to each other in the primary structure often form adjacent strands. Thus, while a β-strand is a secondary structure element because of its geometrically regular features, a β-sheet can be thought of as a tertiary structural feature because it is intrinsically non-local. This example illustrates that the distinctions between secondary and tertiary structural features are not entirely clear.

Yet another secondary structure, the collagen *triple superhelix structure*, can be formed only from polypeptide chains with every third amino acid being glycine. However, even when forming a hairpin from contiguous chain segments, linearly distant residues are brought into proximity at the N- and C- terminal ends of the hairpin.

Assuming all peptide bonds occur in *trans* conformations ($\omega_i = 180°$), the values of the remaining dihedral backbone angles are $\phi_i = 57°$, $\psi_i = 47°$ for the α-helix, $\phi_i = 119°$, $\psi_i = 113°$ for the parallel β-pleated sheet and $\phi_i = 139°$, $\psi_i = 135°$ for the anti-parallel β-pleated sheet. These positions lie within two distinct sterically allowed regions of the angles ϕ and ψ and coincide almost exactly with two main minima of the potential energy as a function of these angles though it is energetically profitable for planar β-sheet structures to be slightly twisted. It follows that systems of hydrogen bonds in the α-helix and β-sheet structures considerably stabilize local backbone conformations and this is why fragments of both secondary structures are so abundant in protein macromolecules.

So-called *turn structures* are also classified as secondary structural elements, but unlike helices and strands, they do not have a repeating, regular geometry. Rather, they can have well-defined spatial dispositions defined by certain values of ϕ and ψ - angles that often require specific residue types and/or sequences, as well as fixed hydrogen bonding patterns. Most turns are local in the primary structure, but omega loops can have a large number of intervening residues lacking defined geometries, with the turn being defined by the conformations of residues that form the constriction that gives this turn its name (Ω). Turns are essential for allowing the polypeptide chain to fold back upon itself to form tertiary interactions. Such interactions are generally long-range, and result in compaction of the protein into a globular, often approximately spherical form. The turn regions are generally located on the outside of the globular structure, with helices and/or sheets forming its core. Turns on the surface of proteins have a wide range of dynamics, from quite mobile in cases where they form few interactions with the underlying protein surface to quite fixed due to extensive tertiary contacts.

One of the things that helps determine the native conformation of a protein is the side chains of all the amino acids involved. Since some amino acid side chains are hydrophobic while others are hydrophilic, all the hydrophobic side chains try to concentrate in the center of the molecule, away from the aqueous environment, while the hydrophilic side chains are attracted to the outside of the molecule in order to lower the energy by making contact with the liquid environment. Additionally, some of the hydrophilic side chains have groups of atoms attached that make them acidic, while others have groups attached that make them basic. Side chains with acidic ends are attracted to side chains with basic ends, and can form ionic bonds. Thus, the side chains interacting with each other help to hold the protein in its native conformation. The side chains project outwardly from both α-helical and β-strand structures, and are available for interactions with other surfaces through hydrophobic contacts and various kinds of bonding interactions to form the tertiary structure. In a helix, the side chains project radially outward, and in a strand successive side chains project alternately up and down. Rotational direction about bonds in the side chains project alternately up and down. Rotations about bonds in the side chain are also restricted. The same steric hindrance that limits the backbone conformation also limits the side chain conformation about the C_α-C_β bond to preferred rotamers defined by rotation angle χ_1 and are restricted by side chain packing in the tertiary structure.

If secondary structural elements result from steric restrictions in ϕ, ψ space, it is less obvious why *tertiary structures* form. Proteins with highly organized tertiary structures generally have a well-developed core of hydrophobic residues contributed from most or all of the secondary structure elements in the chain. Thus, secondary and tertiary structures are intimately interconnected. These buried residues do not form merely a liquid-like oily interior, but rather are usually well-packed, with extensive rotamer restrictions. In aqueous solvents, the hydrophobic effects drive the chain toward compaction to relieve unfavourable solvation of these exposed side chains, but compaction and internal organization are entropically costly due to loss of chain flexibility, and these competing effects nearly cancel each other energetically. This argument could be extended to explain the sensitivity of enzymes to temperature denaturing (i.e., tolerances of only 2-3 degrees C) due to competing internal energy and entropic effects. On the other hand, upon compaction, bonding interactions with solvent molecules are replaced by intra-molecular partners, with a likely net gain in favourable energetic contributions due to several effects. Hydrogen bonding is favoured within secondary structures because these are partially pre-organized by ϕ, ψ restrictions into configurations that permit bonding at little additional entropic cost. In the case of β-sheet information, an additional favourable effect may result when two β-strands are brought into contact.

The view presented above suggests that protein secondary and tertiary structures are interdependent and it is possible that this interdependence is the molecular origin of the protein structural stability, whereby protein secondary and tertiary structures are lost in an all-or-none manner upon changes in the environment that disfavor the folded state, such as higher temperature or solvent additives.

The highest level of protein structural organization is the *quaternary structure*. The subunits that associate may be identical or not, and their organization may or may not be symmetric. In general, quaternary structure results from association of independent tertiary structural units through surface interactions, such as the formation of the hemoglobin tetramer from myoglobin-like monomers. Thus, subunit assembly is a necessary step in tertiary structure formation. The co-dependence of tertiary and quaternary structures parallels the co-dependence between secondary and tertiary structures, and suggest that the distinction among these levels of the protein structure organizational hierarchy may be misplaced. In the past, there have been comparisons made between crystal solids and proteins; however, as Fig. 2.43 shows, such analogies are not well-founded on the physical properties of proteins.

2.7.3   The Process of Protein Folding

The spontaneous act of protein folding is remarkable in that the complex motion of the protein's structural elements, i.e., amino acids, transfers the one-dimensional sequence of data into a three-dimensional object and the process is a result of thermally activated Brownian motion. This process resembles a phase transition such as crumpling but is much more complex and hence harder to model. The net stabilization of the native state conformation of a protein results from the balance of large forces that favor both folding and unfolding. For a hypothetical protein, the energetic contributions to native state stabilization may be distributed as shown in Table 2.17. Thus the net free energy of folding is 10 kcal/mol.

The overall tendency of the side chains of non-polar amino acid residues is to locate in the interior of a protein an effect known as the hydrophobic effect. This is a combined effect of the hydrophobicity of some side chains of amino acids and of the structure of water. Water molecules form hydrogen bonds among themselves and form hexagonal structures on the surface of proteins. The order of hydrophobicity and hydrophilicity for the amino acids is:

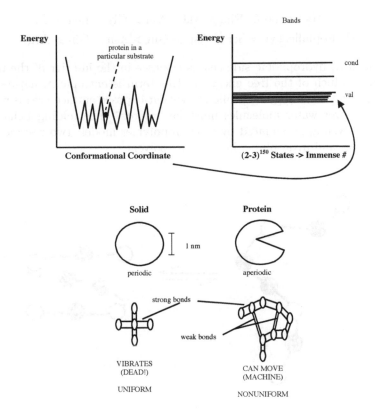

FIGURE 2.43: A schematic comparison between a solid and a protein.

TABLE 2.17: Energy Contributions to Native State Stabilization

Folding	Unfolding
hydrophobic collapse	conformational entropy
intramolecular H-bonding	H-bonding to solvent (water)
van der Waals interactions	
Contributions to the Free Energy of the reaction, U→N:	
-200 kcal/mole	+190 kcal/mole

Hydrophobic:Phe > Ala > Val > Gly > Leu > Cys
Hydrophilic:Tyr > Ser > Asp > Glu > Asn > Gln > Arg

Consequently, hydrophobic side chains coalesce in the interior of the protein structure. Much of the free energy in this term is entropic in nature. The molecular explanation goes as follows: solvent-exposed surface area is reduced and thus fewer water molecules must be ordered. The folding behavior of proteins is well approximated by a heteropolymer model: two residue types, hydrophobic and hydrophilic, in a "poor" solvent.

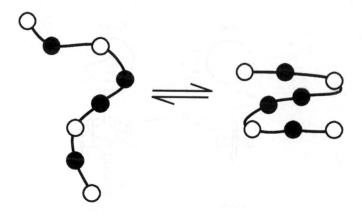

FIGURE 2.44: A heteropolymer model composed of two residue types, hydrophobic and hydrophilic, in a "poor" solvent.

Note that conformational entropy is meant to define the entropy associated with the multiplicity of conformational states of the disordered polypeptide chain.

To consider how long it would take a 100-residue polypeptide to complete a random search for the native state we assume there are three possible conformational states for each residue and that it takes 10^{-13} s to interconvert between each state. For the 100-residue polypeptide, there are 3^{100} [or 5×10^{47}] possible conformational states. Assuming a single unique native state conformation, it would take $(5 \times 10^{47})(10^{-13})$ s $= 5 \times 10^{34}$ sec or 1.6×10^{27} yr. This absurd result often referred as the *Levinthal Paradox* clearly shows that protein folding does not occur by random search.

A protein can assume a very large number of related but distinct conformational states whose distribution can be described by a so-called energy

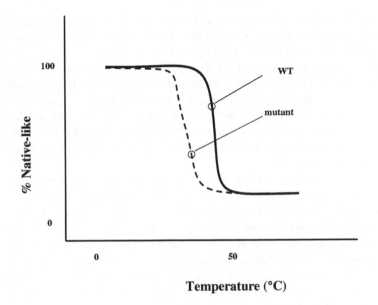

FIGURE 2.45: The percentage of native-like character as temperature rises.

landscape. Each substate is a valley in a (3N-6) dimensional hyperspace where N is the number of atoms forming the protein. The energy barriers between different substates range from about 0.2 kJ/mol to about 70 kJ/mol. It is believed that this energy landscape has a hierarchical structure with different tiers having widely separated barrier heights. Recent in vitro studies demonstrate that protein folding follows paths characterized by retention of partially correct intermediates [241].

The efficiency of protein folding can be compromised by aggregation of folding intermediates that have exposed hydrophobic surfaces. Such aggregates are essentially irreversible. Molecular chaperone proteins bind reversibly to folding intermediates, prevent aggregation and promote their passage down the productive folding path. Molecular chaperones also are known as heat shock proteins because they are synthesized in much greater amounts by cells subjected to a wide variety of stresses, including elevated temperature and oxidative stress.

Specialized enzymes catalyze key steps in protein folding; for example peptidylprolyl cis/trans isomerase (PPIase) catalyzes the cis/trans isomerization of X-pro peptide bonds, where X is any amino acid. Steric interference between neighboring residues in the amino acid sequence is an example of a local interaction. The conformational behavior of an unfolded polypeptide is dominated by local interactions, whereas long-range interactions stabilize secondary and tertiary structure in native proteins. For alpha helices, stabilizing H-bonds occur through residues only a few positions away. In an unfolded polypeptide, a given peptide bond is much more likely to be trans (96%) than cis (4%); except in the case of an X-pro peptide bond where the likelihood of the cis conformer is significantly increased (20%). This relatively high probability of cis presents a significant barrier to the folding of a protein for which the native secondary and tertiary structure demands the trans conformation. Futhermore, some proteins contain a cis X-pro bond in the native structure, so molecules with a trans X-pro bond must interconvert to cis in order for the protein to complete folding.

Denaturation refers to when a protein loses its native conformation. There are several possible agents that can denature proteins including solvent temperature, pH, salt concentration, or hydrophobicity. In a hydrophobic solvent, the amino acids with hydrophobic side chains try to move to the outside of the molecule, and all those with hydrophilic side chains cluster in the centre of the molecule. If a protein remains water-soluble when denatured, it can return to its native conformation if placed back into a "normal" environment. Proteins are chemically and biologically stable unless they are deliberately depolymerized. The decomposition of a polypeptide chain into individual amino acids can be facilitated by hydrolytic enzymes. The thermal melting behavior of protein structure indicates a phase transition to the unfolded state, a characteristic of highly cooperative systems. A typical point mutation (a change of one base in the codon triplet) causes little or no change in the structure of the protein, but reduces protein stability as indicated by the lower melting temperature. This is not a general case. Effects of point mutations depend on the DNA reading frame, the type of substitution (if any) that occurs, and the importance of the particular residue to the structure of the protein, i.e., sickle cell anemia is the result of a point mutation (as previously mentioned). Many amino acid substitutions in protein sequences have no detectable effect on protein function. Mutations causing functional defects usually occur in an enzyme active site or interfere with intermolecular assembly (see Fig. 2.43).

2.7.4　Electrophoresis of Proteins

The term electrophoresis describes the transport of charged macromolecules, e.g., proteins or DNA, through an electrolyte due to an applied voltage. It is a basic laboratory tool in biochemistry to segregate macromolecules according to size. Electrophoresis separates molecules according to their charge:size ratio. In biological applications electrophoresis is used to characterize various objects ranging from bacterial cells to viruses to globular proteins to DNA and has found important clinical applications such as the separation of proteins in blood plasma in order to detect abnormal patterns, and blood composition. The net charge on a protein may vary from -100 to +100 elementary charges and is largely determined by the pH-value of the solution in which a given protein is suspended. A charged protein with a total electric charge Q

is subjected to the electric force $F_e = QE$ where E is the applied electric field and the drag force $F_D = -6\pi\eta av$ where η is the viscosity of the medium, v the velocity of the migrating protein and a its radius. There is an additional complication due to the presence of counter-ions in the solution which shield the protein by surrounding it as if a cloud were created. The shielding "cloud" moves in the opposite direction under the influence of the field E. Without detailed exposition we simply summarize the results as follows. The mobility of a protein (velocity v divided by the electric field intensity) is proportional to the net charge on the protein, inversely proportional to the viscosity, the radius squared and the ionic strength of the solution. For example, a protein of charge $10e$ placed in an aqueous solution of ionic strength 0.1 mol/L subjected to an electric field of 1 V/cm will attain a drift velocity of approximately 10^{-4} cm/s requiring a time of 3 hours for a displacement over 1 cm. If some components possess no net charge but a net dipole moment, they, too, can be separated by an analogous method called dielectrophoresis in which an electric field gradient is used to provide a force on a dipole moment.

2.7.5 Protein Interaction with Environment

It should be emphasized that a structured water medium surrounds all of the protein within a cell. This is the solvation problem and is a result of the protein's hydrophilic exterior forming hydrogen bonds with water. In addition, ions are formed within the cytoplasm which may influence conduction properties through their interaction with side chains of proteins or even their localization within charged pockets of the protein. Consequently, the study of dry proteins does not give an accurate description of their properties.

The interaction of metal ions and proteins is of great importance in the biophysical chemistry of proteins. Such interactions are involved in conformational changes of macromolecular structure and are the basis of one class of gated ion-channels. They may also affect physical properties and the chemical reactivity of biomolecules *in vivo*. These interactions between metal ions may also have an important role in genetic expression, metalo-enzyme activity and metal-nucleic acid processes. There are several possible modes of association. The most common of all such reactions is simply the association of a proton. This drastically changes the electrostatics of the problem so that at low pH when carbonyl groups are protonated, the structure of the protein differs somewhat from its high pH form. Rather than changing the band gap between valence and conduction states, such interactions with metal ions may allow the protein to donate electrons to the metal ions thereby creating holes in the conduction band. Protons (H^+) seem to accept electrons much more readily than do lithium, sodium and larger metallic ions. This property led to some studies of electronic conduction where protons act like a ferry but carry electrons in only one direction [28].

There are only three significant types of charge carriers: electrons, heavy ions and protons. Electrons have a high mobility and there are many familiar materials capable of supporting their conduction such as metals. Heavy ions can easily carry a charge but are immobile within solids. Protons fill the intermediate case, and while they present many experimental difficulties, they may exist within both liquid and solid media. Although three orders of magnitude heavier than an electron, the bare proton or H^+ ion is much smaller than other possible charge carriers simply because it does not carry any electrons. As a result, it has a higher mobility than other ion species. The suspicion of its involvement in conduction is that it may hop along the outside of a protein becoming localized at negatively charged pockets of the protein. Potential barriers are required at electron injection and ejection points to prevent short-circuiting. Protons might also act as carriers of negatively charged ions or electrons. The injected protons in this mechanism of conduction would come from chemical reactions, redox reactions or a proton reservoir at high chemical potential such as the interior of a mitochondrion.

2.7.6   Electron Transfer in Proteins

There is a body of evidence to explain the mechanism of electron transfer in proteins, through a series of redox centers incorporated into protein structure. This allows for the directionality of electron transfer over distances between 0.3-3.0 nm. The transfer is modeled by electron or nuclear tunneling but we should not exclude the possibility of electron transfer along a chain of conjugated orbitals.

The protein-electron carrier system consists of a protein, possibly localized within a lipid bilayer, which has a single redox center. The redox centers are usually prosthetic groups containing molecules of non-proteinous origin which have conjugated orbital systems and often incorporate metal ions. The function of the prosthetic groups is multi-fold. It ensures the fixation of charges and dipoles in its micro-surroundings and catalyzes electron transfer. Secondly, the orientation of the prosthetic group relative to other proteins may enhance recognition of the redox center. Thirdly, the prosthetic group may act to influence the electronic state of associated proteins or vice versa, and result in a degree of isolation from the polar solvent. Finally, it may alter the carrying concentration through its oxidation or reduction thereby controlling electron transport.

Recent studies on electron transfer through protein redox networks have been performed by Ichinose et al., and are in general agreement that electron transfer in biological systems occurs via a redox scheme but is critical of electron tunneling over distances of 3.0-7.0 nm as some have reported. Instead, Ichinose et al. propose that protein is an insulator at physiological temperature and that electron transport is mediated by protons. In this model, an electron is ferried along the protein backbone by protons such that there is no net movement of protons as they hop back and forth, but they carry electrons in one direction only, thereby resulting in conduction.

One of the assets of the redox scheme is that the reduction and oxidation of the protein can act as a dopant just as in traditional semiconductors, if electron transfer is inhibited. This structural flexibility may be employed by the cell to tune its electrical properties for a desired biological function.

Proteins and glasses are similar in at least one aspect: both possess a large number of iso-energetic minima leading to a huge number of thermally assisted transitions between these degenerate minimum energy states. The frequency of these transitions is lowered with the reduction of the temperature but some motions occur even at 100 mK. There is no clear consensus on the usefulness of this energy landscape in physiology. It should also be mentioned that protein conformational dynamics are subject to solvent and protein-solvent structure should be treated as a single system whose behavior can be controlled by the environment. Fluctuations at equilibrium and relaxations from non-equilibrium states are essential characteristics of protein dynamics. Fluctuations correspond to equilibrium transitions between various conformational substates in the ground state manifold while relaxations are transitions towards the ground state manifold from higher energy states. What is striking about these processes in proteins is the very broad distribution of relaxation times and the hierarchical nature of these processes. While there appears to be agreement regarding the functional importance of these processes, it is still unclear how the protein's structure is related to its energy landscape and relaxation dynamics.

2.8 Structure of Nucleic Acids

Nucleic acids come in two varieties in cells: DNA (deoxyribonucleic acid) and RNA (ribonucleic acid). RNA differs from DNA by having one additional oxygen atom on each sugar and one missing carbon atom on each thymine base. DNA and RNA perform distinct functions in the cell. DNA is more stable and hence better suited for information storage. RNA is less stable and hence more useful in information transfer as a messenger, a translator and a synthetic machine [277].

Nucleic acids are linear polymers of nucleotides linked by phosphodiester bonds $-O - PO_2^- - O-$. The sugar ribose occurs only in the ribonucleic acid (RNA). In the deoxyribonucleic acid (DNA) it is replaced by the sugar deoxyribose. Five carbon atoms in the pentose ring are numbered from 1' to 5' and in deoxyribose the hydroxyl group OH bound to carbon 2' is replaced by hydrogen H . In each nucleic acid chain we distinguish the 5' (phosphate) and the 3' (hydroxyl) termini. Successive amino acids are numbered starting from the 5' terminus. The formation of a phosphodiester bond is an endoergic reaction and needs free energy which is released in the nucleotide triphosphate hydrolysis (Fig. 2.46). The sequence in which individual nucleotides with definitive side chains occur along the nucleic acid chain is strictly fixed and genetically determined. We refer to it as the *primary structure* of nucleic acid.

In DNA there are 4 'canonical' nitrogenous bases, guanine (G) and adenine (A), being derivatives of pirimidine, and cytosine (C) and thymine (T), being derivatives of purine. In RNA thymine is replaced by uracil (U). A crucial discovery in molecular biology occurred in 1953, when Watson and Crick discovered that specific pairs of nitrogenous bases can make unique hydrogen bonds in plane (Fig. 2.47).

Nucleic acids are used to store and transmit genetic information in cells. They are composed of long chains of nucleotides, each of which is composed of a sugar phosphate group and a disk-shaped base group. The sugar-phosphate groups are connected together to form a hydrophilic backbone (phosphates have a negative electric charge) and the mainly hydrophobic bases are located off to the side of the chain. In a water environment one side of the chain is protected by the hydrophilic backbone while the other side is exposed. The edges of these bases exhibit chemical complementarity such that A forms two hydrogen bonds with T and C three hydrogen bonds with G. No other combinations of pairs lead to bond formation between them. The matching patterns of hydrogen bonds allow a second strand of DNA, provided it has the proper base sequence, to nestle up to the first and form a stable complex: the famous double helix wherein bases find protection from the aqueous environment. The origin of these bonds is in the fact that the H^+ atom of adenine is attracted to the O^- atom of thymine, for example. Protein synthesis is directed through complicated cell signaling and feedback pathways, although RNA is less removed from synthesis itself.

The reaction of a sugar with a base releases water (an $-OH$ from the sugar plus an H from the base) and produces a sugar-base combination called a nucleoside. Adding a phosphate to a nucleoside releases a water molecule and produces a nucleotide (see Fig. 2.8).

All living things pass on genetic information from generation to generation. This information is contained in the chromosomes which make up an organism's genes. The chromosomal segment that contains the information needed to produce a particular type of complete protein or RNA molecule is contained within each gene.

In Fig. 2.49 (a) we illustrate a section of a DNA double helix.

The helical arrangement occurs when the chromosome replicates itself just before cell division. The arrangement of A paired to T and G opposite C ensures that genetic information is passed to the next generation accurately. We show this process in Fig. 2.50. The two strands or helices of the DNA separate with the help of enzymes leaving the charged parts of the bases fully exposed [295]. The enzymes also operate via electrostatic forces.

To see how the correct order of bases occurs, we focus our attention on the G molecule indicated by an arrow on the lowest strand in Fig. 2.50. Unattached nucleotide bases of all four kinds move around in the cellular fluid. Of the four bases only one will experience attraction to the G when it comes close to it,

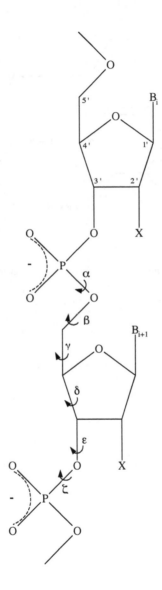

FIGURE 2.46: Notation of angles in nucleic acids.

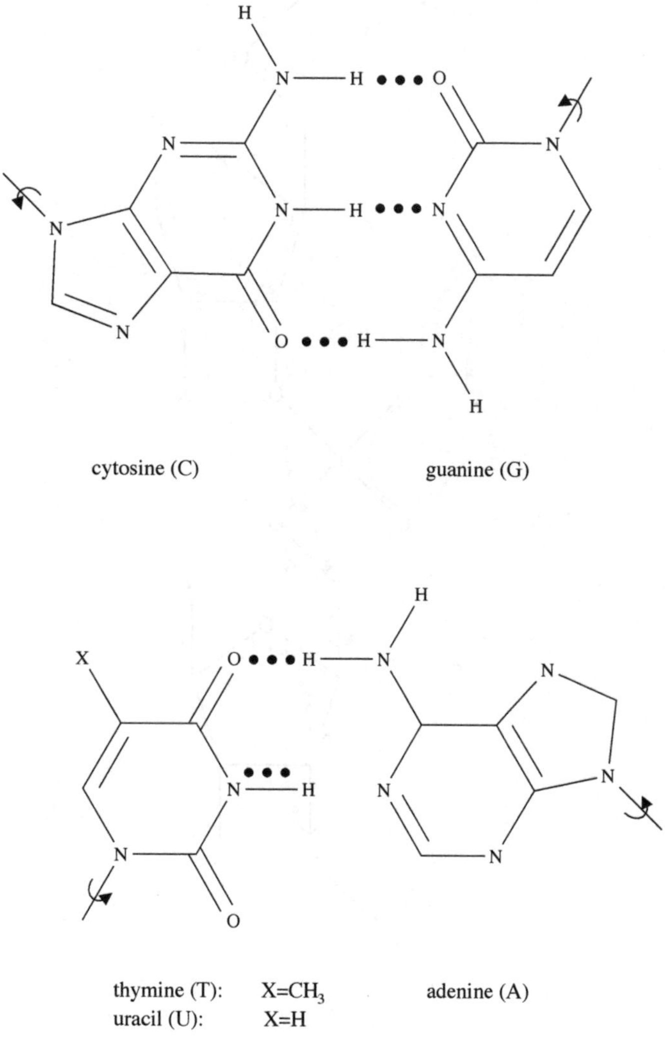

cytosine (C)   guanine (G)

thymine (T):   X=CH₃   adenine (A)
uracil (U):   X=H

FIGURE 2.47:   Canonical nitrogenous bases and Watson-Crick pairing.

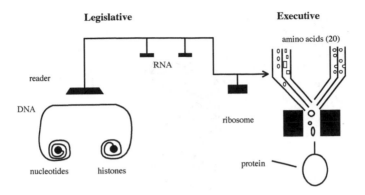

FIGURE 2.48: Each gene contains the information necessary to produce a specific protein.

namely a C. The charges on the other three bases are arranged so they do not get into close proximity of those on G and therefore there will be no significant attractive force exerted on them. These forces decrease rapidly with distance between molecules. Since A, T and C are hardly attracted at all, they will tend to be knocked away by collisions with other molecules before enzymes can attach it to the growing chain. The electrostatic force will often ensure a C remains opposite our G long enough so that an enzyme can attach the C to the growing end of the new chain. We see therefore that electrostatic forces not only hold the two helical chains together but they also operate to select the bases in the right order during replication. In Fig. 2.50 the new number four strand has the same order of bases as the old number 1 strand. Thus the two new helices, 1-3 and 2-4, are identical to the original 1-2 helix. If a T molecule were incorporated in a new chain opposite a G by accident an error would occur. In practice this occurs very infrequently with an error rate of 1 in 10^4. This is the error rate of DNA polymerase, when proofreading is inhibited. Most replicative enzymes have exonuclease (proofreading) activity, and separate DNA repair mechanisms that result in an error rate that is much lower, i.e., in E. coli it is 1 in 10^9 to 1 in 10^{10}. Such an error could result in a mutation resulting in a possible change in some characteristic of the organism. If the organism is to survive this error rate must be low, but for evolution to occur it must be non-zero.

The process of DNA replication is often portrayed as occurring in a clockwork-like fashion, i.e., as if each molecule knew its role and went to its designated place. This is not the case. The forces of attraction between electric charges are rather weak and only become significant when the molecules come close to one another. If the shapes of the molecules are not perfect there is almost no electrostatic attraction. This is why there are few errors.

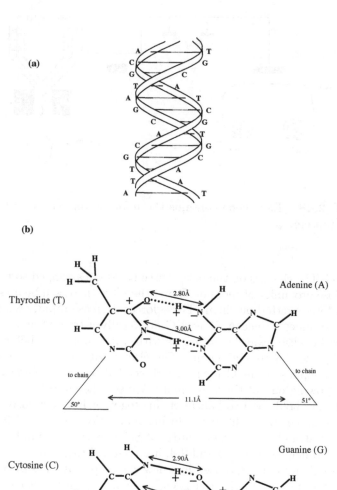

FIGURE 2.49:   (a) A section of DNA double helix and (b) a close-up view showing how A and T and C and G are always paired (the distance unit is $1\text{Å} = 10^{-10}$ m.)

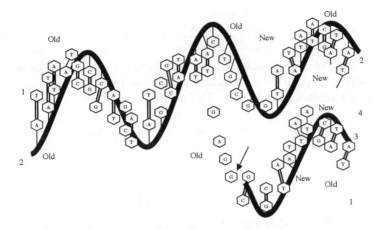

FIGURE 2.50: The replication of DNA.

2.8.1 The Electrostatic Potential of DNA

DNA is a highly charged molecule and it carries two fully ionized monovalent groups on its outer surface per base pair. To model DNA we assume the charge per unit length is λ which is about one negative charge per 1.7Å of length. For simplicity, we assume that DNA is a long charged cylindrical molecule. The diameter of this cylinder is about 20Å. The electrical field of a long strand of DNA in water will clearly depend on the perpendicular distance, r, from the strand and on λ. The electric field lines must point inward because DNA is negatively charged and because we have used a cylinder to model it they must also have cylindrical symmetry. Denoting the electric field, at a distance r from the cylinder, by $E(r)$, we observe that it is inversely proportional to the dielectric constant, ε , and inversely proportional to some power, say r^α, of r, hence we write

$$E(r) = \text{const}\frac{\lambda}{4\pi\varepsilon\varepsilon_0 r^\alpha} \qquad (2.28)$$

In vacuum, the electric field of a monopole of charge Q, a distance R away, is $Q/4\pi\varepsilon\varepsilon_0 R$ so as λ is a charge per unit length, $\alpha = 1$, hence,

$$E(r) = 2\frac{\lambda}{4\pi\varepsilon\varepsilon_0 r} \qquad (2.29)$$

Since $E(r) = -dV/dr$, we have an equivalent form of (Eq. 2.29) as

$$\frac{dV(r)}{dr} = -\frac{\lambda}{4\pi\varepsilon\varepsilon_0 r} \qquad (2.30)$$

Integrating (Eq. 2.30) we see that

$$V(r) = -\frac{\lambda}{4\pi\varepsilon\varepsilon_0} \ln r + k_0 \qquad (2.31)$$

In (Eq. 2.31) k_0 is a constant which is conveniently chosen to be zero.

Now suppose a positive monovalent ion moves in from a distance of one micron to the surface of the DNA. The work done will be

$$\Delta W = |e|V(r = 1\mu m) - |e|V(r = 10\text{Å}) \qquad (2.32)$$

Hence

$$\Delta W = -\frac{\lambda}{4\pi\varepsilon\varepsilon_0}|e| \ln(\frac{10^{-6}}{10^{-9}}) \qquad (2.33)$$

since $1\text{Å} = 10^{-10}m$. Using the values: $k = (4\pi\varepsilon\varepsilon_0)^{-1} = 8.99 \times 10^9 Nm^2/C^2$; $\varepsilon = 80.4$; $\lambda = -|e|/1.7 \times 10^{-10}m$; $e = 1.6 \times 10^{-19}$; in Eq. (2.26), we find that exceeds by about ten times the thermal energy $k_B T$ where k_B is Boltzmann's constant and T is the absolute temperature. This estimate suggests that a monovalent ion stays close to the surface of DNA and we expect DNA to be surrounded by a cloud of Na^+ ions.

2.8.2 DNA: Information and Damage

Genetic coding can be discussed using Information Theory. Deoxyribose nucleic acid (DNA), in the living cell, is the primary genetic material and therefore specifies the information which can be passed on from one generation of cells to the next. It is also intimately involved with the synthesis of new material. The so-called genetic code is a detailed prescription regarding the synthesis of proteins according to the algorithm given in Table 2.18.

For a long time it was not known how information from DNA was transferred to other parts of the cell so that the production of protein can take place. We also saw earlier that there are only four distinct nucleotides in DNA. Thus, discovering that any one of these is in a specific location along a strand increases the information by $\log_2 4 = 2$ bits. As there are twenty amino acids, the identification of one particular amino acid requires $\log_2 20 = 4.22$ bits of information. We therefore conclude that coding must be done by at least three nucleotides arranged in order.

In a cell the DNA is too valuable to do the work of synthesizing proteins directly. First and foremost, huge sections of DNA are 'introns', which is basically 'filler' between coding segments. RNA is processed to have these removed. Another major one is mRNA's shorter lifetime: synthesis of a particular protein is halted when the cytoplasmic supply of the corresponding mRNA degrades. This is an extremely important homeostatic mechanism. The production of messenger RNA (mRNA) molecules from coded DNA is called transcription. The production of proteins is directed by the supply of mRNA, which carries the genetic message from the chromosomes to the

TABLE 2.18: The Genetic Code

3' Terminal Base (first)	U	C	A	G	5' Terminal Base (third)
	Phe	Ser	Tyr	Cys	U
U	Phe	Ser	Tyr	Cys	C
	Leu	Ser	Stop	Stop	A
	Leu	Ser	Stop	Trp	G
	Leu	Pro	His	Arg	U
C	Leu	Pro	His	Arg	C
	Leu	Pro	Gln	Arg	A
	Leu	Pro	Gln	Arg	G
	Ile	Thr	Asn	Ser	U
A	Ile	Thr	Asn	Ser	C
	Ile	Thr	Lys	Arg	A
	Met	Thr	Lys	Arg	G
	Val	Ala	Asp	Gly	U
G	Val	Ala	Asp	Gly	C
	Val	Ala	Glu	Gly	A
	Val	Ala	Glu	Gly	G

ribosomes. Four nucleotides can be coded onto the messenger RNA by the DNA strand namely adenine (A), cytosine (C), guanine (G) and uracil (U). To identify the groups of nucleotides that code for specific amino acids, scientists produced synthetic messenger RNA and examined the resulting synthesized protein. What was found was that if the RNA contains only uracil, for instance, the protein is made up only of the amino acid phenylalinine. As expected the code is based on combinations of three nucleotides (called a codon) and is degenerate, in that one amino acid can be coded for by more than one codon.There are three of the 64 possible combinations which do not code for any amino acid and are believed to act as terminators. When the ribosome reaches this portion of the messenger RNA chain, the growth of the protein is halted. The genetic code is universal and applies to all organisms. If a mutation or error occurs in the DNA, all copies will carry it and incorrect synthesis will take place at the ribosomes. As an example, GAA and GAG code for glutamic acid and GUA and GUG for valine. If, as in this example, an error of U for A occurs in the appropriate portion of a strand of DNA which provides information for the production of hemoglobin, then a valine residue replaces that of glutamic acid. The resulting hemoglobin molecule is called sickle-cell hemoglobin, the name arising from the shape of the red blood cells. Those people with this disease suffer severely from anemia. Many other hereditary diseases are very likely to be caused by minor coding errors of this sort.

Electromagnetic radiation from the ultraviolet range (200-350 nm) to high energy gamma rays can be very harmful to living systems. Cellular injury, mutation and lethality have been demonstrated to result from high intensity UV exposure. One effect seen in samples of DNA extracted from UV irradiated cells is the damage in the form of bond breaks called local denaturation and entanglement with proteins called protein cross-linking (see Fig. 2.51).

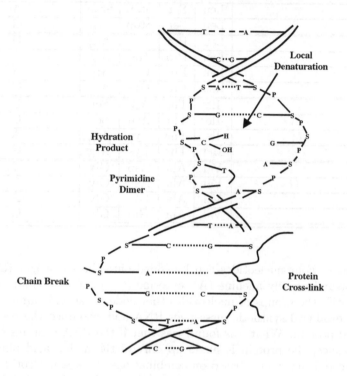

FIGURE 2.51: Schematic illustration of the various defects found in DNA that has been irradiated with ultraviolet light.

2.8.3   Fluorescence in Biomolecules

In many plant materials the phenomenon of fluorescence has been known for more than a century, e.g., in quinone. For many years fluorescence methods have been used to check the presence of trace elements, drugs and vitamins. However, it is surprising that fluorescent assays have only recently been used extensively for identification and location studies, particularly in view of the sensitivity of fluorescence detection and that, under suitable chemical treatment, almost all molecules are capable of fluorescing. This method affords many of the possibilities which radioactive techniques also offer but without any of the associated hazards. In many radioactive assays, studies "in vivo" require laboriously synthesized substances. The same investigations can be undertaken more simply and with equal accuracy using fluorescence. When it is combined with techniques for separation like electrophoresis, the method can detect impurities at concentrations of less than 1 in 10^{10}.

When a substance to be studied is in solution, the container, the solvent and other substances may also fluoresce. Clearly, no solvent should be used which fluoresces in the same general region and the light used as the stimulant should be strictly monochromatic. In some elaborate studies it may be necessary to use a spectrometer to analyze the fluorescent light and the wavelengths of the various components measured before identification becomes possible. In biochemistry, because of its sensitivity, it is used to investigate the location of enzymes and co-enzymes in organisms, to follow the fate of a drug and its metabolic products, and as a marker in genetic experiments. Fluorescence can also be used to determine the amino acid sequence in proteins.

In order to understand fluorescence we must augment our elementary theory of quantum energy levels for electrons in atoms by vibrational degrees of freedom which are present in the energy structure of molecules. Vibrational energy levels arise because the molecular bond acts like a system of two masses connected by an elastic spring which has a natural frequency of vibration ω . The energies are quantized according to the formula

$$E_n = \hbar\omega(n + \frac{1}{2})$$ (2.34)

This is illustrated in the diagram in Fig. 2.52.

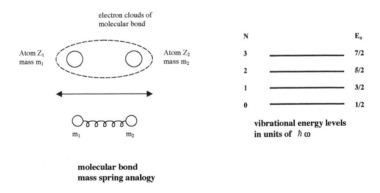

FIGURE 2.52: Quantization of vibration energy levels.

Transitions between vibrational energy levels can be induced by photons provided $\Delta n = \pm 1$. It should be mentioned that each electronic state may have a number of vibrational states. At room temperature most of the molecules will occupy the ground electronic state and its lowest vibrational energy level.

If a molecule does find itself in a high vibrational level it will, after several collisions, drop to a lower level and give up its excess energy to other molecules in its vicinity. This process is called vibrational relaxation. A molecule in an excited electronic and vibrational state will undergo vibrational relaxation followed by an emission of a photon of energy $h\nu'$ which takes it down to the ground state. When the emitted photon is in the visible range the process is called fluorescence. The cycle of excitation by a photon of energy $h\nu$, vibrational relaxation and eventual fluorescence typically takes on the order of 10^{-7} s. Fig. 2.53 illustrates the quantum mechanical origin of fluorescence.

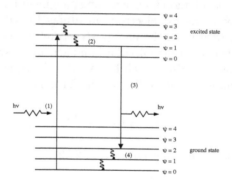

FIGURE 2.53:   Quantum mechanical origin of fluorescence.

When we reincorporate the presence of spin in the electronic states, we find that the ground state is usually a so- called "singlet" state, i.e., it has a net spin of zero for the two electrons occupying it. Some excited states are "singlet" states also. Some other excited states may involve a net spin of unity, which is called a "triplet" state. We show the distinction in Fig. 2.54. This gives rise to the phenomenon of phosphorescence in which the return of the photo-excited electron to the ground state involves an intermediate transition to an excited triplet state. The distinguishing feature of phosphorescence is that it takes place, at times, even several seconds, after the photon absorption process. In addition the phosphorescent photon has an energy h which is always lower than the absorbed photon $h\nu''$ or the fluorescent photon $h\nu'$ (see Fig. 2.55 for illustration).

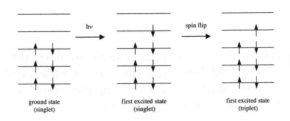

ground state
(singlet)

first excited state
(singlet)

first excited state
(triplet)

FIGURE 2.54: Singlet and triplet states.

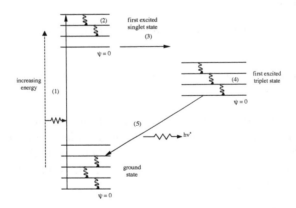

FIGURE 2.55: Quantum mechanical origin of phosphorescence.

Chapter 2 Questions and Problems

QUESTION 2.1 How are myosin and kinesin similar? What effect would each of the following: increasing the concentration of ATP, increasing the concentration of ADP and increasing the temperature, have on the speed of myosin and kinesin movement?

QUESTION 2.2 Put the following interactions in order from generally strongest to weakest: hydrogen bond, covalent bond, van der Waals, dipole-dipole, ion-ion, ion-dipole. In what ways does each of these depend on distance, geometry and solvent?

QUESTION 2.3 Reactions catalyzed by protein enzymes generally have an optimal pH. Levels both below or above this optimum tend to denature the enzyme and thus slow the reaction.

   (a) What types of bonds important to protein structure would be affected by pH? Explain.

   (b) Why might one enzyme be extremely sensitive to pH while another is not?

QUESTION 2.4 What is bioluminescence? Is it the same thing as fluorescence?

QUESTION 2.5 Discuss how plasmas and associated theory could impact on biophysics.

QUESTION 2.6 Recently trans fatty acids have become a health concern in the public eye. How do trans fatty acids differ from their cis counterparts? How does this difference change their physical properties? Why would this change be a health concern and why would the food industry be interested in such a product?

QUESTION 2.7 The following protein sequence is for human insulin:

Gly Ile Val Glu Gln Cys Cys Thr Ser Ile Cys Ser Leu Tyr Gln Leu Glu Asn Tyr Cys Asn

Provide a genetic code that would represent this protein. How many different codes could produce this same sequence?

QUESTION 2.8 What is the most likely charge of insulin at pH 7.0?

QUESTION 2.9 DNA has an average diameter of 2 nm, and its phosphate backbone has a charge of $-1.6 \times 10^{-9}C$ under physiological conditions. Calculate the repulsive force between phosphate groups in DNA.

QUESTION 2.10 In the discussion of dispersion interactions it was stated its magnitude of potential energy follows $1/r^6$. Show this.

QUESTION 2.11 In Section 2.7.4 the displacement of a protein by 1 cm was estimated to require 3 hours. Using this same example calculate the diffusion time.

QUESTION 2.12 If divalent ions are used instead of monovalent ions in Eq. 2.33 how does the work compare to the thermal energy? What does this suggest about the population of divalent ions in the neighbourhood of DNA?

PROBLEM 2.13 Dr. Paul Davies in his paper entitled "Quantum Fluctuations and Life" suggests that quantum processes may be important to understanding biology. He states that the quantum uncertainty principle places a limit on molecular processes such as protein folding and therefore may be a solution to the Levinthal Paradox. As an example consider the energy-time uncertainty relation. This uncertainty sets a limit on the operation of quantum clocks such that for a clock of mass M, and size L:

$$T < \frac{ML^2}{\hbar}$$

where T is the time for the process, and $\hbar$ is the reduced Planck's constant.

(a) For a small protein consisting of a chain of 80 amino acids, each with an average mass, m_{av}, of 100 daltons and an average length, l_{av}, of 0.4 nanometers, find the time limit on the process. How does this value compare with the typical time for the folding of small proteins (typically on the order of 10^{-6} to 10^{-3} seconds)?

(b) Taking into consideration the way in which larger proteins fold why does the above relation not hold for large proteins?

(d) Explain DNA replication and mutation from the point of view of quantum processes.

PROBLEM 2.14 During thermal collisions with water, myosin molecules can get bent. If it takes 5×10^{-20} J of work to accomplish this bending, use Boltzmann's Law to calculate the ratio of bent to relaxed myosin molecules at room temperature.

PROBLEM 2.15 *Condensation reactions.* [after Campbell] The molecular formula for glucose is $C_6H_{12}O_6$. What would be the molecular formula for a polymer made by linking ten glucose together by condensation reactions?

PROBLEM 2.16 *Randomness of the Genetic Code.* Evolution occurs as a result of changing DNA content. DNA may change due to errors during replication, effects of radiation, or viruses. In this problem we look at how a simple base mutation to DNA may affect a protein.

(a) There are 64 different codons. Of the $(64)(9)=576$ possible single base mutations, how many will cause a change in the coded amino acid?

(b) Substituting one kind of amino acid for another randomly may or may not seriously affect the structure and function of a protein. Sometimes, it is hard to predict if a given substitution will make much difference, but in general, changes to the structure are less severe if the side chains of the new and old amino acid are similar. Divide the codons into five groups: those that code for the four types of amino acids, and stop. How many of the possible single base mutations will change the code from one group to another? How many will not change the group? (Show how you arrived at your answer.)

(c) Next, imagine a completely different (unspecified) system of coding the amino acids in which the 20 amino acids and stop were each specified by 3 codons (resulting in 63 possible codons). In this system, the assignment of codons to amino acids is completely random. What is the probability that changing one codon into another will result in a different amino acid? What is the probability that such a change will result in an amino acid from a different group?

(d) Comment on the answers of (c) in comparison to those of (a) and (b). Is the genetic code random?

PROBLEM 2.17 Consider a protein 300 amino acids long, in which each amino acid has just 3 available conformations.

a) Find the total number of conformations available to the protein.

At body temperature, one can imagine searching 10^{13} conformations per second.

b) Find the total time required for this protein to search all of its available conformations. Compare your answer to the age of the universe ($=10^{15}$s).

This question should convince you that proteins must not fold by randomly searching through conformational space.

PROBLEM 2.18 A 50 micron strand of DNA is stretched by an optical trap from equilibrium until its ends are separated by 20 microns.

(a) What is the work done by the trap on the DNA strand?

(b) What is the work done by the DNA strand on the trap?

(c) How much force must the trap exert on the DNA strand to maintain the 20 micron spacing?

PROBLEM 2.19 Two DNA chains of length 5 microns in ion free water run parallel with a separation of 50 Å. What is the repulsive force between the chains?

PROBLEM 2.20 What is the dipole-dipole force between 2 water molecules separated by a distance of 10 Angstrom if one of the two dipoles is perpendicular to the line of separation and one parallel? Is the force repulsive or attractive? The dipole moment of water is $P = 6.2 \times 10^{-30} Cm$.

PROBLEM 2.21 Give an expression for the electrostatic force between two $100kd$ proteins in a 2 molar salt solution, one with a positive charge $Q = 10e$ and one with a negative charge of $-Q$. Compute the electrostatic force at a spacing of 10 and 100 Å. Compare with the force between the proteins in salt-free water and in a vacuum.

Chapter 2 References

1. Alberts, B., Bray, D., Lewis, J., Raff, M., Roberts, K., and Watson, J.D. *Molecular Biology of the Cell.* Garland Publishing, London, 1994.

2. Albrecht-Buehler, G. http://www.basic.nwu.edu/g-buehler/cellint.htm.

3. Alt, W. and Dembo, M. *Mathematical Biosciences* **156**, 207-228. 1999.

4. Amos, L.A. *Trends Cell Biol.* **5**, 48-51 1995.

5. Amos, L.A. and Amos, W.B. *Molecules of the Cytoskeleton.* Macmillan Press, London, 1991.

6. Amos, L.A. and Cross, R.A. *Curr. Opin. in Struct. Biol.* **7**(2), 1997.

7. Amos, L.A. and Hirose, K. *Curr. Opin. in Cell Biol.* **9**, 4-11, 1997.

8. Applequist, D.E., Depuy, C.H., and Rinehart, K.L. *Introduction to Organic Chemistry*, John Wiley and Sons, New York, 1982.

9. Astumian, R.D., and Bier, M. *Physical Review Letters* **72**, 1766, 1994.

10. Bairoch, A. and Apweiler, R. *Nucleic Acids Res.* **26**, 38-42, 1998.

11. Bayley, P.M., Schilstra, M.J., and Martin, S.R. *J. Cell Sci.*, **95**, 33 1990.

12. Benedek, G. B. and Villars, F. M. H. *Physics With Illustrative Examples from Medicine and Biology*, Springer, Berlin, 2000.

13. Bergethon, P.R. and Simons, E.R. *Biophysical Chemistry. Molecules to Membranes*, Springer-Verlag, Berlin, 1990.

14. Boal, D. *Mechanics of the Cell*, Cambridge University Press, Cambridge, 2001.

15. Bras, W. *An X-ray Fibre Diffraction Study of Magnetically-Aligned Microtubules in Solution*, PhD Thesis, Liverpool John Moores University, October 1995.

16. Bras, W. *PhD Thesis*, University of Amsterdam.

17. Brown, J.A. and Tuszyński, J.A. *Physical Review E* **56**, 5834-5840, 1997.

18. Bruinsma, R., *Physics, 6A and 6B*, International Thomson Publishing, 1998.

19. Carlier, M.F., Melki, R., Pantaloni, D., Hill, T.L., and Chen, Y. *Proc. Natl. Acad. Sci. USA* **84**, 5257-5261, 1987.

20. Case, R.B., Pierce, D.W., Hom-Booher, N., Hart, C.L. and Vale, R.D. *Cell* **90**, 959-966, 1997.

21. Chen, C.S., Mrksich, M., Huang, S., Whitesides, G.M., and Ingber, D.E. *Science* **276**, 1425-1428, 1997.

22. Cole, K.S. *Trans. Faraday Soc.*, **33**: 966, 1937.

23. de Gennes, P.-G. *Introduction to Polymer Dynamics.* Cambridge University Press, Cambridge, 1990.

24. De Robertis, E.D.P. and De Robertis, E.M.F. *Cell and Molecular Biology*, Saunders College, Philadelphia, 1980.

25. Derenyi, I. and Vicsek, T. *Proc. Natl. Acad. Sci. USA* **93**, 6775, 1996.

26. Doering, C.R., Horsthemke, W., and Riordan, J. *Physical Review Letters* **72**, 2984, 1994.

27. Dustin, P. *Microtubules.* Springer-Verlag, Berlin, 1984.

28. Edelstein-Keshet, L. *Eur. Biophys. J.* **27**, 521-531, 1998.

29. Fletterick, R.J. *Nature* **395**, 813-816, 1998.

30. Flory, P.J. *Statistical Mechanics of Chain Molecules.* Wiley, New York, 1969.

31. Flyvbjerg, H., Holy, T.E., and Leibler, S. *Phys. Rev. Lett.* **73**, 2372, 1994.

32. Flyvbjerg, H., Holy, T.E., and Leibler, S. *Phys. Rev. E* **54**, 5538-5560, 1996.

33. Frey, E., Kroy, K., and Wilhelm, J. *Adv. Struct. Biol.* **5**, 135-168, 1998.

34. Frieden, C. and Goddette, D. *Biochemistry* **22**, 5836-5843, 1983.

35. Froehlich, H. *Adv. In Electronics and Electron Physics*, **53**: 85-152, 1980.

36. Fygenson, D.K., Braun, E., and Libchaber, A. *Phys. Rev. D* **50**, 1579-1588, 1994.

37. Geeves, M. A. and Holmes, K. C. *Annu. Rev. Biochem.* **68** 687-728, 1999.

38. Gennis, R. B., *Biomembranes: Molecular Structure and Function*, Springer, 1989.

39. Gittes, F., Mickey, E., and Nettleton, J. *J. Cell Biol.* **120**, 923-934, 1993.

40. Hays, T.S. and Salmon, E.D. *J. Cell Biol.* **110**, 391-404, 1990.

41. Hess, B. and Mikhailov, A. *Science* **264**, 223, 1994.

42. Horio, T. and Hotani, H. *Nature London* **321**, 605-607, 1986.

43. Houchmandzadeh, B. and Vallade, M. *Phys. Rev. E* **6320**, 53, 1996.

44. Howard, J. *Mechanics of Motor Proteins and the Cytoskeleton*, Sinauer Associates Inc., Sunderland, Mass, 2001.

45. Huxley, H. E. *Science* **164**, 1356-1366, 1969.

46. Hyman, A.A., Chrétien, D., Arnal, I., and Wade, R.H. *J. Cell. Biol.* **128**, 117-125, 1995.

47. Hyman, A.A., Salser, S., Dreschel, D.N., Unwin, N., and Mitchison, T.J. *Molec. Biol. Cell* **3**, 1155-1167, 1992.

48. Ingber, D.E. *Ann. Rev. Physiology* **59**, 575-599, 1997.

49. Ingber, D.E. *J. Cell Sci.* **104**, 613, 1993.

50. Janmey, P. A., Hvidt, S., Käs, J., Lerche, D., Maggs, A., Sackmann, E., Schliwa, M. and Stossel, T.P. *J. Biol. Chem.* **269**, 32503-32513, 1994.

51. Jülicher, F., Adjari, A., and Prost, J. *Rev. Mod. Phys.* **69**, 1269-1281, 1997.

52. King, R.W.P. and Wu, T.T. *Phys. Rev. E* **58**, 2363-2369, 1998.

53. Klotz, I.M. 1993. *Protein Sci* **2**: 1992-1999.

54. Kozielski, F., Sack, S., Marx, A., Thormahlen, M., Schonbrunn, E., Biou, V., Thompson, A., Mandelkow, E.-M. and Mandelkow, E. *Cell* **91**, 985-994, 1997.

55. Kraulis, J. per., *Journal of Applied Crystallography* **24**, 946-950, 1991.

56. Kull, F.J., Sablin, E.P., Lau, R., Fletterick, R.J., and Vale, R.D. *Nature* **380**, 550-555, 1996.

57. Ledbetter, M.C. and Porter, K.R. *J. Cell Biol.* **19**, 239-250, 1963.

58. Leibler, S. and Huse, D.A. *J. Cell Biol.* **121**, 1357-1368, 1993.

59. Lowe, J. and Amos, L. A. *Nature* **391**, 203-206, 1998.

60. Lu, Q., Moore, G.D., Walss, C., and Luduena, R.F. *Adv. Struct. Biol.* **5**, 203-227, 1998.

61. Luby-Phelps, K. *Curr. Opin. Cell Biol.* **6**, 3-9 1994.

62. Mader, Sylvia. *Biology*, 6th edition. William C. Brown, Dubuque, IA, 1996.

63. Mandelkow, E.-M. and Mandelkow, E. *Cell Motil. and Cytoskel.* **22**, 235-244, 1992.

64. Mandelkow, E.M., Mandelkow, E., and Milligan, R. *J. Cell Biol.* **114**, 977-991, 1991.

65. Maniotis, A. and Ingber, D.E. *Science*, 1997.

66. Marx, A. and Mandelkow, E. *Eur. Biophys. J.* **22**, 405, 1994.

67. Mickey, B. and Howard, J. *J. Cell Biol.* **130**, 909-917, 1995.

68. Mitchison, J.M. *Biology of the Cell Cycle.* Cambridge University Press, Cambridge, 1973.

69. Mitchison, T. and Kirschner, M. *Nature London.* **312**, 237-242, 1984.

70. Némethy, G., Scheraga, H.A. *J Phys Chem* **66**: 1773-1789, 1962.

71. Nicholls, A., Sharp, K., and Honig, B. *Structure, Function and Genetics* **11**, 281, 1991.

72. Nicklas, R.B. and Ward, S.C. *J. Cell Biol.* **126**, 1241, 1994.

73. Nicklas, R.B., Ward, S.C., and Gorbsky, G.J. *J. Cell Biol.* **130**, 929, 1995.

74. Nogales, E., Wolf, S.G., and Downing, K.H. *Nature London.* **391**, 199-203, 1998.

75. Oosawa, F. and S. Asakura. *Thermodynamics of the Polymerization of Protein.* Academic Press, London; New York, 1975.

76. Owicki, J.C., Springgate, M.W., and McConnell, H.M. *Proc. Nat. Acad. Sci USA* **75**, 1616, 1978.

77. Peyrard, M. ed. *Nonlinear Excitations in Biomolecules.* Springer-Verlag, Berlin, 1995.

78. Pink, D. A. *Theoretical Models of Monolayers, Bilayers and Biological Membranes in Biomembrane Structure and Function* Ed. D. Chapman. McMillan Press, London, 319-354, 1984.

79. Rieder, C. L. and E. D. Salmon. *J. Cell Biology.* **24**: 223-233, 1994.

80. Risken, H. *The Fokker-Planck Equation.* Springer-Verlag, Berlin, 1989.

81. Rolfe, D.F.S. and Brown, G.C. *Physiol. Rev.* **77**, 731-758, 1997.

82. Ruppel, K.M., Lorenz, M., and Spudich, J.A. *Curr. Opin. Struct. Bio.* **5**, 181-186, 1995.

83. Saenger, W. *Principles of Nucleic Acid Structure.* Springer-Verlag, New York, 1984.

84. Schulz, G.E. and Schirmer, R.H. *Principles of Protein Structure.* Springer-Verlag, Berlin, 1979.

85. Semënov, M.V. *J. Theor. Biol.* **179**, 91-117, 1996.

86. Sept, D. *Models of Assembly and Disassembly of Individual Microtubules and their Ensembles,* PhD thesis, University of Alberta, 1997.

87. Sept, D., Limbach, H.-J., Bolterauer, H., and Tuszyński, J.A. *J. Theor. Biol.* **197**, 77-88, 1999.

88. Sinden, R. *DNA Structure and Function,* Academic Press, San Diego, 1990.

89. Singer, S.J. and Nicolson, G.L. *Science* **175**, 720, 1972.

90. Small, J. V., Stradal, T., Vignal, E., and Rottner, K. *Trends in Cell Biol.* **12**: 112-120, 2002.

91. Stryer, L. *Biochemistry.* W.H. Freeman and Co., San Francisco, 1981.

92. Svoboda, K., and Block, S.M. *Cell* **77**, 773, 1994.

93. Tabony, J. and Job, D. *Nature London.* **346**, 448-451, 1990.

94. Tanford, C., *The Hydrophobic Effect: Formation of Micelles and Biological Membranes,* 2nd ed., Wiley, 1980.

95. Tobacman, L. S. and Korn, E. D. *J. Biol. Chem.* **258**, 3207-3214, 1983.

96. Tran, P.T., Walker, R.A., and Salmon, E.D. *J. Cell Biol.* **138**, 105-117, 1997.

97. Vassilev, P.M., Dronzine, R.T., Vassileva, M.P., and Georgiev, G.A. *Biosci. Rep.* **2**, 1025-1029, 1982.

98. Vater, W., Stracke, R., Böhm, K.J., Speicher, C., Weber, P., and Unger, E. *The Sixth Foresight Conference on Molecular Nanotechnology.* Santa Clara, CA (USA) 1998.

99. Vogel, G. *Science* **279**: 1633-1634, 1998.

100. Volkenstein, M.V. *General Biophysics.* Academic Press, San Diego, 1983.

101. Zuckermann, M. J., Georgallas, A., and Pink, D. A. *Can. J. Phys.,* **63**, 1228-1234, 1985.

Chapter 3

What Is a Biological Cell?

3.1 Cytoplasm: structural elements, physical properties, cellular water and its characteristics, biological ferroelectricity

The fluid contents of a cell are known as the cytoplasm. The cytoplasm provides the medium in which fundamental biophysical processes such as cellular respiration take place. Its properties are somewhat different than those of dilute aqueous solutions. The contents must be accurately known for in vitro studies of enzymatic reactions, protein synthesis and other cellular activities. Typical constituents of the cytoplasm are listed in Tables 3.1 (ionic) and 3.2 (bio-molecular). Most of the trace ions in Table 3.1 are positively charged; however the cytoplasm cannot have any net electrical charge, thus the difference is made up of the other constituents such as proteins, bicarbonate (HCO^{3-}), phosphate (PO_4^{3-}) and other ions which are for the most part negatively charged, a few of which are significantly electronegative.

TABLE 3.1: Major Components of Cytoplasm [167] in a Typical Mammalian Cell

Ions	Concentration	Non-Ionic Constituents	
K^+	140 mM	protein	200-300 mg/mL
Na^+	10 mM	actin	2-8 mg/mL
Cl^-	10 mM	Tubulin	4 mg/mL
Ca^{2+}	0.1 μ M	pH	7.2
Mg^{2+}	0.5 mM	(specific tissues may differ)	

Most cells maintain a neutral pH and their dry matter is composed of at least 50% protein (see Table 3.2). The remaining dry material consists of nucleic acids, trace ions, lipids and carbohydrates. A few metallic ions are found which are required for incorporation into metallo-proteins but these ions such as iron (II) (Fe^{2+}) are typically found in nano-molar concentrations.

There is experimental evidence for the existence of two phases of the cy-

TABLE 3.2: Percentage Content and Molecular Numbers of Key Cellular Components

Molecule	percent content in dry weight	number of molecules
DNA	5%	2×10^4
RNA	10%	4×10^4
Lipids	10%	4.5×10^6
Polysaccharides	5%	10^6
Proteins	70%	4.7×10^6

toplasm. These are the so-called liquid and solid phases, sol and gel, respectively. In the solid phase, the major constituents of the cell are rendered immobile while in the liquid phase, the cytoplasm's viscosity does not differ significantly from water [129]. Diffusion in the cytoplasm is affected mainly by macromolecular crowding. In the solid phase, diffusion is slowed by a factor of three relative to diffusive movement in water. Such properties of the cytoplasm seem to be regulated in some sense by the cytoskeleton, but the manner by which this regulation is accomplished is unclear. It is believed that it involves the tangling and detangling of a mesh of various protein filaments. The important point is that once the cell has acted to organize itself, the transition to a solid phase can allow it to expend relatively minimal energy to maintain its organization [39]. Contrary to early perceptions, the cytoplasm is not a viscous soup-like amorphous substance but a highly organized, multi-component, dynamic network of interconnected protein polymers suspended in a dielectrically polar liquid medium. This point of view will become quite clear when we discuss the cytoskeleton in the following section.

3.1.1 Osmotic Pressure of Cells

As shown in Table 3.1 a variety of solute molecules are contained within cells (see also Fig. 3.1). The cellular fluid (cytosol) has a chemical composition of 140 mM K^+, 12 mM Na^+, 4 mM Cl^- and 148 mM A^- where 1 mM stands for a concentration of 10^{-3} mol/liter. The symbol A stands for protein. Cell walls are semipermeable membranes and permit the transport of water but not of solute molecules. We can apply the osmotic pressure concept to cells, but because of the content of the cellular fluid, we need to find the osmotic pressure of a mixture of solute molecules. We use Dalton's Law to determine the osmotic pressure inside a cell. A mixture of chemicals, with concentrations c_1, c_2, c_3, dissolved in water has a total osmotic pressure equal to the sum of the partial osmotic pressures, Π , of each chemical. Thus

$$\Pi = \Pi_1 + \Pi_2 + \Pi_3 + \ldots = RT(c_1 + c_2 + c_3 + \ldots) \qquad (3.1)$$

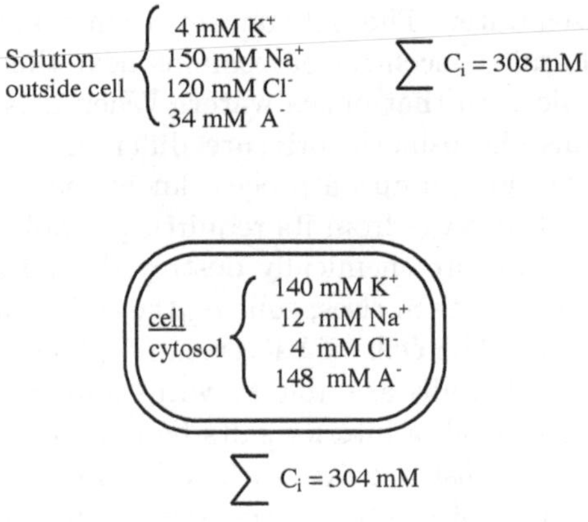

$$
\text{Solution} \atop \text{outside cell}
\left\{
\begin{array}{l}
4\,\text{mM K}^+ \\
150\,\text{mM Na}^+ \\
120\,\text{mM Cl}^- \\
34\,\text{mM A}^-
\end{array}
\right.
\qquad \sum C_i = 308\,\text{mM}
$$

$$
\text{cell} \atop \text{cytosol}
\left\{
\begin{array}{l}
140\,\text{mM K}^+ \\
12\,\text{mM Na}^+ \\
4\,\text{mM Cl}^- \\
148\,\text{mM A}^-
\end{array}
\right.
$$

$$
\sum C_i = 304\,\text{mM}
$$

FIGURE 3.1: Solute molecules inside and outside of a cell.

The total osmotic pressure inside a cell, Π_{in}, is therefore

$$
\Pi_{in} = RT \frac{(140 + 12 + 4 + 148) \times 10^{-3} mol}{1 liter} \times \frac{1 liter}{10^{-3} m^3} = 7.8 \times 10^4 Pa \quad (3.2)
$$

where we used the concentrations given above and a temperature of $T = 310K$ since the gas constant is $R = 8.31 J/mol K$. Cell walls would be expected to burst under such large pressures. However, they do not because the exterior fluid also exerts an osmotic pressure in the opposite direction. The cell exterior is composed of 4mM K^+, 150 mM Na^+, 120 mM Cl^- and 34 mM A-. As a consequence the total osmotic pressure of the cell exterior, Π_{out} , is given by

$$
\Pi_{out} = RT \frac{(4 + 150 + 120 + 34) \times 10^{-3} mol}{1 liter} \times \frac{1 liter}{10^{-3} m^3} = 7.9 \times 10^4 Pa \quad (3.3)
$$

Π_{out} is again a large osmotic pressure but because Π_{in} and Π_{out} are quite close in values, the osmotic pressure difference between the exterior and interior part of the cell is very small, as it is the net pressure exerted on the cell wall. For fragile animal cells, it therefore becomes vitally important to keep their interior and exterior osmotic pressures closely matched. The cell therefore has a sophisticated control mechanism to do this.

3.1.2 Osmotic Work

If two solutions have the same osmotic pressure, we call them iso-osmotic. However, if the pressures are different, the one at higher pressure is called hypertonic and the one at lower pressure is called hypotonic. When cells are placed in a solution and neither swell nor shrink we call the solution isotonic. In the tissues of most marine invertebrates the total osmotic concentration is close to that of the sea water. The salt concentration of sea water is about 500 mM. As long as the salt concentration remains near this value the blood of many crabs is isotonic with that of sea water. When it is outside this range, the system maintains the osmotic pressure difference across its membrane through the activity of ion pumps, a process known as osmoregulation.

The cell begins to drift away from its required physiological cytosolic composition if the ion pumps are chemically destroyed. Across the cell wall the osmotic pressure difference then rises, causing the cell to swell, become turgid and eventually explode. The cells of bacteria and plants are not osmotically regulated since their cell walls are able to withstand pressures in the range of 1 to 10 atm. The removal of interior salts by teleost fish and the importation of salt by freshwater fish require work to be done. The minimum work performed when n moles of solute are transferred from one solution with a concentration c_1 to a solution with concentration c_2 is given by

$$W = nRT \ln \frac{c_1}{c_2} \tag{3.4}$$

where c_1 might be the salt concentration in the tissue of a fish and c_2 the salt concentration in sea water. In this case $c_2 > c_1$ so the osmotic work done by the sea water should be negative - the physical reason for this being that energy is required to move salt molecules from a solution of low concentration to one of high concentration. The actual work done by the cells of a fish in excreting salt will be the additive inverse of that in (Eq.3.4) above.

Osmotic pressure is also used by the cells of plants and, in particular, trees. Tree roots have a high osmotic pressure inside them which leads to absorption of water from the soil. A key role is also played, it is believed, by osmotic pressure in the growth of plants. The openings on the surfaces of cell leaves, called stomata, are bordered by guard cells that can regulate their internal pressure by controlling the potassium concentration. Water absorption causes these cells to swell under osmotic pressure and the stomata are closed. Contained within the cytoplasm are the components of the cytoskeleton and certain smaller compartments known as organelles which are specialized to perform their respective functions. We discuss the components of the cytoskeleton in the section that follows.

3.2   Cytoskeleton: microtubules, actin filaments, intermediate filaments actin, gelsolin and other proteins participating in cytoskeletal organization

3.2.1   The Cytoskeleton

One of the most important issues of molecular biophysics is the modelling of the complex behavior of the cell's cytoskeleton. Interiors of living cells are structurally organized by the cytoskeleton networks of filamentous protein polymers: microtubules, actin (microfilaments) and intermediate filaments with motor proteins providing force and directionality needed for transport processes. Unlike the hardened skeleton that supports mammals such as humans, the cytoskeleton is a dynamic structure that undergoes continuous reorganization. The cytoskeletal network of filaments has the responsibility of defining the cell shape [83], protecting the cell from changes in osmotic pressure, organizing its contents, providing cellular motility and finally is responsible for separating chromosomes during mitosis.

The cytoskeleton is unique to eukaryotic cells. It is a dynamic three-dimensional structure that fills the cytoplasm. This structure acts as both muscle and skeleton, for movement and stability. The long fibers of the cytoskeleton are polymers composed of protein subunits. The key proteins of the cytoskeleton are listed below [5]:

- Tubulin and microtubules

- Actin and actin filaments

- Intermediate filaments

- Motor proteins (families thereof):

- kinesin

- myosin

- dynein

- Specialized molecules in the two phases of the cytoplasm are :

- Gel promoters

- Sol promoters

- Gelsolin

- Cross linkers (integrin, talin)

In addition, we list some cross-linking, sequestering and severing molecules, namely for actin: a-actin, filamin, cap-Z; for MT's: microtubule associated proteins (MAP's), for example tau or MAP2 which are empirically important in neurons.

For example, in the epithelial (skin) cells of the intestine (see Fig. 3.2); all three types of fibers (microfilaments, microtubules and intermediate filaments) are present. Microfilaments project into the villi, giving shape to the cell surface. Microtubules grow out of the centrosome to the cell periphery. Intermediate filaments connect adjacent cells through desmosomes.

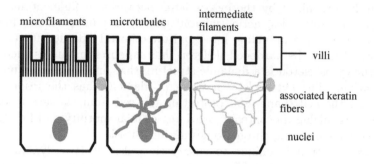

FIGURE 3.2:   Cytoskeletal components of intestinal epithelial cells.

3.2.2   Biopolymers of the Cytoskeleton

With the exception of the cellulose fibers of the plant cell wall which are polysaccharides, the filaments of importance to the cell are all made up of protein polymers [103]. Some cells, such as the auditory outer hair cells, contain strings of the protein spectrin. The extracellular matrix of connective tissue is traversed by fibres of collagen, a family of proteins exhibiting a variety of forms. Collagen, present in connective tissues such as tendon, is organized hierarchically into ropes and sheets. Type I collagen, one of the more common varieties, produces a single strand with a molecular mass of 100 kilodaltons. Three such strands organize into tropocollagen macromolecules that are 300 nm long and 1.5 nm in diameter. Furthermore, a collagen fibril whose diameter is 10-300 nm contains numerous strands of tropocollagen. These fibrils are commonly further organized into a parallel formation referred to as a collagen fiber. In the cytoskeleton of the erythrocyte, two dimers of spectrin are connected end-to-end to form a tetramer that has a length of approximately 200 nm and whose monomer units have a mass of between 240 and 260 kilodaltons. In Table 3.3 we have summarized some physical characteristics of the key cellular polymers [103].

It can be readily concluded based on the above that the persistence length increases with the mass density. This can be deduced from the expression for the persistence length, i.e., the distance over which a polymer maintains its direction in space, of a uniform cylindrical rod as a function of its radius R where:

$$\xi_p = \beta Y I \frac{\pi R^4}{4} \tag{3.5}$$

from which one finds that, for example, for microtubules and actin their Young modulus ranges between 10^9 and $2 \times 10^9 N/m^2$, a value comparable to collagen but much smaller than that for steel.

TABLE 3.3:   The Key Biomolecular Polymers in a Cell

Polymer	Unit	Linear Density (Da/nm)	Persistence Length (nm)
Spectrin	2-stranded filament	4,600	10-20
IF	4-stranded protofilament	4,000	
Collagen (type I)	3-stranded filament	1,000	
DNA	double helix	1,900	51-55
F-actin	filament	16,000	10,000-20,000
Microtubules	13 protofilaments	160,000	2×10^6-6×10^6

3.2.3   Tubulin

The tubulin that polymerizes to form MTs is actually a heterodimer of α-tubulin and β-tubulin. These two proteins are highly homologous and have 3D structures which are nearly identical. Although the similarity of α-tubulin and β-tubulin had long been suspected, the fact that tubulin has resisted crystallization for about 20 years prevented confirmation of this hypothesis until fairly recently. Nogales et al. [245] were able to perform cryo-electron crystallography to 3.7 $\mathring{A}$ resolution on sheets of tubulin formed in the presence of zinc ion. Fig. 3.3 has been produced from the Nogales data, via the protein data bank [25; PDB entry: 1tub], using MOLSCRIPT [179] and it makes the similarities between the two proteins clear. Each is composed of a peptide sequence more than 400 members long which is highly conserved between species. The amino acid sequences for these proteins may be compared given the data in Table 2.16 which lists the conventional one and three letter codes for the 20 naturally-occurring amino acids. Codes should be read from left to right and spaces are inserted each 10 residues for clarity.

The alpha and beta tubulin monomer each consist of two beta-sheets flanked by alpha helices to each side. Each monomer can be divided into three domains: an N-terminal nucleotide-binding domain, an intermediate domain that, in beta-tubulin, contains the taxol-binding site and a C-terminal domain that is thought to bind to motor proteins and other MAPs. The monomer structures are strikingly similar to the crystal structure of the bacterial division protein [197], which has been determined to 2.8$\mathring{A}$ resolution. FtsZ can form tubules in vitro and has been proposed to be a prokaryotic predecessor of the eukaryotic tubulins. The tubulins and FtsZ share a core structure of 10 beta-strands surrounded by 10 alpha-helices.

In Fig. 3.4 we have shown the electrostatic charge distribution in vacuum for the tubulin heterodimer that includes the two C-termini. Note that it is highly negative.

Based on the charge of the amino acid, the amino acids may be classified into 3 groups: those with a positive charge, those with a negative charge and the neutral residues. The size of the residue and its ability to react with other amino acids will affect protein folding and consequently function. The

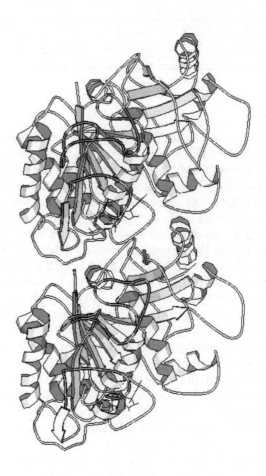

FIGURE 3.3:   A diagram of the tubulin molecule produced from the Nogales et al. [245] electron crystallography data shows the similarity between the α-subunit (upper half) and β-subunit (lower half). The stick outlines near the base of each subunit indicate the location of GTP when bound.

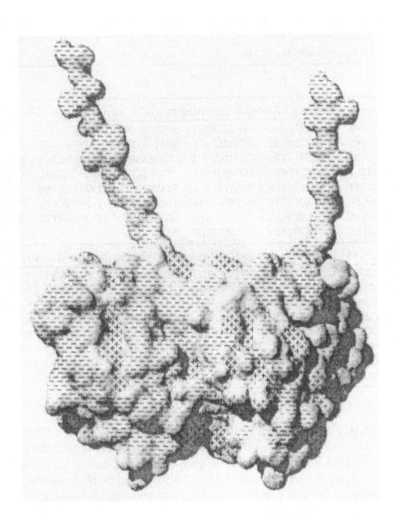

FIGURE 3.4: A map of the electrostatic charge distribution on an alpha-tubulin monomer microtubule.

sequences of a few representatives samples of tubulin have been retrieved from the Swiss-Prot protein sequence database [25] and are shown in Table 2.16 in chapter 2.

TABLE 3.4: The amino acid sequence of human α_1- and β_1- tubulin shows a high degree of homology. The sequence is given along with the total number of amino acids and the molecular weight of the molecule.

Human α_1-tubulin amino-acid sequence (451 amino acids, 50157 Da)					
MRECISIHVG	QAGVQIGNAC	WELYCLEHGI	QPDGQMPSDK	TIGGGDDSFN	TFFSETGAGK
HVPRAVFVDL	EPTVIDEVRT	GTYRQLFHPE	QLITGKEDAA	NNYARGHYTI	GKEIIDLVLD
RIRKLADQCT	RLQGFLVFHS	FGGGTGSGFT	SLLMERLSVD	YGKKSKLEFS	IYPAPQVSTA
VVEPYNSILT	THTTLEHSDC	AFMVDNEAIY	DICRRNLDIE	RPTYTNLNRL	IGQIVSSITA
SLRFDGALNV	DLTEFQTNLV	PYPRIHFPLA	TYAPVISAEK	AYHEQLSVAE	ITNACFEPAN
QMVKCDPGHG	KYMACCLLYR	GDVVPKDVNA	AIATIKTKRT	IQFVDWCPTG	FKVGINYQPP
TVVPGGDLAK	VQRAVCMLSN	TTAIAEAWAR	LDHKFDLMYA	KRAFVHWYVG	EGMEEGEFSE
AREDMAALEK	DYEEVGVHSV	EGEGEEEGEE	Y		

Human β_1-tubulin amino-acid sequence (444 amino acids, 49759 Da)					
MREIVHIQAG	QCGNQIGAKF	WEVISDEHGI	DPTGTYHGDS	DLQLDRISVY	YNEATGGKYV
PRAILVDLEP	GTMDSVRSGP	FGQIFRPDNF	VFGQSGAGNN	WAKGHYTEGA	ELVDSVLDVV
RKEAESCDCL	QGFQLTHSLG	GGTGSGMGTL	LISKIREEYP	DRIMNTFSVV	PSPKVSDTVV
EPYNATLSVH	QLVENTDETY	CIDNEALYDI	CFRTLRLTTP	TYGDLNHLVS	GTMECVTTCL
RFPGQLNADL	RKLAVNMVPF	PRLHFFMPGF	APLTSRGSQQ	YRALTVPDLT	QQVFDAKNMM
AACDPRHGRY	LTVAAVFRGR	MSMKEVDEQM	LNVQNKNSSY	FVEWIPNNVK	TAVCDIPPRG
LKMAVTFIGN	STAIQELFKR	ISEQFTAMFR	RKAFLHWYTG	EGMDEMEFTE	AESNMNDLVS
EYQQYQDATA	EEEEDFGEEA	EEEA			

In humans, no less than six α isotypes and ten β isotypes, in addition to the well-known γ tubulin, have been found so far [200]. Although the sequence of amino acids is highly conserved overall, certain regions of α_I-tubulin show divergence from α_{II}-tubulin and so on. Recent studies have shown that the differences in α-tubulin are more subtle than those in β-tubulin. Table 3.5 gives a comparison between the main β-tubulin isotypes in cows. The location of cells expressing that particular variant of tubulin are given along with the homology in percent with β_1 which is derived from a comparison of primary sequences. Tubulin isotypes and the relationship between their structure and function present an intriguing scientific puzzle. For example, β_I has been largely conserved in evolution; in β_{IV}, divergences have been conserved for over 0.5 billion years. Correlations between distribution and function indicate that β_{II}, and β_{VI} are often found in blood cells, β_{III} is frequent in tumors, a β_{III} dimer is more dynamic than β_{II} or a β_{IV}, while β_{II} and β_{IV} crosslink easily. The mitotic spindle contains $\beta_I, \beta_{II}, \beta_{IV}$; the nucleus: β_{II} and γ forms nucleation sites for microtubule assembly.

TABLE 3.5: Localization and Homology of Bovine β-Tubulin [200]

Isotype	Localization	Homology	Abundance in Brain (%)
β_1	everywhere, thymus	100.0	3
β_2	brain	95.0	58
β_3	brain, testis, tumours	91.4	25
β_4	brain, retina, trachea	97.0	11

There exist at present several different studies that confirm the existence of different conformational states of tubulin; however there is little quantitative information on the structural changes. The first indication that there was more than a single conformational state of tubulin came simply from observing the assembly of MT's. Tubulin bound to GTP, or assembly-ready tubulin, binds together and forms straight protofilaments. However, following polymerization and GTP hydrolysis GDP-bound tubulin forms curved protofilaments that sometimes close up on themselves to produce oligomer rings. Another manifestation of the different conformations came when Hyman et al. [122] measured the energy released from a slowly hydrolyzable analog of GTP known as GMPCPP. The energy released when the analog bound to tubulin was reduced compared to the quantity of energy released by the free molecule. The speculation was that the difference must go into changing the conformation of the tubulin dimer. In addition, when GMPCPP was bound to tubulin in a MT, the energy release was further diminished.

3.2.4 Microtubules

The thickest and perhaps the most multifunctional of all filaments are microtubules, which are polymers comprised of the protein called tubulin that we discussed above. They are found in nearly all eukaryotic cells [69]. The elementary building block of microtubules is an alpha-beta heterodimer whose dimensions are 4 by 5 by 8 nm that assembles into a cylindrical structure that typically has 13 protofilaments (see Fig. 3.5). The outer diameter of a microtubule is 25 nm and the inner diameter is 15 nm [13]. The monomer mass is approximately 55 kilodaltons. MTs are larger and more rigid than the actin microfilaments (MFs) or intermediate filaments (IFs), and thus they serve as the major architectural strut of the cytoplasm.

Microtubules act as a scaffold to determine cell shape, and provide a set of "tracks" for cell organelles and vesicles to move on. They are involved in a number of specific cellular functions such as: (a) organelle and particle transport inside cells (e.g., nerve axon) through the use of motor proteins, (b) signal transduction, (c) when arranged in geometric patterns inside flagella and cilia, they are used for locomotion or cell motility, (d) cell division. During cell division they form mitotic spindles which are indispensable for chromosome segregation, (e) organization of cell compartments (e.g., positioning of ER, Golgi and mitochondria). MTs perform these tasks by careful control over its assembly and disassembly (MT dynamics) and by interactions with microtubule associated proteins (MAPs).

Ledbetter and Porter [186] were the first to describe these tubules found within the cytoplasm whose structure has been well-established by light microscopy, immunofluoresence and cryo-electron microscopy. The tube is composed of strongly bound linear polymers, known as protofilaments, that are connected via weaker lateral bonds to form a sheet that is wrapped up to form a cylinder. While the electron crystallography of Nogales et al. [245] has shown that the α- and β-monomers are nearly identical, this small difference on the monomer level allows the possibility of several lattice types. In particular, the so-called MT A and B lattices have been identified (see Fig. 3.5). Moving around the MT in a left-handed sense, protofilaments of the A lattice exhibit a vertical shift of 4.92 nm upwards relative to their neighbors. In the B lattice, this offset is only 0.92 nm because α and β and monomers have switched roles in alternating protofilaments. This change results in the development of a structural discontinuity in the B lattice known as the seam. In addition to lattice variation, it is also known from experimental observation that the number of protofilaments may differ from one MT to another. Although 13 protofilaments is by far the most common situation *in vivo*, Chrétien et al. have observed that the protofilament number need not be conserved along the length of an individual MT. This leads to the emergence of structural defects in the form of excess or missing protofilaments in a single MT.

Microtubule formation takes place in relation to changes in cell shape, movement and in preparation for mitosis. It occurs under the following conditions:

- Critical concentration of tubulin   1 mg/ml

- Optimal pH = 6.9

- Strict ionic requirements (Mg, Ca, Li,...)

- Assembly can be inhibited by reduced temperatures ($7°C$)

The onset of assembly crucially depends on: (a) the temperature range [90], (b) the concentration of tubulin in the cytoplasm, (c) the supply of biochemical energy in the form of GTP.

It is important to note that tubulin contains two bound GTP molecules one of which is exchangeable and binds to the beta tubulin molecules while the other one that binds to alpha tubulin is fixed. Dimers polymerize head to tail into protofilaments (pf's), and each MT polymer is polarized. The plus end is

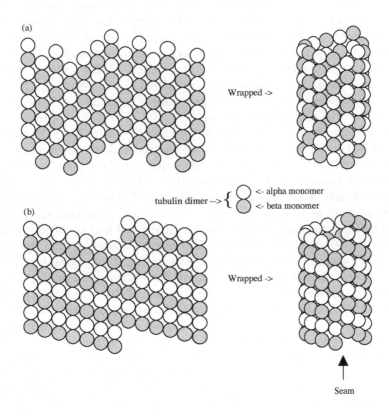

FIGURE 3.5: The 13A and 13B MT lattices: (a) in the A lattice, perfect helical spirals are formed, (b) in the B lattice, there is a structural discontinuity known as the seam.

a fast assembly end; the minus end is a slow assembly end [271]. Organization of the MTs in vivo is also polarized; minus ends are bound to a microtubule organizing center (MTOC) while plus ends extend outward to cell periphery.

The first stage of MT formation is called "nucleation". The process requires tubulin, Mg^{++} and GTP and proceeds fastest at 37 C. This stage is relatively slow until the microtubule is initially formed. During nucleation, an alpha and a beta tubulin molecule join to form a heterodimer. Then these attach to other dimers to form oligomers which elongate to form protofilaments. When a tubulin molecule adds to the microtubule, the GTP is hydrolyzed to GDP. Eventually the oligomers join to form the ringed microtubule. The second phase, called "elongation", proceeds much more rapidly. MT lengths may range from hundreds of nm to micrometers. MT's undergo irregular periods of growth, rapid shortening (catastrophe) and regrowth (rescue *in vitro*). Microtubules may vary in their rate of assembly and disassembly depending on the prevailing conditions.

The assembly dynamics at each end of the MT differ also. The so-called plus end is between three and five times as dynamic as the negative end. That is to say that both the growth and shortening of the positive end of the MT occur at rates at least three times those of the negative end. Tubulin half-life is nearly a full day; however, the half life of a given microtubule may be as small as 10 minutes. Thus, they are in a continual state of flux. This is believed to result from the needs of the cell and is called "dynamic instability" [195, 113]. This behavior is the consequence of GTP hydrolysis during MT polymerization. GTP caps on growing MTs are stable [178], but GDP caps are unstable leading to catastrophic events and a dramatic loss of polymerized tubulin. Tests have shown that microtubules will form normally with non-hydrolyzable GTP analog molecules attached. However, they will not be able to depolymerize. Thus, the normal role of GTP hydrolysis may be to promote the constant growth of microtubules as they are needed by a cell. Mathematical models of the dynamic instability use either the Monte Carlo approach [31] or the master equation formalism (see Appendix D for general exposition and Flyvbjerg et al. [104; 81] for details of application to microtubules). Both polymerized and unpolymerized monomers bind either GTP or GDP. Upon assembly, the energy of hydrolysis from GTP to GDP is conferred to the tubulin subunits but the reason for this is unknown. We believe that at least part of this relatively large amount of energy is stored in an MT in the form of stacking fault energy which, when a critical amount is exceeded, can be released in the form of an earthquake-like collapse of the entire MT structure. This hypothesis would be consistent with the observation by Hyman et al. [121] of a structural change accompanying GTP hydrolysis.

Above a tubulin concentration threshold, MT ensembles show a quasi-periodic, regular pattern of damped oscillations [36]. This indicates that interactions between individual microtubules must play a crucial role in affecting this drastic change in behavior. What is intriguing is how the stochastic individual behavior may result in smooth collective oscillations observed to take place at high tubulin concentrations [179]. The answer may simply be given by the application of statistical mechanics to ensemble averaging. A complete explanation of the above requires an application of the master equation formalism or, alternatively, the use of chemical kinetics [184, 116; 252]. Furthermore, it has been shown that microtubules will self-organize *in vitro*, leading to spatial pattern formation [265]. A mathematical description of this process can be provided in terms of reaction-diffusion equations [136].

The measured flexural rigidity for MT's corresponds to a Young's modulus of 1.4 GPa in normal MTs and can be raised by more than a factor of two to 3.4 GPa when hydrolysis is prevented. In an independent study, a Young's modulus of 4.6 GPa was derived from the buckling of microtubules that required the application of a 10 pN force. The Young's modulus of F-actin has also been measured and is of the same order of magnitude though conflicting measurements make actin both more and less rigid than MTs. This number is sufficiently large to imply that the cell must rely on depolymerization rather than deformation to effect a shape change. This measured value of Young's modulus indicates that bending of a protofilament into an arc with a radius of curvature of 20 nm, as observed by Mandelkow et al., would require about 0.14 eV/dimer (3.2 kcal/mol) which is slightly less than the energy of GTP hydrolysis, 0.22 eV (5.1 kcal/mol). This is believed to explain the difference observed in the free energy release when free floating GTP is hydrolyzed compared to the hydrolysis of GTP bound to a MT.

Normal cell organization prevents the existence of free MT's since MT's become attached to centrosomes or organized within minutes. MTOC's contain MAP's that bind minus ends of MT's and nucleate MT assembly. Within the cell body, the majority of the MT's emanate from a centriole. The negative ends of MT's are anchored at these microtubule organizing centers. The MTs in situ are interconnected and intraconnected by microtubule associated proteins (MAP's). MAP's have a stabilizing effect on the dynamics of MT's as mentioned above. At the base of cilia and flagella, MTOC's are called basal bodies or centrioles. In interphase cells, the MTOC is found in the centrosome (cell center) which contains two centrioles plus pericentriolar material.

In the cell itself, microtubules are formed in an area near the nucleus called the "aster". Usually the minus end is the anchor point. There are regulatory processes that appear to control MT assembly in a cell. Microtubule growth is promoted in a dividing or moving cell. Meanwhile, microtubule growth is repressed in a stable, polarized cell such as a neuron. The cell can provide a GTP cap on the growing end of a microtubule to regulate further growth. This happens when the tubulin molecules are added faster than the GTP can be hydrolyzed. Thus, the microtubule becomes stable and does not depolymerize. Another way of capping a microtubule is to put a structure at its end, such as a cell membrane.

Note that assembling microtubules exert pushing forces on chromosomes during mitosis. The force that a single microtubule can generate was measured by attaching microtubules to a substrate at one end and causing them to push against a rigid barrier at the other end. The subsequent buckling of the microtubules was analyzed to determine both the force on each microtubule end and the growth velocity.

Finally, it should be mentioned that MT's and their individual dimers possess a net charge of approximately -50 e at neutral pH and a dipole moment on the order of 1800 debye. Thus, MT's are electrets, i.e., oriented assemblies of dipoles which are predicted to have piezoelectric properties due to the coupling between elastic and electric degrees of freedom of the protein [46]. Interestingly, the predicted dipole orientation of tubulin units in a MT is that of an almost perpendicular direction to the surface of the cylinder resulting in a net near zero value of the dipole moment for the microtubule as a whole. It has been recently emphasized that microtubule networks play an essential role in cellular self-organization phenomena which include pattern formation resulting from reaction-diffusion instabilities in the mechanisms of cytoskeletal self-organization. We believe that electrostatic interactions are crucial in the mechanism of self-organization of the cytoskeleton that is so prominent in mitosis.

Microtubule associated proteins (MAP's) are tissue and cell type specific. Several classes of MAP's with different functions have been identified. They are high molecular weight proteins (200-300 K) or the so-called tau (20-60 k) proteins. General classes are minus-end binding (i.e., MTOC), plus-end binding (e.g., the kinetochore of mitotic chromosomes [135], polymer severing, polymer stabilizing and cross-linking (i.e., MAP-2 and tau in neurons), and motor proteins (the dyneins and the kinesins, see [15] that are discussed separately later on). One of their domains binds to tubulin polymers or to unpolymerized tubulin. This speeds up polymerization, facilitates assembly and stabilizes the microtubules. The other end binds to vesicles or granules. MAP's may vary with the cell type. The best examples of MAP's are found in neurons where diverse patterns of MAP attachments to MT's can be found. Furthermore, it is believed that some of these MAP's may bind to special sites on the alpha tubulin that form after it is assembled into a microtubule. These are sites where a specific molecule is acetylated or the tyrosine residue is removed from the carboxy terminal. They are important marker sites for stabilized microtubules, because they disappear when microtubules are depolymerized. In summary, MAPs accelerate polymerization, serve as "motors" for vesicles and granules and essentially control cell compartmentalization.

3.2.5 Actin Filaments

Microfilaments have the smallest diameter of all the cytoskeletal filaments and are the most common. They are composed of a contractile protein known as actin and consequently are often referred to as actin filaments. Actin is the most abundant cellular protein. Microfilaments are fine, thread-like protein fibers, 3-6 nm in diameter (see Fig. 3.6). Microfilaments' association with the protein myosin is responsible for muscle contraction. Microfilaments can also carry out cellular movements including gliding, contraction and cytokinesis.

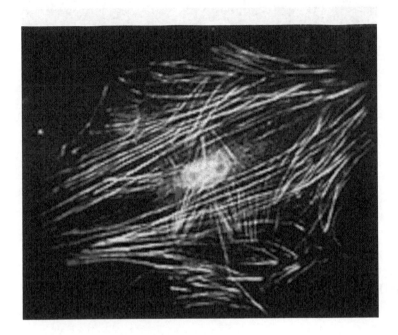

FIGURE 3.6:   Microfilaments with fluorescent label.

Actin is one of the most abundant proteins in eukaryotic cells. G- and F-actin are reversible structural states of actin. Filamentous actin (F-actin) is a polymer of the protein known as globular actin (G-actin). These strands are capable of forming both stable and labile components within cells. F-actin is the main component of thin filaments in muscle sarcomere, but it can also be found in non-muscle cells. Actin is often associated with other proteins in the cytoskeleton, providing it with a dynamic component [84]. Actin is a structural protein that has 375 residues (see Fig. 3.7) which has been highly conserved in evolution and whose globular monomer form has a weight of 42 kilodaltons. It binds ATP and Mg^{2+}.

G-actin monomers assemble into F-actin filaments of two-stranded geometry (see Fig. 3.8) with a mass density three times as high as that of spectrin filaments. Polymerization of actin [209] has two different rates at the poles of nucleating filament barbed and pointed ends. ATP actin binds faster to the barbed end. ATP is then hydrolyzed to ADP after incorporation.

The length distributions of actin filaments are exponential with a mean of about 7 μm. The polymerization of F-actin from G-actin is a largely mono-tonic process that is dependent on ATP. The F-actin assembles primarily

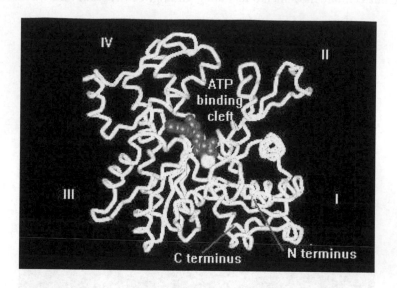

FIGURE 3.7: Structure of the G-actin monomer.

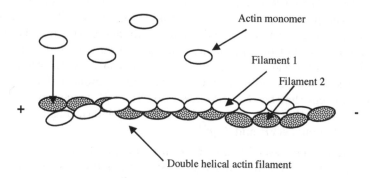

FIGURE 3.8: Formation of the F-actin polymer.

through a standard nucleation/elongation mechanism to a saturation value which depends on pH [269]. However, this length is independent of the initial concentration of actin monomer, an observation inconsistent with a simple nucleation-elongation mechanism. With the addition of physically reasonable rates of filament annealing and fragmenting, a nucleation-elongation mechanism can reproduce the observed average length of filaments in two types of experiments: (1) filaments formed from a wide range of highly purified actin monomer concentrations, and (2) filaments formed from 24 mM actin over a range of CapZ concentrations. Once assembled, microfilaments have a diameter of about 8 nm (Fig. 3.9).

In the cell, F-actin can be found exhibiting characteristic forms as follows:

- thin filaments in striated muscle

- individual filaments in cortical cytoplasm

- bundled filaments (stress fibrils) in interphase fibroblasts

- short actin bundles in microvilli

- filament bundles ending in focal contacts.

- filaments running parallel with the membrane at the membrane-cytoplasm interface

8 nm

FIGURE 3.9: The double stranded helix of an actin filament. Each of the F-actin strands has a polarity indicated by the arrows (each sphere represents a monomer of G-actin).

Microfilaments are found linked together by actin-associated proteins and congregate into one of three major forms (Fig. 3.10). The polymerization dynamics and filament organization of actin into spatial structures has been modelled, for example, by Edelstein-Keshet [92]. In the configuration of parallel strands, microfilaments often form the core of microvilli, while in an anti-parallel arrangement, actin may act in conjunction with myosin to bring about muscle contraction in the presence of ATP [118]. Microfilaments are often found with the lattice configuration near the leading edge of growing or motile cells where they provide greater stability to the newly formed region. New actin filaments are nucleated at the leading edge of the cell's growth and trailing microfilaments are disassembled.

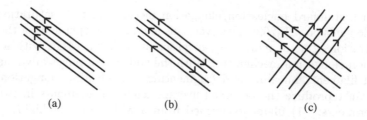

(a)   (b)   (c)

FIGURE 3.10:   The three main arrangements of microfilament aggregation which form with the aid of actin-crosslinking proteins: (a) parallel bundles, (b) contractile bundles and (c) gel or lattice like arrangements.

3.2.6   Actin Binding Proteins

Actin filaments are known to interact with at least 48 other proteins. Cytosolic myosin motors drive along actin filaments. Actinin interacts with actin and a variety of other proteins including integrins. Spectrin is another multifunctional actin-binding protein that binds actin filaments to transmembrane proteins. Talin is a 230 kD protein concentrated at regions where actin filaments attach to and transmit tension across the cytoplasmic membrane to the extracellular matrix. It is absent at cell-cell interaction sites. It is also thought to interact with the integrins and vinculin, a 117 kdal protein found at all cell-cell and cell-matrix interaction sites. These are focal contacts (adherens junctions), dynamic structures related to cell adhesion and movement. Tensin is also found in close proximity to actin and integrins. Profilin is a small actin-binding protein with high expression in cytoplasm of non-muscle cells. PIP and PIP2 cause dissociation of complexes and are found to bind at the barbed end of the filaments. Actinin is involved in non-muscle cells where some actin bundles interact with the membrane. Gelsolin caps fast-growing ends of filaments. Gelsolin may also be involved in filament disassembly. It is a representative of a number of proteins forming a family believed to act in regulating actin polymerization. For example, cofilin participates in depolymerization of actin filaments further back from the leading edge within the lamellipodium. A summary of the actin-binding proteins is given in Table 3.6.

At the leading edge of a moving cell is the lamellipodium. The forward extension of a lamellipodium is driven by actin polymerization. Lamellipodia contain an extensively branched network of actin filaments, with their plus ends oriented toward the plasma membrane [298]. Several proteins already discussed above participate with actin in generating the forward movement of the lamellipodium in the so-called ameboid movement.

For example, profilin, which promotes ADP/ATP exchanges by G-actin, to yield the ATP-bound form with the ability to polymerize, is located at the leading edge of an advancing cell. Arp2/3, which binds to the sides of actin fil-

TABLE 3.6: Summary of Actin-Binding Proteins

Function	Protein type(s)
bind monomeric actin and prevent polymerization	profilins
cap one end of the filament	capZ36
sever filaments and bind to barbed ends	gelsolin
bind laterally along the filament	tropomyosin
move along the filaments	myosins
connect filaments and network bundles	α-actinin synapsin villin

aments and nucleates the growth of new filaments, is associated with the actin filament network in lamellipodia. Various actin cross-linking proteins stabilize the actin network in lamellipodia. Newly formed actin filaments are stable, as an advancing lamellipodium moves past them, until they disassemble further back from the edge.

3.2.7 Intermediate Filaments

The cytoskeleton of all eukaryotic cells contains intermediate filaments (IF's) which are thinner than MT's, thicker than actin and more stable than both. These are microscopic ropes with protein subunits in the 10-15 nm range. Intermediate filaments have a fairly complex hierarchical organization in which two protein building blocks generate a dimer through an anti-parallel association of two molecules in a coiled-coil configuration. Pairs of dimers form linear protofilaments 2-3 nm in width. The IF itself is a bundle of 8 protofilaments forming a cylinder of 10 nm diameter. IF's form by association of helix rod domains. Parallel unstaggered two-stranded helical coiled coils of 44-54 nm size depending on IF are formed. Zipper structure of hydrophobic residues is evident. IF aggregation is achieved when two dimers form a tetramer (see Fig. 3.11). Protofilaments form longitudinal extension of tetramers. A number of possible topologies of filaments have been found, among them: parallel, antiparallel, staggered and registered. A significant role has been observed of end domains in IF structures. Tail domains may protrude from the main body of a filament making it accessible for interactions with other parts of the cytoplasm. The pool of unpolymerized IF subunits is insignificant since newly synthesized proteins polymerize very quickly. Filaments appear to require no cofactors or energy input for their assembly.

There are a variety of IF's including vimentin, desmins, nuclear lamins and the keratins of epithelial cells. IF's appear to provide resilience, tensile strength and plasticity to the cell and may provide the main cytoskeleton form with complex arrays in cells. These molecules are quite unlike the globular molecules, tubulin and G-actin since they form non-polar structures. The overlap between dimers allows for the filaments to stretch and, as a result, intermediate filaments are able to withstand large stresses without breaking. An extensive network of intermediate filaments surrounds the nucleus forming what is known as the nuclear envelope and it extends to desmosomes in the plasma membrane. The formation and disassembly of the nuclear lamina is regulated by phosphorylation. This allows for the elimination of the nuclear envelope prior to mitosis. The filaments also extend out into the cell periphery where they act to maintain cell integrity and may connect with the cell membrane. Specific cells, such as neurons, have their own distinctive intermediate filaments, known as neurofilaments.

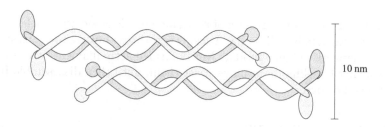

10 nm

FIGURE 3.11:　Schematic of an intermediate filament. The filamentous protein tetramer is non-polar and composed of two coiled-coil dimers.

Their tissue specificity is very high. IF's may vary substantially in sequence since there are at least 40 genes encoding for them. IF expression closely correlates to the behavior and role of the cell. IF's of different sequences have the same overall structure with a rod domain which is a long central domain with a coiled coil. Below we list the key types of the IF's and briefly describe their properties:

a) **Type I Type II keratins** are expressed in epithelia, sheet tissues that delimit functional components which are expressed in epithelia. Keratin filaments are found in the cytoplasm and loop into desmosomal plaques. They maintain the physical integrity of the cell.

b) **Type III vimentin** can form homopolymers expressed by fibroblasts, hemopoetic and endothelial cells in embryogenesis formed in primary mesoderm and endoderm cells.

c) **Desmin IF type III** is found in smooth, striated and cardiac muscle

d) **Type IV Neurofilaments:** (NF-L, NF -M, NF-H) are heavily phosphorylated, especially NF-M and NF-H. Charged tails point out from the filament and may space filaments and regulate the axonal diameter.

e) **Type V lamins:** form a network rather than individual filaments. Significant differences exist in both rod and tail domain structures. Nuclear lamin proteins provide an envelope of the insoluble cytoskeletal material below the nuclear membrane. Lamin disassembly at mitosis is triggered by phosphorylation.

IF control is thought to be produced by phosphorylation which alters fibril assembly/solubility. Other known post-translational modifications include the glycosylation of keratins and lamins. There is a general response of IF's to the cell cycle, movement, differentiation, etc. Mutations of IF's cause serious pathologies, e.g., the seven keratin diseases known, of which the most common is the epidermolysis bullosa simplex in which basal epidermis cells detach from underlying connective tissue resulting in severe blistering.

Single-stranded DNA and RNA associate readily with many IF structures. Binding is predominantly through the H1 domain and it interferes with fibril assembly. DNA-lamin interactions are thought to be essential for the integrity of the nuclear chromatin. **IF associated proteins** are: Plectin (300Kd) which associates with several cytoskeletal components and links to membranes directly, Lamin receptors and lamin-associated proteins.

3.3   Tubulin Isotype Homology Modelling - from [58, 327]

In addition to α and β tubulin, MTs often require the presence of additional proteins, Microtubule Associated Proteins (MAPs), for correct assembly. Further, three other tubulins are required for efficient microtubule assembly *in vivo*. The first, γ tubulin, while not essential for MT assembly, is found at microtubule-organizing centers (MOCs) and nucleates MTs and establishes their polarity [247]. Two additional widespread tubulins, δ and ε, have uncertain roles in MT assembly, although proposed models describe possible roles [158].

Like α, β and γ tubulins, δ and ε tubulins exist in many eukaryotic organisms as a family of related proteins referred to as isotypes or isoforms. An apparent conservation of the three-dimensional pattern of conserved amino acids has been observed in several families of structurally homologous proteins [223]. This conservation points to the existence of a folding nucleus, suggesting that buried residues within a protein tend to evolve more slowly than those on the surface. Interestingly, mapping evolutionary rates onto the sequences of α and β tubulin produced the opposite pattern [271]. Here, the relative rate of sequence divergence at sites within the core of each tubulin isotype seems to occur more frequently than at sites on the protein surface. This phenomenon is most likely due to the considerable number of intermolecular interactions that occur between tubulin monomers when assembled into MTs as well as their interactions with other proteins and ligands. In general, surface residues within the tubulin superfamily seem to be conserved; however variation does seem to be clustered within the protein surface that comprises the longitudinal interface between protofilaments within the MT. This implies that inter-dimer interactions between protofilaments may be key to understanding the stability properties that each isotype confers to the MT. In humans, seven α and eight β tubulin isotypes have been identified [199, 268, 204]. At the molecular level, the role and interactions of tubulin with other proteins is complex and appears to involve structural variations between the α and β isotypes [264]. The extensive distribution of α and β tubulin isotypes may provide a link between the structure of tubulin and their influence on the physical properties of MTs. Experimental evidence indicates MT dynamics is altered by inclusion of different tubulin isotypes [251]. This implies that when the equilibrium of MT assembly and disassembly is disrupted, the cell can respond by producing a specific tubulin isotype mixture to restore a normal balance. Small differences in the structures and therefore chemical affinities may translate into significant increases in the growth and catastrophe rates for MTs. Differences in observed MT assembly and disassembly rates could also be due to changes in the electrostatics of tubulin, in particular, due to its net charge and, when screened by counter-ions, due to its dipole moment and orientation. Tubulin electrostatics, although significantly screened by counter-ions, does affect MT assembly, influencing dimer-dimer interactions over relatively short distances, less than 5 nm [26]. Sept et al. [290] demonstrated by differing longitudinal offsets of protofilaments relative to each other that the two primary types of MT lattices, i.e., type A and type B, closely correspond to the calculated local energy minima. As one of the applications of our electrostatic calculations, we estimate in this paper the strengths of the dipole-dipole interactions within MT lattices composed of different human tubulin isotypes. It is clear that cellular tubulin isotype levels change in response to external conditions. An example is overexpression of β tubulin isotype III (β_{III}) after exposure to agents like paclitaxel [262, 194, 134, 233]. It is of great interest to determine what makes this particular isotype special with respect to drug resistance. It is equally intriguing to find out why different isotypes exhibit different dynamic instability rates. For illustration purposes, Table 3.7 lists an overview of the β tubulin distribution in normal human cells. Current anti-tubulin drugs bind with only slight preferences for one isotype over another [169, 30]. For example, vinca alkaloids bind best to β_{II} [169], suggesting why they are useful in treating leukemias and Hodgkin's lymphoma, since in these cancers lymphocytes express β_{II} while normal ones do not [133]. Key is that cancerous cells do not express the same composition of tubulin isotypes as the non-cancerous cells from which they derive [285]. Therefore, a drug highly specific for an isotype in cancerous cells could selectively injure those cells, while doing less damage to healthy cells. Consequently, knowledge of tubulin isotype structural and functional properties is extremely valuable, as we might exploit their differences in rational drug design.

The tubulin sequences generally contain more acidic than basic residues and accordingly have a negative overall electrostatic charge. In an overly simplistic analysis we could expect that these negatively charged objects would repel each other, preventing MT formation. Reality is more complex, how-

TABLE 3.7: Tissue Distribution of β Tubulin Isotypes in Normal Cells

Isoform	Organ Expression	Cellular Expression
β_I	most tissues	most cells
β_{II}	brain, nerves, muscle; rare elsewhere	particular cell types
β_{III}	brain	neurons only
	testis	Sertoli cells
	colon (very slight amounts)	epithelial cells only
β_{IVa}	brain only	neurons and glia
β_{IVb}	most tissues (not as widespread as I)	ciliated cells, lower in others
β_V	unknown	unknown
β_{VI}	blood, bone marrow, spleen	erythroid cells, platelets
β_{VII}	brain	unknown

ever, as these systems are electrostatically virtually neutral in ionic solutions, and at long ranges counterions in the solvent annul tubulin's net negative charges. Experimentally, tubulin has been found to have an unscreened negative charge of approximately 0.19 e per dimer which was found at pH 6.8 in the presence of 1-5 V/cm electric fields [303]. At short ranges the charge is less fully screened and other solvent-related screening and dielectric effects may be relevant. However, in these cases, some regions of interaction will be non-trivially closer together than others, and close charges may dominate more distant charges in the electrostatic field. The extreme of this occurs between monomers, where contacting regions have interactions strong enough to form a protein-protein bond. The net charge or monopole is perhaps the simplest approximation to the complicated electrostatic field around tubulin and dominates the multipole expansion in vacuum. Slightly more complex is the electrostatic dipole moment, which has opposing positive and negative directions along a central axis, but is radially symmetric about this axis. Like the monopole, the dipole does not adequately represent the complex field near tubulin's surface, and is subject to attenuation by solvent mediated effects at longer ranges, but having both a position and an orientation, a dipole gives more complex interactions and a better approximation in an aqueous environment [326]. With specific charges assigned to particular locations of the protein's atoms in space, we can calculate a dipole moment about a suitably chosen origin of the co-ordinate system. If the protein has zero net charge, this dipole moment is obviously independent of the choice of origin. However, when this is not the case, care must be taken in offsetting the net charge and its fictitious contribution to the dipole moment if the coordinate system's origin is not centered properly. In this work we assigned charges to atomic coordinates based on the amino acid of the protein. It is important to note that no absolute value can be stated for the charge of the entire atom as it depends on its surrounding and that many such models of atomic charge have been developed for molecular simulations. The dipole moment then calculated is only an estimate, and results from different choices of the charge model generally differ albeit not significantly.

The Uniprot Knowledgebase, which consists of the Swiss-Prot release 42.1 and TrEMBL release 25.12 (March 15, 2004) [18], was downloaded and filtered for amino acid sequences that were annotated with the term "tubulin", which produced a set of 2294 entries. This set was then filtered to remove entries labeled as fragmentary, producing a subset of 652 sequences. Using default parameters within Clustal W (version 1.82) [137] these were aligned and using cluster analysis assembled into a guide tree (dendrogram). Many of the sequences were not tubulin, but unrelated proteins with tubulin in their name (e.g., tubulin tyrosine ligase or tubulin-folding cofactor D). Because these sequences differ from those of tubulins, they cluster separately in the dendrogram, providing a clear delineation between 557 tubulin-like and 95 non-tubulin sequences. A second dendrogram guide tree was then produced using the tubulin-like sequences. This second tree identified three major branches, with 190 α tubulins, 285 β tubulins and 82 γ, δ and ε tubulin sequences, respectively. Following classification of the tubulin sequences comparative models of each were created using the PDB with 1JFF and 1SA0 $\alpha\beta$ -dimer structures as preferred templates. These are accurate and representative structures of the two crystallization approaches used. In 1SA0 the colchicine molecule is at opposite ends of the two dimers and each dimer has a different interaction with the stathmin-like domain. As the 1JFF dimer is in a different crystal packing state from those of 1SA0, all three dimers have a distinct environment. After the models were built, the physical properties of the tubulins inspected were estimated from these models using analysis tools within Gromacs v. 3.3 [191]. The models were imported using pdb2gmx with the GROMOS96 43a1 force field [331] providing atomic charges, size and hydrogen atom locations. As a side-effect, the net charge of the sequence is reported. Connolly solvent accessible surfaces [72] were then calculated with g_{sas}. After using editconf to shift the origin to the geometric centre, which is expected to be a good approximation of the protein centre-of-mass, $g_{dipoles}$ was used to calculate a dipole moment for each structure investigated. The dipole moment components were used to find the resultant magnitude and orientation, so that use of these properties could be made for the calculation of dipole-dipole energies when the various tubulin dimers have been embedded into an MT structure. Additionally, we have identified the net charges on the most variable and perhaps most functionally important structure of tubulin, its carboxy-terminal domain. The numerous tubulin sequences provide an opportunity to create a library of tubulin models, from which we can calculate physical characteristics that can be correlated with biological roles in microtubules. These characteristics include volume, surface area, net charge and dipole moments. We present our results below.

3.3.1 Solvent Accessible Surface Area

The solvent accessible surface (SAS) areas are largely the same between the α and β tubulins with mean values of approximately 220 nm^2, and standard deviations of $10nm^2$ and $30nm^2$, respectively. This difference in variation is visible in Figs. 3.12 and 3.13 where surface areas for the α and β tubulin families are compared with the estimated dipole moments, respectively. Little correlation is evident between these two physical properties.

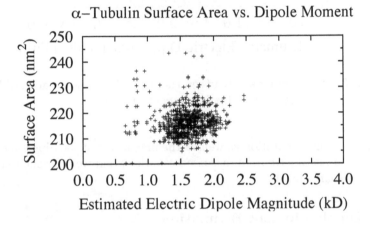

FIGURE 3.12: Comparison between the α tubulin's solvent accessible surface areas and their estimated dipole moment magnitudes.

3.3.2 Net Charge

Figs. 3.14 and 3.15 plot a distribution of the values of the net electrostatic charge and the dipole moments of the tubulin models. The β tubulins are on average more highly charged with an average value of approximately $-24e$ compared to the $-22e$ average for the α tubulins, although the β tubulins have a slightly broader distribution (standard deviation of $3.0e$ versus $2.6e$). There appears to be little if any correlation between the dipole moment and the net charge as might be expected. Further, it should be kept in mind that the electrostatic charges are greatly weakened in solution due to the presence of counter-ions as we mentioned earlier. On the other hand, the values of the dipole moment may be strongly affected by the hydration shell of water molecules on the tubulin surface. However, of great interest in connection with polymerization of tubulin into microtubules and with protein-protein and drug-protein binding is the actual distribution of charges on the surface

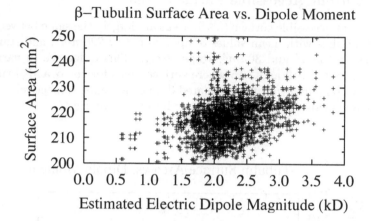

FIGURE 3.13:   Comparison between the β tubulin's solvent accessible surface areas and their estimated dipole moment magnitudes.

of tubulin and the orientation as well as the magnitude of the dipole moment. Some of these issues have been discussed elsewhere and we refer the reader to the original source [275] .

3.3.3   Dipole Moment Estimation

As seen in Fig. 3.16 and 3.17, α tubulins have dipole moments calculated about their centers in the range of 1,000-2,000 D, while the β tubulins' dipole moments are on average stronger with values ranging between 2,000-3,000 D. These values will be used below in this paper to evaluate the strength of dipole-dipole interactions within an MT.

3.3.4   Human Repeats and Dipole-Dipole Interactions

The 100 repeats of the human tubulin sequences allow measurements of the reproducibility of the obtained results. In particular the dipole moment estimates have roughly a 300 D standard deviation, and the surface areas a 10 and 30 nm^2 standard deviation for the two tubulin subfamilies, respectively. These values suggest that much of variation between tubulin sequences may be due to the computational artifacts related to Modeller variability and not any necessarily due to the intrinsic differences arising at the sequences level. To describe the geometry of a microtubule, we use a cylindrical coordinate system. We follow the MT description given by [190], and place the centers of each dimer at a radius of 112 $\mathring{A}$ and in 13 protofilament-like rows at the 13 angular multiples of $(360/13)°$. Along these rows, the centers are every 81.2 $\mathring{A}$. The interaction energy between two dipoles $\vec{p_1}$ and $\vec{p_2}$ separated by $\vec{r}$ is

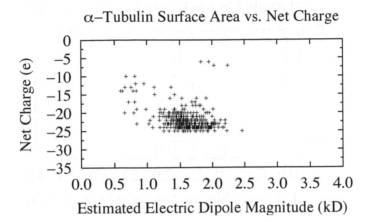

FIGURE 3.14: Comparison between the α tubulin's dipole moments and their net monomer charges.

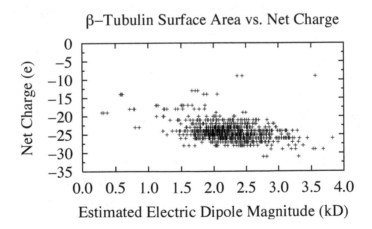

FIGURE 3.15: Comparison between the β tubulin's dipole moments and their net monomer charges.

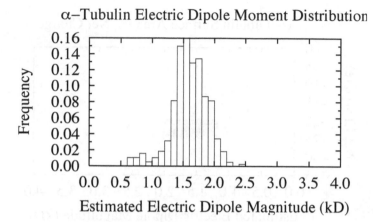

FIGURE 3.16: A histogram of dipole moment magnitudes estimated for α tubulin sequences. The two magnitudes resulting from the use of two models have been averaged, and this mean value sorted.

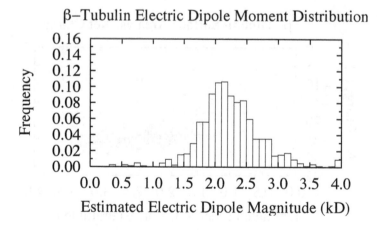

FIGURE 3.17: A histogram of dipole moment magnitudes estimated for β tubulin sequences. The three magnitudes from the three models has been averaged, and this mean sorted.

given by the equation

$$U = \frac{1}{4\pi\varepsilon\varepsilon_0}\left(\frac{\vec{p_1}\cdot\vec{p_2}}{r^3} - 3\frac{(\vec{p_1}\cdot\vec{r})(\vec{p_2}\cdot\vec{r})}{r^5}\right) \tag{3.6}$$

A dipole at each dimer center may be described as $\vec{p} = (p_r, p_\theta, p_z)$ with a radial, tangential and longitudinal component. The interaction energy of two dipoles of known positions may be rewritten as

$$4\pi\varepsilon\varepsilon_0 U = c_1 p_{r1}p_{r2} + c_2 p_{r1}p_{\theta2} + c_3 p_{r1}p_{z2} + c_4 p_{\theta1}p_{r2} + c_5 p_{\theta1}p_{\theta2}$$
$$+ c_6 p_{\theta1}p_{z2} + c_7 p_{z1}p_{r2} + c_8 p_{z1}p_{\theta2} + c_9 p_{z1}p_{z2} \tag{3.7}$$

where c_i are known geometry-dependent constants depending on $\vec{r}$ and rotation of the coordinate system components between the dipole locations. To study the interaction energy between a reference dipole and those of the adjacent protofilament as the adjacent protofilament is displaced along its length, we take several dipoles from the adjacent protofilament near the reference and use a computer to evaluate the c_i between adjacent dipoles and the reference. By having all the adjacent dipoles identical, and assuming $4\pi\varepsilon\varepsilon_0$ uniform, the c_i can be summed to obtain the same form as in Eq. 3.7. We considered the adjacent section of length $2l$, expecting that considering longer sections would approach a limit, as more separated dipoles have weaker interactions. We placed the reference dipole at $z = 0\mathring{A}$ and the adjacent dipoles at $z(81.2\mathring{A})(i+d)$ where i is an integer between $-l$ and $+l$, and the displacement, d, varied from 0 to 1. At $d = 0$, we found that a limit indeed is rapidly approached in our calculations, with little difference in the summed c_i between $l \geq 10$, and $l = 10000$. For $l = 1500$ and $d = 0, 0.01, 0.02, \ldots 1.00$ we have calculated the summed values of the coefficients c_i. These vary nearly sinusoidally with displacement, see Fig. 3.18. Fits to three low-frequency Fourier components, $c_i(d) = A_i + \sum_{k=1}^{3} B_{ki}\cos kd + C_{ki}\sin kd$, give coefficients where either the sine or the cosine series is small ($\sim 10^6$) relative to the other. In the significant series the second-order terms are all less than 5 % of the first-order terms, and the third-order terms less than 3 % of the second-order terms. In only two of the nine cases are the constant terms significantly non-zero. Thus we take only the first-order and constant terms and get the approximations $c_i(d) = A_i + B_i \cos d$ or $c_i(d) = C_i \sin d$. Repeating the above with separate dipoles for the α and β monomers can be quickly done. We place α dipoles at the locations discussed above obtaining the same result. A row of β dipoles placed midway between the dipoles of the α row has the same form but an additional displacement of one-half must be included in the calculations. Similarly, a reference β dipole may be placed at $z = 40.6\mathring{A}$ which is the same as subtracting one-half from the displacements. A similar β-β interaction requires no adjustment of the offsets. By summing these four contributions a single sum is again obtained. For simplicity we use only one α and one β monomers. Then, due to symmetry many coefficients cancel and the result

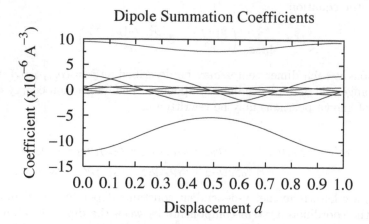

FIGURE 3.18: The nine coefficients c_i as functions of the displacement, d. Note the high level of symmetry, and that two of the curves are identical.

may be rewritten as $4\pi\varepsilon\varepsilon_0 U = A + B\cos d$ where both A and B have the form:

$$c_1(p_{r\alpha}^2 + p_{r\beta}^2) + c_2(p_{\theta\alpha}^2 + p_{\theta\beta}^2) + c_3(p_{z\alpha}^2 + p_{z\beta}^2) + c_4(p_{r\alpha} + p_{r\beta})$$
$$+ c_5(p_{\theta\alpha} + p_{\theta\beta}) + c_6(p_{z\alpha} + p_{z\beta}) \quad (3.8)$$

for appropriate choices of the coefficients c_i. We considered all the human $\alpha\beta$ tubulin pairs in our set of replicate runs, and substituted each calculated dipole moment into the resulting energy formula. As A, the magnitude of the oscillation is typically small compared to B, the offset term; we now ignore the sinusoidal term. The results show a great deal of similarity, with the five α monomers being largely indistinguishable, as are many of the β monomers. Only the Q99867 sequence is significantly different. The distribution of interaction energies for each β tubulin sequence is shown in Fig. 3.19. A similar calculation can be done within a protofilament by omitting the self-interaction from the summation. Since displacement makes no sense in this case, only a single case needs to be considered, avoiding the sinusoidal term from the outset. Again, only Q99867 stands out (see Fig. 3.20 for illustration). In a recent paper, Schoutens [294] estimated the magnitude of dipole-dipole interactions in an MT structure and concluded that it has a destabilizing effect which can contribute to as much as 50-60% of the elastic energy of the lattice. Since the absolute values of the dipole-dipole energy based on our calculations depend on the dielectric constant of the protein and the medium, whose values in turn change depending on the conditions of the solution (pH, ionic concentrations, etc.), it is rather complicated to give a precise number. On the other hand, we see that there is some dependence on the isotype chosen to form an MT lattice which may at least partially explain the differences in the tubulin isotypes' dynamics.

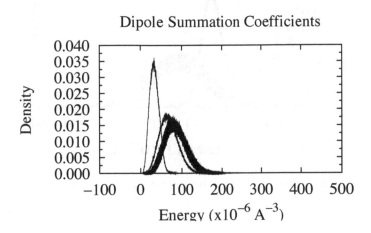

FIGURE 3.19: The per dimer interaction energies with an adjacent protofilament in a microtubule.

Structures of α and β tubulin are known to be similar, being nearly indistinguishable at approximately 6 Å, with only 40 % amino acid identity [190]. Since sequences within the α and β tubulin groups are more similar to each other than to those sequences belonging of the other group, it is reasonable to believe that any given sequence should have a structure very similar to that of another member of its group. Further support for this comes from the published structures of [245] and [198] which are of a porcine sequence, but fit to structural data from inhomogeneous bovine samples. Accordingly, crystallographic structures can be used as a framework to produce model structures of different sequences with some confidence. In practice, regions of consecutive amino acids absent in the templates produce the largest variabilities in the models. In one case (1JFF A chain) this was large enough that we find the models to be implausible. We believe that this core library of over 500 α and β tubulins is likely large enough to sample the diversity of tubulin variation, therefore no attempt to systematically add to this set from other data sources has been made here, although select sequences of interest from other sources have also been studied separately. Indeed our results suggest that variation from the model generation process is the dominant source of variation that we see. Isotype composition has a demonstrable effect on microtubule assembly kinetics [251]. This could be due to changes in the electrostatics of tubulin, which although significantly screened by counter-

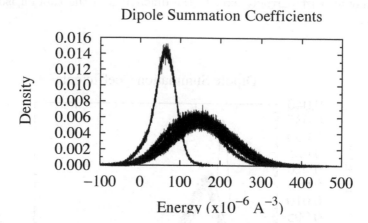

FIGURE 3.20: The per dimer interaction energy within a microtubule protofilament.

ions does affect microtubule assembly by influencing dimer-dimer interactions over relatively short distances (approximately 5 nm) as well as the kinetics of assembly. These short-range interactions have been studied by calculating the energy of protofilament-protofilament interactions [290]. Through this work, it was concluded that the two types of microtubule lattices (A and B lattices differ in the relative positions of the protofilaments relative to each other) correspond to the local energy minima. The dipole moment of tubulin could also play a significant role in microtubule assembly and in other processes. For example, this could be instrumental in the docking process of molecules to tubulin and in the proper steric configuration of a tubulin dimer as it approaches a microtubule for binding. An isolated (in vacuum) tubulin dimer has an electric field dominated by its net charge, and has a nearly spherical isopotential surface.

In contrast, a dimer, surrounded by water molecules and counter-ions, as is physiologically relevant, has an isopotential surface with two lobes much like the dumbbell shape of a mathematical dipole. In a microtubule, individual dimers have near parallel dipole moments relative to their nearest neighbors. The greater the individual dipole of each of its units, the less stable the microtubule will generally be since these near parallel dipole-dipole interactions are repulsive, pushing apart the microtubule. Note the strength of the interaction is proportional to the square of the dipole moment; microtubules of tubulin units with larger dipole moments should be more prone to disassembly catastrophes compared to those containing lower dipole moment tubulins. For cells expressing more than one isotype, one can conceive of microtubule dynamics regulated by altering relative amounts of different isotypes according to their dipole moments. However, these effects are moderated by local interactions between two dimers, which produces a more complicated electrostatic field than the assembly of dipoles described here would suggest; for details see Baker et al. [26].

3.3.5 Motor Proteins

Two of the principal motor proteins that attach to MTs are kinesin and dynein. While kinesin moves towards the plus end of a MT, dynein is negative end directed. Each of these proteins consists of a globular head region and an extended coiled-coil tail section as shown in Fig. 3.21. The study of motor proteins including myosin, which has a similar structure, has shown that while the long tail is able to increase the force generated by the molecular motor, the essential components for force generation are located within the globular head. Models of motor protein movement can essentially be divided into two different mechanisms, those with diffusion and those with a *power stroke* [187]. The efficient propagation of these proteins, often with pairs such as kinesin and dynein moving in opposite directions simultaneously and seemingly avoiding collisions, led to the proposal that they may be directed by electrostatic interactions with the MT [46] and it is interesting to note that the binding of kinesin to MTs has since been shown to be primarily electrostatic [350].

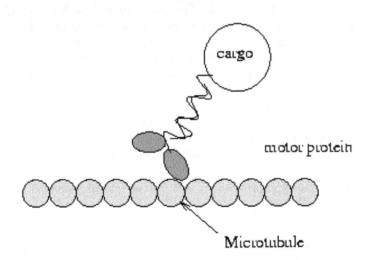

FIGURE 3.21: A motor protein is shown 'walking' along a MT protofilament.

The motion of the motor protein in either model is accomplished through the hydrolysis of ATP. The diffusion models require an oscillating potential that is presumably driven by a conformation change of the motor-MT bond. Activation of the complex by ATP leads to a potential that is relatively flat. Diffusion occurs in this state and once the potential reverts to its asymmetric form, the geometry of the potential is such that forward propagation of the motor protein is favored. A review of such schemes may be found thanks to Jülicher et al. [165]. In the power stroke models by contrast, it is the motor protein whose structure changes. Sometimes such models are envisioned as models of proteins walking since one imagines the protein to stretch and bind at a second location before relaxing to its original conformation when the 'back leg' releases its grip on the MT to start the process anew once additional ATP has arrived. The motor protein has two or more distinct states where at least one conformational change occurs and is driven by ATP hydrolysis as has been experimentally demonstrated for myosin by Spudich et al. [275]. Phosphorylation of the motor protein such as in the case of myosin may lead to subsequent conformational changes. Thus the protein may be viewed as walking along the MT powered by ATP. It is interesting to note that the use of GTP to control MT dynamics and ATP to control motor protein motion along MTs allows a cell to have control over both the cars (motor proteins) and the track (MTs) individually.

M. Buttiker [53] and R. Landauer [184] proposed a model for molecular motors which results in uni-directional motion of a molecule along a quasi-one-dimensional protein fiber. This model was essentially a generalization of the thermal ratchet model due to R. Feynman [100]. The key ingredient of this mechanism is an inhomogeneous temperature distribution with the periodicity of the filamentous structure to which the motor binds. However, it is well known from the theory of thermal conductivity that temperature variations over tens of nanometers are equilibrated on time scales of microseconds. Therefore, models of this type appear unrealistic. Recent attempts at explaining the motor protein behavior use isothermal ratchets in which the motor is subjected to an external potential that is periodic and asymmetric. In addition, a fluctuating force F(t) is acting on the motor. A recent review article on this topic by F. Jülicher [165] distinguishes three classes of such models, namely:

(i) A Langevin-based approach [84] with the equation of motion given by

$$\xi \frac{dx}{dt} = -\partial_x W(x) + F(t) \tag{3.9}$$

where x is the direction along the protofilament axis, ξ is the friction coefficient, $W(x)$ is the potential due to the fiber and $F(t)$ is the fluctuating force whose time average is zero, i.e.:

$$< F(t) >= 0 \tag{3.10}$$

This fluctuating force is typically related to the kinetics of ATP binding to the motor protein and a subsequent hydrolysis of ATP into ADP that provides the excess energy allowing the protein to unbind from the filament.

(ii) An approach in which the potential fluctuates in time [21]

$$\xi \frac{dx}{dt} = -\partial_x W(x,t) + f(t) \tag{3.11}$$

where the Gaussian (white) noise term $f(t)$ satisfies the fluctuation-dissipation theorem and the typical conditions are imposed:

$$< f(t) >= 0 \tag{3.12}$$

$$< f(t)f(t') >= 2\xi T \delta(t - t') \tag{3.13}$$

(iii) A generalized model with several internal states of the particle which is described by the Langevin equation that depends on the state i=1,...N, i.e.,

$$\xi_i \frac{dx}{dt} = -\partial_x W_i(x) + f_i(t) \tag{3.14}$$

The Gaussian noise term also satisfies the fluctuation-dissipation theorem via:

$$< f_i(t) >= 0 \tag{3.15}$$

$$< f_i(t)f_j(t') >= 2\xi T \delta(t - t') \delta_{ij} \tag{3.16}$$

A more convenient and elegant approach is through the Fokker-Planck formalism [267] for the probability distribution function $P(x,t)$ that describes the motion of the motor in a statistical manner. Here, $D^{(1)}$ is the drift coefficient and $D^{(2)}$ the diffusion coefficient.

$$\frac{\partial P}{\partial t} = \left[-\frac{\partial}{\partial x} D^{(1)}(x) + -\frac{\partial^2}{\partial x^2} D^{(2)}(x) \right] P(x,t) \tag{3.17}$$

So far, models of the above types have been fairly successful in representing the gross features of the experimental data obtained [306]. However, the use of a hypothetical potential and a number of arbitrary parameters casts some doubt on the usefulness or even correctness of such models. So, can this model of motor protein movement be successfully applied to the problem of chromosome separation? The prevailing model of how the MTs that form the mitotic spindle separate chromosomes is that motor proteins attach to the chromosomes and the distal ends of spindle MTs. The force generation is thereby developed locally by the motor proteins and cannot depend on the length of the MT. However, the force attracting the chromosome to the pole has been shown to increase as the spindle shortens [135]. Only a repulsive force from the pole or generated by the tubulin subunits of the microtubule would solve this difficulty. However, if it is the motor proteins are responsible for the force generation, it must be explained how it is that the chromosome movement does not require ATP [301]. Therefore if the motor proteins are important for more than simply binding the chromosome, this has not been shown to date and the relationship between force and the length of the spindle MTs must still be explained.

3.4   Anisotropic Elastic Properties of Microtubules - from ref [325]

As discussed earlier, the cytoskeleton is composed of three different types of filaments organized in networks: microfilaments (MF), intermediate filaments (IF) and microtubules (MT). Each of them has specific physical properties and structures suitable for their role in the cell. For example, the two-dimensional arrangement of MFs in contractile fibers, so-called stress fibers, appears to form cable-like structures involved in the maintenance of the cell shape and transduction pathways. F-actin can support large stresses without a great deal of deformation and it ruptures at approximatively $3.5N/m^2$ [161]. IFs have a rope-like structure composed of fibrous proteins consisting of two coiled coils and are mainly involved in the maintenance of cell shape and integrity. Ma *et al.* [205] shows that IFs resist high applied pressures by increasing their stiffness. They can withstand higher stresses than the other two components without damage [161].

By biological standards, MTs are rigid polymers with a large persistence length of $6mm$ [39]. From Jamney's experiments [161], MTs suffer a larger strain for a small stress compared to either MFs or IFs. The rupture stress for MTs is very small and typically is about $0.4 - 0.5N/m^2$ [161]. The lateral contacts between tubulin dimers in neighboring protofilaments have a decisive role for MT stability, rigidity and architecture [221]. Tubulin dimers are relatively strongly bound in the longitudinal direction (along protofilaments), while the lateral interaction between protofilaments is much weaker [174].

There have been a number of experimental studies in recent years dealing with the various aspects of the elasticity of MTs. On the other hand, theoretical effort in this area has been much more limited. For example Jánosi *et al.* [163] have studied the elastic properties of MT tips and walls and the various shapes observed from electron micrographs have been shown to be consistent with their simple mechanical model. However, their model deals chiefly with the geometrical characteristics of MTs and not with their anisotropic properties which is the focus of our effort. The limited flexibility of inter-protofilament bonds in MTs, assembled from pure tubulin, has also been investigated by Chrétien *et al.* [65] via moiré patterns in cytoskeletal micrographs. The position of tubulin subunits and their arrangement on the MT surface enables the moiré period to be predicted.

In this section we discuss the elastic properties of MTs in particular and try to utilize the experimental data found in the literature for our theoretical estimations. Our starting point is a review of the information found in the literature regarding the results of experiment and theoretical models of elastic properties of MTs. Using recently published information regarding dimer-dimer interactions and the molecular geometry of the MT we provide estimates of the elastic moduli accounting for the anisotropy of the MT filament.

The response of a cylinder of length L and cross-sectional area A, to an extensional force (stretching), F, is described by *Young's modulus*, $\mathcal{Y}$, and the relationship between these physical quantities is given by Hooke's Law,

$$\frac{F}{A} = \mathcal{Y}\frac{\Delta L}{L}, \tag{3.18}$$

where ΔL is the small length change and the relative extension, $\frac{\Delta L}{L}$, is the unidirectional strain along the direction of the force applied to the cylinder.

Another elastic property is characterized by the *shear or twisting modulus*, G, that is expressed, for a homogeneous isotropic material, by [174]

$$G = \frac{\mathcal{Y}}{2(1+\nu)} \tag{3.19}$$

where ν is Poisson's ratio, representing the relative magnitude of transverse to longitudinal strain and its value typically lies in the range $0 < \nu < 0.5$.

Another indicator of the elastic property of a solid is the parameter called *flexural rigidity*, κ_f, that determines the resistance of a filament to a bending force. The higher the flexural rigidity, the greater the resistance to bending. The flexural rigidity is completely determined by the properties of the bonds between the atoms within each protein subunit and the properties of the bonds that hold the subunits together in MTs. For isotropic and homogeneous materials, the flexural rigidity, κ_f, can be represented as the product of two terms (characterizing material properties and their shape), by

$$\kappa_f = \mathcal{Y}\mathcal{I}_y \tag{3.20}$$

where $\mathcal{I}_y$ is a shape-dependent parameter called the second moment of inertia. For a hollow cylinder [145,39] it is given by $\mathcal{I}_y = \pi(R_o^4 - R_i^4)/4$, where R_o and R_i are the outer and the inner radii, respectively. However, for a hollow cylinder made up of n cylindrical protofilaments where each protofilament has radius r, $\mathcal{I}_y$ is given by $\mathcal{I}_y = \left(\frac{2}{\pi^2}n^3 + n\right)\frac{\pi}{4}r^4$ [145].

The *persistence length*, ξ_p, is another index which describes the filament's resistance to thermal forces. Roughly speaking this length is the length over which a filament appears straight and can be expressed by

$$\xi_p = \frac{\kappa_f}{k_B T} = \frac{\mathcal{Y}\mathcal{I}_y}{k_B T} \tag{3.21}$$

where k_B is Boltzmann's constant and T is the temperature in Kelvin. The more flexible the filament, the smaller the persistence length. Polymers for which persistence length and contour length are similar are called semiflexible polymers.

In Table 3.8 we list the different values found in the literature for the above defined parameters. From Table 3.8 we observe that the shear and Young's moduli are significantly different. This observation describes an anisotropic material. Hence, MTs are inhomogeneous and anisotropic. The bending deformation and stretching of filaments are governed by Young's modulus and the sliding between filaments is governed by the shear modulus [174]. Note that the values of the shear modulus, G, seem to be strongly dependent on the experimental conditions and show little consistency as we will attempt to elucidate in Section 3.4.

TABLE 3.8: Experimental Data for Elastic Properties of MTs (and Other Proteins) from the Literature

Young's Modulus, $\mathcal{Y}$	$(1.2 - 2.7) \times 10^9 N/m^2$	[145,314,343,302]
	$(1 - 5) \times 10^8 N/m^2$	[174,297a]
Shear modulus, G	$1.4 N/m^2$	[281]
	$10^3 N/m^2$ between MTs	[314]
	$34 N/m^2$ gel (concentration-dependent)	[161]
	$1.4 \times 10^6 N/m^2$ at $25°C$ (temp.-dependent)	[174]
Flexural rigidity, κ_f	$(16 - 30) \times 10^{-24} N.m^2$	[94,145,238,190,62,115,121]
	$(2.9 - 5.1) \times 10^{-24} N.m^2$	[99]
Poisson's ratio, ν	0.3 for macromolecules	[297b,297a]
	0.4 for MFs	[343]
Persistence length, ξ_p	$(1 - 6.3) \times 10^{-3} m$	[39,94,115]

Using the refined tubulin crystal structure produced by Löwe et al. [198], Sept et al. [289, 290] calculated the protofilament-protofilament energy as a function of the shift along the protofilament axis. Two protofilaments were constructed similar to Fig. 3.22 and translated relative to each other along the MT axis. Using Poisson-Boltzmann calculations and a surface area term the interaction energy was computed over a 80Å translational in 2Å steps. The data published in this work indicate the presence of two stable equilibrium positions that correspond to lattice types A and B.

In fact, a recent paper by VanBuren et al. [330] estimated the values of lateral and longitudinal bond energies in an MT structure using a stochastic model of assembly dynamics. Their analysis predicts the lateral bond energy to be in the range of -2.2 to -5.7 kT (which compares to -9.1×10^{-21} to -2.4×10^{-20} J) while the longitudinal bond energy is given as -6.8 to -9.4 kT (-2.8×10^{-20} to -3.9×10^{-20} J). However, without the knowledge of the potential's positional dependence these values cannot provide information on the corresponding elastic coefficients.

Using this potential map (Sept et al., Figure 1[290]) we have evaluated the

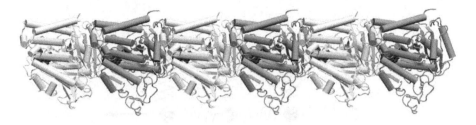

FIGURE 3.22: MT protofilament. α and β monomers are colored gray and white respectively.

corresponding elastic coefficient in a harmonic approximation around the potential minimum. We found the value $k \approx 4N/m$ which will be used in the next section. Note that an early phenomenological estimate of the spring constant k [227] was approximately 40 times smaller and predicted incorrectly that the rigidity of MTs is 100 times lower than that of actin. Unfortunately, still no results are available at present which would correspond to the elastic coefficients along different directions and not just as the result of protofilament-protofilament shifting along the protofilament axis as demonstrated in Sept et al. [289, 290]. While this calculation imposes a computational challenge, it is only a matter of time before these values are obtained in a subsequent calculation of MTs anisotropic elastic properties.

Recently, De Pablo et al. have estimated experimentally by radial indentation of MTs with a scanning force microscope tip a spring constant value related to radial interactions between neighboring protofilaments [81]. Indentations, induced by the nanometer sized tip positioned on the top of an MT, result in a linear elastic response with the spring constant $k = 0.1N/m$. Tubulin dimers are relatively strongly bound in the longitudinal direction (along protofilaments), while the lateral interaction between protofilaments is much weaker [174,246,290]. Hence if we define κ_p the elastic constant between the two dimers along a protofilament, and κ_l the elastic constant between the two dimers belonging to adjacent protofilaments, then $\kappa_p > \kappa_l = k$.

We will now make estimates of the anisotropic elastic moduli measured for individual MTs based on the internal geometry of the MT and the molecular forces acting between individual dimers. To give a simple conceptual example we follow [145] and imagine a perfect cubic lattice connected by springs, see Fig. 3.23.

Suppose a force F is applied along the line of springs as indicated in Fig. 3.23. Each bond experiences this force which is perpendicular to one side of the lattice. Each spring will be extended by a small distance Δr, from its equilibrium state r_0, according to the application of Hooke's law to

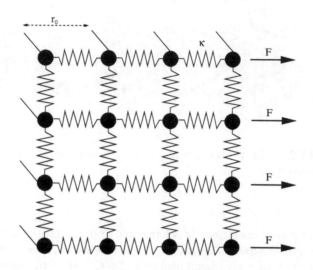

FIGURE 3.23: A cubic lattice of particles distant of r_0 when at equilibrium state, connected by elastic springs with spring constants κ and subjected to a force F [145].

a spring, i.e., $F = \kappa \Delta r$. Dividing this relation through by r_0^2 we obtain

$$\frac{F}{r_0^2} = \frac{\kappa}{r_0} \frac{\Delta r}{r_0}. \tag{3.22}$$

The left-hand side of Eq. (3.22) represents the force per unit area or stress whereas $\frac{\Delta r}{r_0}$ on the right-hand side, deformation per unit length, describes the strain involved. From Eq. (3.18), we therefore obtain Young's modulus of this material as $\mathcal{Y} = \frac{\kappa}{r_0}$. In essence the value of Young's modulus is proportional to the spring constant which depends on the strength of the intermolecular forces. It is inversely proportional to the equilibrium separation between neighboring molecules.

Using the same method, we estimate below the effective elastic moduli for a longitudinal, a lateral and a shear deformation (Fig. 3.24).

3.4.0.1 Effective Young's modulus due to longitudinal compression (directed parallel to the protofilament of an MT):

When a compressive force, F, is applied as shown in *1)* of Fig. 3.24, there will be a small displacement, Δr_p, of the distance between two dimers of a particular protofilament given by

$$F = \kappa^{\parallel} \Delta r_p \tag{3.23}$$

where $\kappa^{\parallel}$ is the elastic constant between the two dimers along a protofilament (this was referred to earlier as κ_p). Dividing both sides of Eq. (3.23) by the

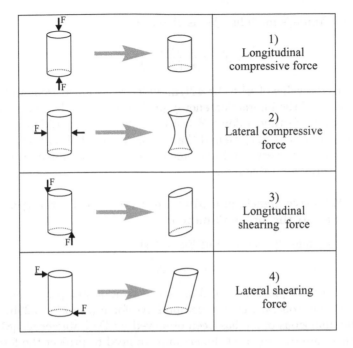

FIGURE 3.24: Different types of forces applied to a cylinder and resulting deformations. *1)* A compressive force applied longitudinally to the tip of the filament. *2)* A compressive force applied laterally to the wall of a filament. *3)* A longitudinal shearing force. *4)* A lateral shearing force.

area, A, over which the force is applied we obtain

$$\frac{F}{A} = \frac{\kappa^{\parallel}}{A} \Delta r_p. \tag{3.24}$$

Suppose N is the number of protofilaments around the MT, then

$$A = \frac{1}{N} \pi (R_o^2 - R_i^2) \tag{3.25}$$

where R_o and R_i are the outer and the inner radii of the MT, respectively. Hence from Eqs. (3.23)–(3.25) we obtain

$$\left(\frac{F}{A}\right) = \frac{N \kappa^{\parallel} l}{\pi (R_o^2 - R_i^2)} \left(\frac{\Delta r_p}{l}\right). \tag{3.26}$$

The left-hand side of Eq. (3.26) is the stress applied and the term in the brackets on the right, assuming l is the equilibrium distance between dimers, is the strain or incremental displacement per unit length. Hence, from Eq. (3.18) the effective Young's modulus, $\mathcal{Y}_{\|}$, is given by

$$\mathcal{Y}_{\|} = \frac{N\kappa^{\|}l}{\pi(R_o^2 - R_i^2)}. \tag{3.27}$$

We estimate the value of $\kappa^{\|}$ to be $4\,\mathrm{N/m}$ which is obtained from an harmonic approximation of the interaction energy profile published by Sept et al. (see Fig. 1 in [290]). Further, taking $N = 13$, $l = 8nm$, $R_o = 12.5nm$, and $R_i = 7.5nm$, and from Eq. (3.27) we find that our estimate of $\mathcal{Y}_{\|}$ is $1.32 \times 10^9 N/m^2$ for [290] which is in very good agreement with the experimental data found in the literature (see Table 3.8).

3.4.0.2 Effective elastic modulus due to a lateral force (perpendicular to the protofilament):

In this case (namely, entry *2*) of Fig. 3.24)

$$F = \kappa_\perp \Delta r, \tag{3.28}$$

where Δr is the displacement of dimers in two adjacent protofilaments and $\kappa_\perp$ is the elastic constant for the compressive force (see Fig. 3.25b), the experimental analogous of $\kappa_\perp$ has been observed by De Pablo et al. [81].

Since there are two adjacent dimers and we need to project the force along the line joining the two dimers, we multiply both sides of Eq. (3.28) by $2\sin\theta$, where θ is the angle subtended by 2 adjacent dimers at the center of the MT (see Fig. 3.25b). We then divide the result by an appropriate area (see Fig. 3.25a), A, given by:

$$A = \pi R_o l, \tag{3.29}$$

l being the distance between two neighboring layers (Fig. 3.25a) calculated as a center-to-center distance. Thus

$$\frac{2F\sin\theta}{A} = \frac{2\kappa_\perp \sin\theta}{\pi l}\left(\frac{\Delta r}{R_o}\right). \tag{3.30}$$

The part on the left-hand side of Eq. (3.30), $2F/A$, represents the applied stress, and putting the left-hand $\sin\theta$ to the right-hand side gives $(\Delta r/R_o \sin\theta)$ as the resultant strain where $R_o \sin\theta$ is the equilibrium separation of one pair of dimers. Hence the effective lateral elastic modulus, $\mathcal{Y}_\perp$, is estimated to be given by

$$\mathcal{Y}_\perp = \frac{2\kappa_\perp \sin\theta}{\pi l}. \tag{3.31}$$

If we use the value of $\kappa_\perp$ as the curently available estimate for the elastic coefficient by [81], i.e., $\kappa_\perp = 0.1 N/m$, $\theta = \frac{\pi}{6}$, and $l = 8\,\mathrm{nm}$ then the corresponding value of the Young's modulus is found to be $\mathcal{Y}_\perp = 4 \times 10^6 N/m^2$. At any rate, due to the presence of the $\sin\theta$ factor in Eq. (3.31) and a small value of θ ($\theta = \frac{2\pi}{N-1}$ where N is the number of protofilaments) we expect the Young's modulus to be anisotropic for the parallel and perpendicular directions along which it may differ by as much as an order of magnitude.

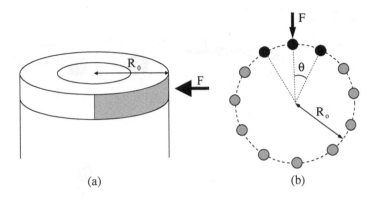

(a)   (b)

FIGURE 3.25: *a)* Side elevation of an MT showing the area in grey to which the force F is applied. *b)* Cross-section of an MT showing the direction of the applied force.

3.4.0.3 Effective elastic modulus due to a longitudinal shearing force:

The cross-sectional area of the MT is πR_o^2, whereas the length of the section of an MT (see Fig. 3.26a) is $L = nl$ where n is the number of layers of dimers and l the distance between two neighboring layers. The net macroscopic displacement, Δx (see Fig. 3.26a), is obviously given by $\Delta x = n\delta x$ where δx (see Fig. 3.26b) is the microscopic lateral displacement due to the shearing force. With f_r denoting the inter-dimer force (see Fig. 3.26b) between a displaced dimer, A', in Figure 3.26a and a relatively stationary one, C', we have

$$f_r = \kappa_s(l' - l) \tag{3.32}$$

where κ_s is the microscopic elastic constant and $l' - l$ denotes the change in the interdimer distance (i.e., the difference in length between $B'C' = l$ and $A'C' = l'$).

We denote the restoring force across the section of the MT by f_{rh} (see Fig. 3.26b) which, by vectorial addition, is related to the restoring force between dimers A' and C', f_r, by

$$f_{rh} = f_r \sin \alpha. \tag{3.33}$$

If N is the number of protofilaments around the perimeter of the MT we have

$$\begin{aligned}
F = N f_{rh} &= N f_r \sin \alpha && \text{from Eq.(3.33)}\\
&= N\kappa_s(l' - l) \sin \alpha && \text{from Eq.(3.32)} \qquad (3.34)\\
&= N\kappa_s l(\tfrac{1}{\cos \alpha} - 1) \sin \alpha,
\end{aligned}$$

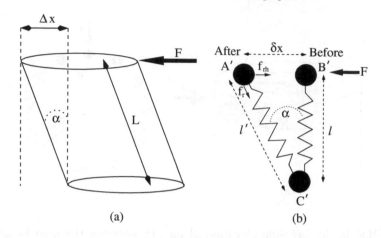

FIGURE 3.26: *a)* Section of an MT of length L, subjected to a lateral shearing force, F. *b)* Movement of a single dimer as a result of the force F indicating the undisturbed position (before, B') and the position following displacement (after, A').

since it follows from Fig. 3.26b that $l = l' \cos \alpha$. Rearranging Eq. (3.34),

$$F = N\kappa_s l(1 - \cos \alpha) \tan \alpha \qquad (3.35)$$

and, using the fact that the cross-sectional area is $A = \pi R_o^2$ and $\tan \alpha = \frac{\delta x}{l} = \frac{\Delta x}{L}$, we have

$$\frac{F}{A} = \frac{N}{\pi R_o^2} \kappa_s l(1 - \cos \alpha) \frac{\Delta x}{L} = G\left(\frac{\Delta x}{L}\right) \qquad (3.36)$$

where G is the shearing modulus. Hence we deduce that the shearing modulus can be finally expressed by

$$G = \frac{N}{\pi R_o^2} \kappa_s l(1 - \cos \alpha) \simeq \frac{N}{\pi R_o^2} \kappa_s l \frac{\alpha^2}{2} \qquad (3.37)$$

where α is the angle of deformation in radians. We have used the fact that α is small by expanding the cosine. The angle α can lie between the following limits:

$$0.001 \leq \alpha \text{ in radians } \leq 0.1 \qquad (3.38)$$

From Eq. (3.37) we see that the shear modulus, G, varies greatly with possible values of α and hence also with the size and magnitude of the displacement. Taking typical values, for example $N = 13$, $l = 8nm$, $R_o = 12.5nm$ and $\kappa_s \approx 0.5N/m$, we find G may vary between the following estimated limits

$$53N.m^{-2} \leq G \leq 0.5 \times 10^6 N.m^{-2} \qquad (3.39)$$

Note that the large variation in the value of G found in Eq. (3.39) may provide
an explanation of the large differences seen in the various experimental data
(see Table 3.8). The crucial point to make here is that the deformation angle,
α, enters into the shear modulus formula. Combining Eqs. (3.36) and (3.37)
allows us to calculate the force required for the shearing action as a function
of the angle, α. For example, with the use of typical structural parameters
and for $\alpha = 0.1$ the shearing force amounts to more than $20pN$ which is quite
significant. The case of a lateral shearing force (see entry *4)* in Fig. 3.24) is
analyzed in the same manner but we have not discussed it here due to the
lack of proper experimental techniques for such measurements at present.

3.5   Centrioles, Basal Bodies, Cilia and Flagella

Basal bodies and centrioles consist of a 9-fold arrangement of triplet microtubules. A molecular cartwheel fills the minus end of the cylinder and is
involved in initiating the assembly of the structure. The cylinders, called centrioles (see Fig. 3.27), are always found in pairs orientated at right angles to
each other. Dense clouds of satellite material associated with the outer cylinder surfaces are responsible for the initiation of cytoplasmatic microtubules.

Consequently, centrioles organize the spindle apparatus on which the chromosomes move during mitosis. They also control the direction of cilia or flagella movement. Cilia and Flagella Cellular movement is accomplished by cilia
and flagella. Cilia are hair-like structures that can beat in synchrony causing
the movement of unicellular paramecium. In large multicellular organisms
their role is to move fluid past the cell. Cilia are also found in specialized lin-
ings in eukaryotes. Flagella are whip-like appendages that undulate to move
cells. They are longer than cilia, but have similar internal structures made of
microtubules. Cilia and flagella have the same internal structure. The only
difference is in their length. Prokaryotic and eukaryotic flagella differ greatly.
Both flagella and cilia have a 9 + 2 arrangement of microtubules wherein
the 9 fused pairs of microtubules are on the outside of a cylinder, and the 2
unfused microtubules are in the center. Dynein "arms" attached to the microtubules serve as the molecular motors. Cilia and flagella are organized from
centrioles that move to the cell periphery. These are called "basal bodies".
Note that numerous cilia can be projecting from the cell membrane. Basal
bodies control the direction of movement of the cilia. The difference between
cilia and flagella and centrioles is that they contain 9 sets of triplets and no
doublet in the center. How the triplets in the basal body turn into the cilium
doublet remains a mystery and the cartoon illustration in Fig. 3.28 compares
the cross section of a cilium with that of a centriole. Fig. 3.29 compares the
cross-section of a flagellum to that of a microtubule axoneme doublet.

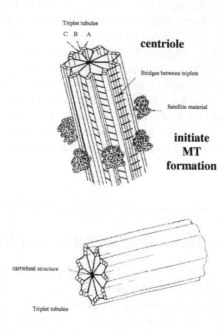

FIGURE 3.27:   The structure of a centriole.

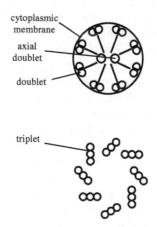

FIGURE 3.28:   Comparison between the structure of a cilium (above) and a centriole (below)

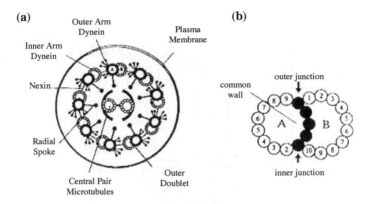

FIGURE 3.29:   A schematic cross-section through a flagellum (a), and the microtubule (MT) doublet (b), following [195].

Fig. 3.30 is a schematic representation of a longitudinal section of a cilium in a paramecium showing the basal body, the transition zone and the proximal part of the axenome. The membrane MT bridges in the ciliary necklace as well as in the plaque area are thought to be connected to the intramembranous particles observed on the ciliary membrane.

Fig. 3.31 shows several snapshots of the beating motion in a cilium and its correlation with calcium waves.

Cilia and flagella move because of the interactions of a set of microtubules inside of them (see Fig. 3.32). Collectively, these are called an "axoneme". Two of these microtubules join to form one doublet in the cilia or flagella. Note that one of the tubules is incomplete. Furthermore, there are important microtubule associated proteins (MAP's) projecting from one of the microtubule subunits. The core doublets are both complete. Extending from the doublets are sets of arms that join neighboring doublets composed of dynein, spaced at 24 nm intervals. Nexin links are spaced along the microtubules to hold them together. Projecting inward are radial spokes that connect with a sheath enclosing the doublets.

The dynein arms have ATPase activity. In the presence of ATP, they can move from one tubulin to another. They enable the tubules to slide along one another so the cilium can bend. The dynein bridges are regulated so that sliding leads to synchronized bending. Because of the nexin and radial spokes, the doublets are held in place so sliding is limited lengthwise. If nexin and the radial spokes are subjected to enzyme digestion, and exposed to ATP, the doublets will continue to slide and telescope up to nine times their length.

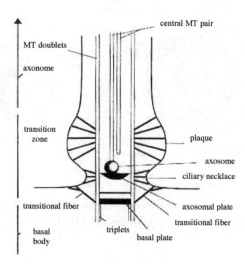

FIGURE 3.30: Schematic representation of a longitudinal section of a cilium.

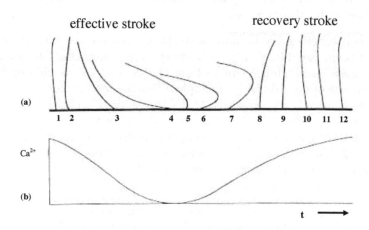

FIGURE 3.31: Different stages during the beating of a cilium (a). The oscillations of Ca^{2+} in the transition zone (b).

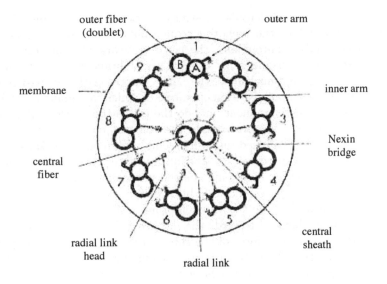

FIGURE 3.32:   An electron micrograph of a cross-section of a cilium following *Molecular Cell Biology* by Lodish et al. [195].

3.6   Networks and Meshworks of Protein Filaments, Stress Fibers and Tensegrity

To understand fully the way living systems form and function (including the mechanism of cell division), we need to uncover the basic principles that guide biological organization. A remarkable number of natural systems are constructed using a common form of architecture known as tensegrity, a term which refers to a system that stabilizes itself mechanically because of the way in which tensional and compressive forces are distributed and balanced within itself. Tensegrity structures are mechanically stable because of the way the entire structure distributes and balances mechanical stresses and not because of the strength of individual members. Since tension is continuously transmitted across all structural members, a global increase in tension is balanced by an increase in compression within members distributed throughout the structure. As described in detail above, the interiors of living cells contain an internal framework called the cytoskeleton, composed of three different types of molecular protein polymers, known as microfilaments, intermediate filaments and microtubules.

Cell shape is regulated by a complex balance of both internal and external forces exerted by the extracellular matrix. This balance has been described by (1993, 1997) in terms of tensegrity. Cells get their shape from tensegrity due not only to the cytoskeleton's three major types of filaments but also from the extracellular matrix, the anchoring scaffolding to which cells are naturally secured in the body. Throughout the cell a network of contractile microfilaments exerts tension and pulls the cell's membrane and all its internal constituents toward the nucleus at the core. Opposing this inward pull are two main types of compressive elements, one of which is outside the cell and the other inside. The component outside the cell is the extracellular matrix while the compressive "girders" inside the cell can be either microtubules or large bundles of cross-linked micro-filaments within the cytoskeleton. The third component of the cytoskeleton, the intermediate filaments, interconnects microtubules, contractile microfilaments, the surface membrane and the cell's nucleus. In a nutshell, contractile actin bundles act as molecular cables. These cables exert a tensile force on the cell membrane and the internal constituents of the cell, pulling them all towards the nucleus. Microtubules act as struts that resist the compressive force of the cables. In many cases, it is important to maintain the cell's shape to preserve its functionality. Chen et al. [63] have shown experimentally how cells switch between genetic programs when forced to grow into specific shapes, while King and Wu [172] have shown on theoretical grounds how the cell geometry changes the susceptibility of the cell to electromagnetic fields.

Although the cytoskeleton is surrounded by membranes and penetrated by viscous fluid, it is this hard-wired network of molecular struts and cables that stabilizes cell shape. The microtubules are compressed, rigid elements. The actin filaments are tensile. Recently, Maniotis and Ingber [180] demonstrated that pulling on receptors at the cell surface produces immediate structural changes deep inside the cell. Thus, cells and nuclei do not behave like viscous containers filled with a cystoplasmic soup. The existence of a tensegrity force balance provides a means to integrate mechanics and biochemistry at the molecular level. The tensegrity model developed by Ingber suggests that the structure of the cell's cytoskeleton can be changed by altering the balance of physical forces transmitted across the cell surface. This finding is important because many of the enzymes and other substances that control protein synthesis, energy conversion and growth in the cell can be bound to the cytoskeleton. Therefore, there can be a dynamical interplay between the cytoskeletal geometry and the kinetics of biochemical reactions including gene activation. Remarkably, by simply modifying their shape, cells can switch between different genetic programs. Cells that spread flat were found to be more likely to divide, whereas round cells activated a death program called celled apoptosis. When cells were neither too extended nor too retracted, they neither divided nor died but, instead, differentiated themselves in a tissue-specific manner. Thus, mechanical restructuring of the cell and cytoskeleton accompanies cellular growth, division or death. Hence, mechanical forces are transmitted over specific molecular paths in living cells. Because a local force can change the shape of an entire tensegrity structure, the binding of a molecule to a protein can cause the different, stiffened helical regions to rearrange their relative positions throughout the length of the protein. Even in vitro, MT's exhibit travelling waves of assembly and disassembly as well as the formation of polygonal networks. Reversing the laboratory conditions and then causing a subsequent assembly stage has revealed the presence of memory effects in these pattern formation phenomena. Table 3.9 summarizes the tension forces produced by tensegrity structures in the cell.

TABLE 3.9: Tension Forces Within the Cell

Tensegrity	Force in pN
Tension generated by mybrofil	10^5 to 10^6
Viscous drag force on *Vorticella*	8.6×10^3
Viscous drag force on a chromosome	0.1
Force needed to stop a chromosome	700

An important force-generating family of proteins has not been included here yet and it deserves special attention due to their dynamic behavior and pervading presence in all living cells. This class is called motor proteins.

3.7 Cell Nucleus and Chromosomes; Their Physical Properties and the Mechanism of Folding and Segregation

The cell nucleus is the headquarters of the cell since it regulates all cell activity by controlling enzymes. The nucleus is also the main library of the cell. It contains the blueprints and instructions that have evolved over one billion years of evolution, which tell the cell how to operate, how to rebuild itself after every cell division, and how to act and interact with other cells in an organism. Its gene control system monitors signals from the nucleus and its outside world, the cytoplasm. It co-ordinates the cell's activities, which include intermediary metabolism, growth, protein synthesis and reproduction (cell division). It seems to be structured in a hierarchy of levels of genomic instructions. The spherical nucleus occupies about 10 percent of a cell's volume, making it the cell's most prominent feature.

3.7.1 Nuclear Chromatin, Chromosomes, Nuclear Lamina

The nucleus consists of a nuclear envelope (the outer membrane) and nucleoplasm. The nucleoplasm contains chromatin and the nucleolus. The nuclear envelope is a double membrane which has four phospholipid layers. It also has large pores through which material passes back and forth. Most of the nuclear material consists of chromatin, the unstructured form of the cell's DNA that organizes to form chromosomes during mitosis or cell division. It consists of DNA looped around histone proteins. The nucleolus is a knot of chromatin. The chromatin is composed of DNA. DNA contains the information for the production of proteins. This information is encoded in the four DNA bases: adenine, thymine, cytocine and guanine. The specific sequence of these bases tells the cell what order to put the amino acids. There are three processes that enable the cell to manufacture protein. Replication allows the nucleus to make exact copies of its DNA. Transcription allows the cell to make RNA working copies of its DNA. In translation, the messenger RNA is used to line up amino acids into a protein molecule.The code is actually translated on structures that are also made in the nucleus, called ribosomes which provide the structural site where the mRNA sits. The amino acids for the proteins are carried to the site by transfer RNAs. Each transfer RNA (tRNA) has a nucleotide triplet that binds to the complementary sequence on the mRNA. The tRNA carries the amino acid at its opposite end.

Note that only the cells of advanced organisms, known as eukaryotes, have a nucleus. Generally there is only one nucleus per cell, but there are exceptions such as slime moulds and the Siphonales group of algae. Simpler single-cell organisms (prokaryotes), like the bacteria and cyanobacteria, do not have a nucleus. In these organisms, all the cell's information and administrative functions are dispersed throughout the cytoplasm.

A double-layered membrane, the nuclear envelope, separates contents of the nucleus from the cellular cytoplasm. The envelope is riddled with holes called nuclear pores that allow specific types and sizes of molecules to pass back and forth between the nucleus and the cytoplasm. It is also attached to a network of tubules, called the endoplasmic reticulum, where protein synthesis occurs. These tubules extend throughout the cell and manufacture the biochemical products that the organism is genetically encoded to produce and in a particular cell its genes can be actively repressed or promoted.

3.7.2   Chromatin/Chromosomes

Packed inside the nucleus of every human cell is nearly 2m of DNA, which is divided into 46 individual molecules, one for each chromosome and each about 4 cm long. Packing all this material into a microscopic cell nucleus is an extraordinary feat of packaging where it is combined with proteins and organized into a precise, compact structure, a dense string-like fiber called chromatin. Each DNA strand wraps around groups of small protein molecules called histones, forming a series of bead-like structures, called nucleosomes, connected by the DNA strand. The uncondensed chromatin has a "beads on a string" appearance. The string of nucleosomes, already compacted by a factor of six, is then coiled into an even denser structure, compacting the DNA by a factor of 40. This compression and structuring of DNA serves several functions. The overall negative charge of the DNA is neutralized by the positive charge of the histone molecules, the DNA takes up much less space, and inactive DNA can be folded into inaccessible locations until it is needed. There are two types of chromatin. Euchromatin is the genetically active portion and is involved in transcribing RNA to produce proteins used in cell function and growth. Heterochromatin contains inactive DNA and is the portion of chromatin that is most condensed.

During the cell cycle, chromatin fibers take on different forms inside the nucleus [193]. During interphase, when the cell is carrying out its normal functions, the chromatin is dispersed throughout the nucleus in a tangle of fibers. This exposes the euchromatin and makes it available for the transcription process. When the cell enters metaphase and prepares to divide, the chromatin changes dramatically. First, the chromatin strands make copies of themselves through the process of DNA replication. Then they are compressed in a 10,000-fold compaction process into chromosomes. As the cell divides, the chromosomes separate, giving each cell a complete copy of the genetic information contained in its chromatin.

Nucleolus - The nucleolus is a membrane-less organelle within the nucleus that manufactures ribosomes, the cell's protein-producing structures. The nucleolus looks like a large dark spot within the nucleus. A nucleus may contain up to four nucleoli, but within each species the number of nucleoli is fixed. After a cell divides, a nucleolus is formed when chromosomes are brought together into nucleolar organizing regions. During cell division, the nucleolus disappears.

Nuclear Envelope - The nuclear envelope is a double-layered membrane that encloses the contents of the nucleus during most of the cell's lifecycle. The space between the layers is called the perinuclear space and appears to connect with the rough endoplasmic reticulum. The envelope is perforated with tiny holes called nuclear pores. These pores regulate the passage of molecules between the nucleus and cytoplasm, permitting some to pass through the membrane, but not others. The inner surface has a protein lining called the nuclear lamina, which binds to chromatin and other nuclear components. During mitosis, the nuclear envelope disintegrates, but reforms as the two cells complete their formation and the chromatin begins to unravel and disperse.

Nuclear Pores - These pores in the nuclear envelope regulate the passage of molecules between the nucleus and cytoplasm, permitting some to pass through the membrane, but not others. Building blocks for building DNA and RNA are allowed into the nucleus as well as molecules that provide the energy for constructing genetic material. The pores are fully permeable to small molecules up to the size of the smallest proteins, but form a barrier keeping most large molecules out of the nucleus. Some larger proteins, such as histones, are given admittance into the nucleus. Each pore is surrounded by an elaborate protein structure called the nuclear pore complex, which probably selects large molecules for entrance into the nucleus.

3.8 Mitochondria and Proton Pumps: Energy Generation and Utilization in the Cell

All processes of life need energy to be sustained, which originates from quanta of visible light, emitted by the sun and absorbed by pigments of photosynthetic units. After this process of molecular excitation, the absorbed energy is accumulated and transmitted to other parts of the cell, rest of the plant, and finally to other organisms which are not able to obtain energy by photosynthesis directly.

3.8.1 Cell Energetics: Chloroplasts and Mitochondria

The energy of molecular excitation is transformed into its chemical form of high-energy compounds. The most common accumulator of chemical energy in the cell is adenosine triphosphate (ATP), formed in the process of photosynthesis and used in nearly all processes of energy conversion in other cells. The hydrolysis of ATP, producing adenosine diphosphate (ADP) and catalyzed by special enzymes, the ATPases, allows the use of this stored energy for ionic pumps, processes of molecular synthesis, production of mechanical energy and many other uses. The amount of energy stored by the ADP $\rightarrow$ ATP reaction in the cell, however, is limited due to osmotic stability. Therefore, other molecules, like sugars and fats, are used for long-term energy storage. The free energy of ATP is used to synthesize these molecules. Subsequently, in the respiratory chain, these molecules are decomposed, and ATP is produced again.

Chloroplasts are double membraned ATP-producing organelles found only in plants. Inside their outer membrane is a set of thin membranes organized into flattened sacs stacked up like coins called thylakoids. The disks contain chlorophyll pigments that absorb solar energy which is the ultimate source of energy for all the plant's needs including manufacturing carbohydrates from carbon dioxide and water [206]. The chloroplasts first convert the solar energy into ATP stored energy, which is then used to manufacture storage carbohydrates which can be converted back into ATP when energy is needed. The chloroplasts also possess an electron transport system for producing ATP. The electrons that enter the system are taken from water. During photosynthesis, carbon dioxide is reduced to a carbohydrate by energy obtained from ATP [206]. Photosynthesizing bacteria (cyanobacteria) use yet another system whereby they do not manufacture chloroplasts but use chlorophyll bound to cytoplasmic thylakoids. The two most common evolutionary theories of the origin of the mitochondria-chloroplast ATP production system are: a) endosymbiosis of mitochondria and chloroplasts from the bacterial membrane system and, b) the gradual evolution of the prokaryote cell membrane system of ATP production into the mitochondria and chloroplast systems. Proponents of endosymbiosis maintain that mitochondria were once free-living bacteria, and that "early in evolution ancestral eukaryotic cells simply ate their future partners" [340].

The **mitochondrion** is the site of aerobic respiration. Most of the key processes of aerobic respiration occur across its inner membrane. The mitochondrion, where ATP is produced, functions to produce an electro-chemical gradient (see Fig. 3.33), by accumulating hydrogen ions between the inner and outer membrane. This electro-chemical energy originates from the estimated 10,000 enzyme chains in the membranous sacks on the mitochondrial walls. As the charge builds up, it produces an electrical potential that releases its energy by causing a flow of hydrogen ions across the inner membrane into the inner chamber. The energy causes an enzyme to be attached to ADP which catalyzes the addition of a third phosphorus to form ATP.

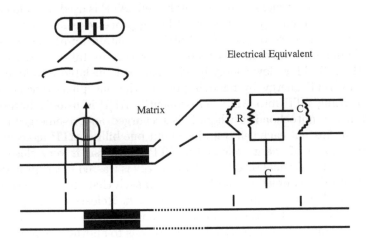

FIGURE 3.33: A diagram showing the mitochondrial membrane and its electrical equivalent.

However, the many contrasts between the prokaryotic and eukaryotic means of producing ATP provide strong evidence against the endosymbiosis theory. No intermediates to bridge these two systems have ever been found and arguments put forth in the theory's support are all highly speculative. In the standard picture of eukaryote evolution, the mitochondrion was a lucky accident. It is proposed that first, the ancestral cell, probably an archaebacterium, acquired the ability to engulf and digest complex molecules. At some point, however, this predatory cell didn't fully digest its prey, and an even more successful cell resulted when an intended meal took up permanent residence and became the mitochondrion.

3.8.2   The Cell as a Machine

In order to function, every machine requires specific parts interconnected in an intelligent fashion in order to perform the desired function. In addition, a steady supply of energy must be provided to convert it, with some level of efficiency, into useful work. Likewise, all biological cells, like machines, must have many well-engineered parts to work. Indeed, cells are constructed from yet smaller machines known as organelles. Cell organelles include mitochondria, Golgi complexes, endoplasmic reticulum and the protein filaments of the cytoskeleton. Even below this level there are machine-like parts of the cell, such as motor proteins and enzymes, that perform specific functions involving energy input and power output, e.g., transport. As mentioned above, a critically important macromolecule is ATP which is a complex nano-machine that serves as the primary energy currency of the cell. ATP is used to build complex molecules, provide energy for nearly all living processes such that it powers virtually every activity of the cell. Nutrients contain numerous low-energy covalent bonds but unfortunately these are not very useful to do most type of work in the cell. Thus, low energy bonds must be translated into high-energy bonds using ATP energy by removing one of the phosphate-oxygen groups, thereby turning ATP into ADP. Subsequently, ADP is usually immediately recycled in the mitochondria where it is recharged and re-emerges again as ATP. At any instant each cell contains about one billion ATP molecules. Because the amount of energy released in ATP hydrolysis is very close to that needed by most biological reactions, little energy is wasted in the process [272]. Generally, ATP is coupled to another reaction such that the two reactions occur nearby utilizing the same enzymatic complex. Release of phosphate from ATP is exothermic while the coupled reaction is endothermic. The terminal phosphate group is then transferred by hydrolysis to another compound, via a process called phosphorylation, producing ADP, phosphate (Pi) and energy. Phosphorylation often takes place in cascades, making it an important signaling mechanism within the cell. An important feature of ATP is that it is not excessively unstable, but instead is designed so that its hydrolysis is slow in the absence of a catalyst. This insures that its stored energy is released only in the presence of an appropriate enzyme.

The *Gibbs free energy* is defined as

$$G = E + pV - TS \tag{3.40}$$

where E is the internal energy, p pressure, V volume, T absolute temperature (K) and S entropy. One of the most important biochemical reactions in the context of energy generation in animals is the oxidation of glucose:

$$C_6H_{12}O_6 \rightarrow 6CO_2 + 6H_2O \tag{3.41}$$

in which 1 mol of glucose produces $\Delta G = +686 kcal$ of the Gibbs free energy, which is the maximum work obtainable from this reaction. In total, 180 g of glucose reacts with 134 ℓ of O_2 to produce 686 kcal of energy, i.e., 5.1 kcal is produced per ℓ of O_2. Work can be obtained indirectly either by: (a) burning the glucose and using the heat released, or (b) using glucose as a step in series of complex reactions releasing work at the end of the series. This latter approach has been utilized in animal cells via the so-called Krebs cycle. In it, for every mole of glucose metabolized, 38 mol of ATP is formed from ADP in the reaction:

$$ADP + \text{phosphate} \rightarrow ATP \qquad (3.42)$$

The overall reaction can be written as

$$glucose + 6O_2 + 38ADP + 38 \text{ phosphate} \rightarrow 38ATP + 6CO_2 + 6H_2O \quad (3.43)$$

and it requires an input of 382 kcal of energy. However, as each ATP hydrolysis reaction, which is

$$ATP \rightarrow ADP + phosphate \qquad (3.44)$$

i.e., inverse to (Eq. 3.42), 8 kcal of energy are made available. Consequently, with 38 ATP molecules one stores 304 kcal for available work, of which only about 50% is converted into useful work (as in muscle contraction) the rest being lost to heat production. Thus, the overall efficiency of biochemically-based molecular "engines" in living cells is on the order of 20%.

Next we examine fat breakdown. We suppose 302 g of fat react with 414 liters of O_2 according to

$$C_3H_5O_3(OC_4H_7)_3 + 18.5O_2 \rightarrow 15CO_2 + 13H_2O \qquad (3.45)$$

when 1941 kilocalories of energy are produced. This then corresponds to 4.7 kilocalories per liter of O_2.

In these two examples the number of kilocalories per liter of O_2 are fairly close. For protein breakdown a similar picture emerges. On average 4.9 kilocalories are produced per liter of O_2.

3.8.3 Active Transport

The discussion of molecular phenomena and biological processes so far has included the general categories of diffusion, osmosis and reverse osmosis, all of which on the cellular level are passive in nature. The driving energy for passive transport comes from molecular kinetic energy or pressure. There is another class of transport phenomena, called active transport, in which the living membrane itself supplies energy (typically via ATP) to cause the transport of substances. Biological organisms sometimes need to transport substances from regions of low concentration to high concentration - the direction opposite to that in osmosis or dialysis. Of course, sufficiently large back pressure causes reverse osmosis or reverse dialysis, but there are known instances in which substances move in the direction that reverse osmosis or reverse dialysis would take them even if existing pressures are insufficient. In these instances, active transport must be taking place, which means that living membranes expend their own energy to transport substances. Active transport can also aid ordinary osmosis or dialysis and explains why some transport proceeds faster than expected from osmosis or dialysis alone.

Active transport is extremely important in nerve cells. Changes in the concentration of electrolytes across nerve cell walls are responsible for nerve impulses. After repeated nerve impulses, significant migration has occurred and active transport "pumps" the electrolytes back to their original concentrations.

3.8.4   Ion Channels and Ion Pumps

The membrane possesses pores or channels that allow a selective passage of metabolites and ions in and out of the cell. They can even drag molecules from an area of low concentration to an area of high concentration working directly against diffusion. An example of this is the sodium/potassium pump. Most of the work done on the transport across membranes is done via ion pumps such as sodium-potassium pumps. The energy required for the functioning of the pump comes from the hydrolysis of ATP in which a phosphorylated protein is identified as an intermediate in the process. The hydrolysis of a phospho-protein usually causes a conformational change that opens a pore that drives the sodium and potassium transport. Some membrane proteins actively use energy from the ATP in the cell to perform mechanical work. Here, the energy of a phosphate is used to exchange sodium atoms for potassium atoms. It can be demonstrated that the free energy change in the hydrolysis of a phospho-protein with a value of 9.3 kJ/mol will drive a concentration gradient of 50:1 uphill. For each ATP molecule hydrolyzed, three sodium ions are pumped out and two potassium ions are pumped in. As shown in Fig. 3.34 ion channels come in three general classes: (a) voltage-gated, (b) ligand-gated and (c) gap junctions. They differ not only in their design geometry but also in the use of physical and chemical mechanisms for the selection of ions for passage. Fig. 3.35 shows schematically how a conformational change in an ion channel leads to the opening and closing of the gate. Fig. 3.36 illustrates the concept of an equivalent electrical circuit for an ion channel which is modulated by a voltage.

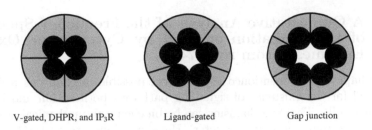

V-gated, DHPR, and IP$_3$R   Ligand-gated   Gap junction

FIGURE 3.34:   The three key types of ion channels.

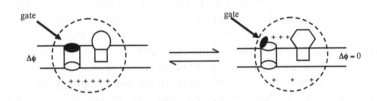

FIGURE 3.35:   The functioning of a gate: modulation of a local electric field by an ion channel.

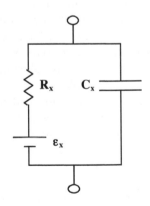

FIGURE 3.36:   Equivalent electrical circuit of an ion channel.

3.9 A Quantitative Analysis of the Frequency Spectrum of the Radiation emitted by Cytochrome Oxidase Enzymes - from ref [274]

In Chapter 1 we mentioned that several possible mechanisms have been identified for the existence of signalling pathways both within and between biological cells. These mechanisms can be biochemical, mechanical, electronic or even electromagnetic in character. Here, we elaborate on the latter type as it pertains to cell-cell communication.

3.9.1 Introduction

In a series of studies spanning a period of some 25 years of research G. Albrecht-Buehler (AB) demonstrated that living cells possess a spatial orientation mechanism located in the centriole [6,7,8]. This is based on an intricate arrangement of microtubule filaments in two sets of nine triplets each of which are perpendicular to each other. This arrangement provides the cell with a primitive "eye" that allows it to locate the position of other cells within a two to three degree accuracy in the azimuthal plane and with respect to the axis perpendicular to it. He further showed that electromagnetic signals are the triggers for the cells' repositioning. It is still largely a mystery how the reception of electromagnetic radiation is accomplished by the centriole. Another mystery related to these observations is the original electromagnetic radiation emitted by a living cell. Using pulsating infra-red signals scattered off plastic beads AB mimicked the effects of the presence of another living cell in the neighbourhood. The question that still remains which we wish to address in this section is the source of infra-red radiation speculated by AB to originate in the mitochondria. Mitochondria are the organelles which produce the energy in the form of ATP molecules each of which carries approximately 0.5eV stored in the high energy phosphate group, part of which is due to the electrostatic repulsion between the oxygens in the phosphate group. The synthesis of ATP in the mitochondria is a multiple step process relying on the transfer of protons across the mitochondria wall and creating a pH gradient. The explanation of how this works brought Mitchell a Nobel Prize in 1971 [224]. At the centre of this mechanism is the functioning of proton pumps embedded in the mitochondrial wall. Broadly speaking, these pumps are intricately built with an active moveable part consisting of an enzyme called cytochrome oxidase that opens and closes a channel through which individual H^+ ions enter the mitochondria. A living cell is a non-equilibrium physical system which requires a continuous energy input to sustain its activities. In animal cells energy production involves breakdown of nutrients and a subsequent production of ATP molecules which are the universal currency of energy in cells. The production of ATP molecules in turn is directly linked to the regular functioning of the mitochondrial pumps. It is therefore logical that one of the signatures of a living cell would be related to the oscillatory (pulsating) effect of mitochondrial proton flow. AB did suggest that the mitochondria are the best candidates for the explanation of infra-red emission effects. He even pointed in the direction of the porphyrin containing proteins, i.e., the cytochromes. It is our intention in this section to provide a quantitative account of how the structure of the porphyrin molecule results in the spectrum of emitted IR radiation observed in AB's experiments. Based on statistical analysis of a culture of 800 cells AB found a response function to the wavelength of electromagnetic radiation found below. The peak in Fig. 3.37 appears to coincide with 850 nm which is in the near IR and in the following pages we propose a detailed quantitative mechanism explaining such an emission spectrum.

3.9.2 The Biochemical Structure and Function of the Cytochromes

The mechanisms involved in the process of proton-gradients are illustrated in Fig. 3.38. to convert the energy of redox reactions of photoredox processes into a movement of protons from the left- to the right-hand side of the membrane; the redox driven protonomotive system is postulated. The separation of protons and anions across the membrane holds the energy and the stored potential energy is utilised in two ways. When there is a right-to-left flow of protons through the ATP, synthetase is coupled to the energy requiring synthesis of ATP from ADP and inorganic phosphate. This latter flow is coupled to active transport of biochemical components. The reverse flow involves the hydrolysis of ATP on the left side accompanied by the left to right proton flow.

ATP synthetases are located asymmetrically in membranes which are driven by the electrochemical potential of the protons and have a H^+/P stochiometry characterising it. The major sub-units are shown in Fig. 3.39: to the left cytochrome c oxidase and to the right the quinol oxidase. In both cases the motif of a single heme plus heme/copper binuclear centre is evident. An analogy may be drawn, despite the fact that the enzymes differ in the nature of their electron entry sites, between electron donation from cytochrome c to the Cu_a/cytochrome a sites and from quinol (QH_2) to heme. The oxidation system in mitochondria is made up principally of the membrane associated with the electron transport chain.

The oxidation of reduced NAD, the reduction of O_2 to water which accompanies it and the storage of energy are the overall function of this system. It is assumed in the chemiosmotic hypothesis that the intermediate form is created by transporting hydrogen ions from the inner mitochondrial matrix to the exterior of the inner mitochondrial membrane. A representation of the process in the mitochondria can be schematically illustrated by the following reactions

(a)

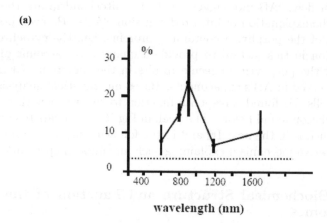

(b)

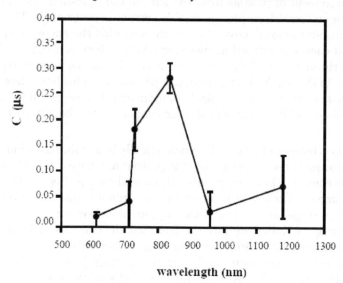

FIGURE 3.37: (a) The percentage of cells that removed the light scattering particle as a function of wavelength. The most "attractive" wavelength was between 800 and 900nm (*http ://www.basic.nwu.edu/g-buehler/irvision.htm*). (b) Spectral sensitivity of destabilisation versus wavelength in nm.

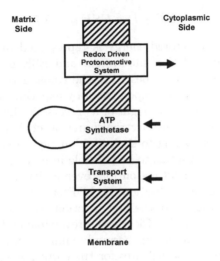

FIGURE 3.38: A model of the transducing components located on the inner side of the mitochondrial membrane.

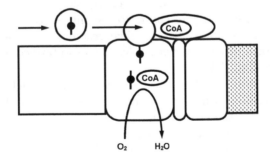

FIGURE 3.39: A schematic diagram of cytochrome c oxidase (cyt.aa3).

$$NADH + H^+ + \frac{1}{2}O_2 \Leftrightarrow NAD^+ + H_2O + ESE$$

$$ADP + P + ESE \Leftrightarrow ATP$$

The electrochemically stored energy is represented by ESE. The reaction in (a) describes the oxidation sub-system and ESE is associated with the membrane. The process in (b) describes the transduction system. Available models have restrictions placed upon them by experimental features. The reactions in (a) and (b) can be thought of as reversible since, in the presence of excess ATP and an appropriate electron source other than water, the reactions can be run backwards leading to such high energy reduced intermediates as NADH. This mechanism seems to be possible in many systems and a normal pathway in others. A second reason for reversibility is that respiratory control of the rate of oxygen utilisation by ADP suggests a near-to-equilibrium distribution of the products of the first reaction. By a utilisation of ESE the reaction is shifted to the right. Clearly the reversible and near-to-equilibrium features suggest that equilibrium thermodynamics can be used. We may assume that for eq. (b) the actual sites for the corresponding processes are the ATP synthetase complexes of the inner mitochondrial membrane.

As a super family of metalloenzymes, the terminal oxidases share the basic function of catalysing, by a 4 electron process, the complete reduction of oxygen to water. For our purposes it is important to emphasise that a proton electrochemical gradient is established across the membrane by the terminal oxidases. This is achieved by taking up the required protons to produce water from one side of the membrane only. These latter oxidases have also been shown to pump protons across the membrane. This occurs in a process which implies a coupling, against the diffusion gradient, between endergonic proton translocation and an exergonic electron transfer process. The terminal oxidases therefore contribute significantly to the maintenance of a transmembrane proton electrochemical gradient. Crystallographic structures of cytochrome oxidase enzymes have been established and we show one example below in Fig. 3.40.

It has been firmly established that there is a near-zero activation energy in oxidation of reduced cytochrome c by ferricyanide. The uptake of an electron at the iron centre or the energy for release may come from the vibrational energy of the protein. Furthermore, for our purposes in this section, it would seem quite probable that conformational mobility of the protein may assist in the electron-transfer process and the control of the gating process itself. The removal of sites for binding ferricyanide and elimination of surface positive charges via modification of lysine residues does not affect the activation energy which remains near zero. The treatment of the protein with 4M guanidine hydrochloride (which relaxes the binding of ferrocyanide to the reduced molecule) does not affect the electron transfer rate. The notion that conforma-

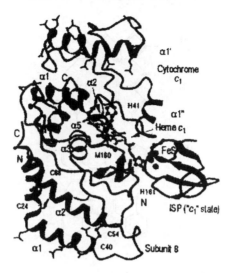

FIGURE 3.40: A view from the mitochondria intermembrane space down the membrane normal: a stereoview of cytochrome c, sub-unit 8 and the metalcluster binding fold of the ISP at the "c_1" state. The disulphide bonds in sub-unit 8 are shown and ligands of heme c_1, His^{161} of ISP.

tional changes are important in the electron transfer function of cytochrome c is supported by the above evidence.

At the centre of the enzyme one always finds an all important heme group which plays a cardinal functional role. The heme group, which is essentially a network of 36 conjugated bonds arranged in a flat disc, is derived from porphyrin molecules. Below we show the basic structure of the heme group in Fig. 3.41(a) as well as a specific example in Fig. 3.41(b). Iwata et al. [160] pointed out that the structure of the cytochrome bc1 complex, especially the iron-sulphur complex, suggests the existence of a new electron transport mechanism of the enzyme. The importance of electro-conformational coupling, i.e., the influence of oscillating electric fields for the enzymes cellular energy production, has been demonstrated by Tsong et al. [320] within a physical framework. The heme group and related molecules often contain a metal ion at their centre. One can imagine that such systems of conjugated bonds will exhibit physical properties which are somewhere "between" those produced by the energy levels of single ions and the continuous energy band structure of solid state crystals. The electronic structure generated by the charge of the metal ion when it interacts with its environment will be considerably modified. Some of the states may lie close to each other not allowing transitions to take place because of selection rules arising from conservation of momentum and spin. It is possible therefore that they may store energy in the form of infra-red photons being well protected from the thermal chaos

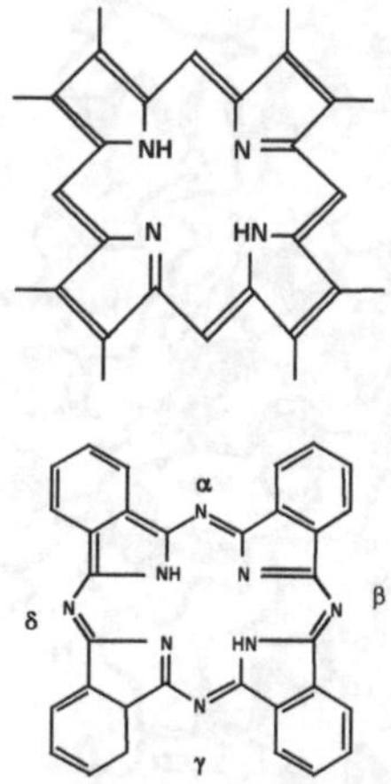

FIGURE 3.41: (a) Basic molecular structure of porphyrins without the side chains being specified. (b) H_2 phthalocyanine (for H_2 porphyrin omit peripheral benzene rings and replace $\alpha, \beta, \gamma, \delta, -N$ by $-CH = $'s).

of the cellular world until a very specific trigger discharges them. These discharges may release photons of higher energy than the single electrons which built up the charge. They may also generate sudden electrical conductivity of the molecule because electrons in higher energy levels may have energies comparable to the energies of the conduction bands of the crystal in which they are embedded. Thus the porphyrin molecule could serve as a powerful amplification mechanism even though the absorbed photons have relatively small energies. In the next section we present a very simplified picture of the lowest states of the divalent ferrous ion in the centre of a heme group.

3.9.3   A Simplified Model Calculation

We now present a simplified model calculation, following from Tuszyński et al. [274]. We consider a structure made up of N carbon atoms in a plane and arranged in a circle with an Fe^{2+} ion at its centre having an incomplete d shell containing six electrons and all other ion states are filled. The presence of a ferrous or ferric ion in the cytochrome molecule is a reasonable presumption since it undergoes a redox reaction by hydrogenase whereby

ferricytochrome $c_3(4Fe^{3+}) + 2H_2 \Leftrightarrow$ ferrocytochrome $c_3(4Fe^{2+}) + 4H^+$

We choose to examine the divalent variety and how its environment modifies its orbital states since its ground term has an orbital angular momentum, L = 2 (a high angular momentum state), whereas the ferric variety is an S-state ion which is not significantly affected by charges in its environment.

If a is the radius of the first co-ordination circle, the carbon atoms will be at positions $\left(x = a\cos\left(\frac{2\pi}{N}m\right), y = a\sin\left(\frac{2\pi}{N}m\right), z = 0\right)$ where the integer, m, takes the values m = $1, 2, \ldots, N$.

We consider one of the 3d electrons for simplicity at the position $x = r\sin\theta\cos\phi$, $y = r\sin\theta\sin\phi$, $z = r\cos\theta$. If $-q|e|$ is the charge on any one carbon atom, the potential energy, V, of one 3d electron will be obtained by writing the squared distance between the carbon atom and a 3d electron as

$$V = \frac{q|e|^2}{\sqrt{r^2 + a^2 - 2ar\sin\theta\left(\cos\left(2\frac{\pi}{N}m\right)\cos\phi + \sin\left(\frac{2\pi}{N}m\right)\sin\phi\right)}} \tag{3.46}$$

$$\left(x - a\cos\left(\frac{2\pi}{N}m\right)\right)^2 + \left(y - a\sin\frac{2\pi}{N}m\right)^2 + z^2 \tag{3.47}$$

Rewriting (eq. 3.46), V becomes

$$V = \frac{q|e|^2}{\sqrt{r^2 + a^2 - 2ar\sin\theta\cos\left(2\frac{\pi}{N}m - \phi\right)}} = \frac{q|e|^2}{\sqrt{1 + \frac{r^2}{a^2} - 2\frac{r}{a}\sin\theta\cos\bar\theta}} \tag{3.48}$$

where

$$\bar\theta = \frac{2\pi}{N}m - \phi \tag{3.49}$$

By assuming $r \ll a$ we may expand the square root so that (eq. 3.48) becomes

$$V = \frac{q|e|}{a}\sum_{n=1}^{\infty}\left(\frac{r}{a}\right)^n P_n(\sin\theta\cos\bar\theta) \tag{3.50}$$

where the P_n are the nth Legendre polynomials. More explicitly, to the $n = 4$ term

$$V = \frac{q|e|^2}{a}\{1 + \frac{r}{a}\sin\theta\cos\bar\theta + \frac{r^2}{a^2}\frac{1}{2}\left[3\sin^2\theta\cos^2\bar\theta - 1\right]$$
$$+ \frac{r^3}{a^3}\frac{1}{2}\left[5\sin^3\theta\cos^3\bar\theta - 3\sin\theta\cos\bar\theta\right] \tag{3.51}$$
$$+ \frac{r^4}{a^4}\left[35\cos^4\bar\theta\sin^4\theta - 3\cos^2\bar\theta\sin^2\theta + 3\right] + \ldots\}$$

Utilising trigonometric multiple angle formulas [125] (eq. 3.51) becomes

$$V = \frac{q|e|^2}{a}\left\{1 + \frac{r}{a}\sin\theta\cos\bar{\theta} + \frac{r^2}{4a^2}\left[3\sin^2\theta(\cos 2\bar{\theta} + 1) - 2\right]\right.$$

$$+ \frac{r^3}{8a^3}\left[5\sin^3\theta(\cos 3\bar{\theta} + 3\cos\bar{\theta}) - 12\sin\theta\cos\bar{\theta}\right] \tag{3.52}$$

$$\left. + \frac{r^4}{64a^4}\left[35\sin^4\theta(\cos 4\bar{\theta} + 4\cos 2\bar{\theta} + 3) - 120\sin^2\theta(\cos 2\bar{\theta} + 1) + 24\right] + \ldots\right\}$$

Summing over all the carbon atoms, i.e., m from 1 to N, the contributions from $\cos\bar{\theta}, \cos 2\bar{\theta}, \cos 3\bar{\theta}$ vanish leaving the potential energy of one electron due to N carbon atoms, $\bar{V}$, as

$$\bar{V} = \frac{q|e|^2}{a}N\left\{1 + \frac{r^2}{4a^2}[3\sin^2\theta - 2] + \frac{r^4}{64a^2}[105\sin^4\theta - 120\sin^2\theta + 24] + \ldots\right\} \tag{3.53}$$

When $\bar{V}$ is summed over all 6 electrons, retaining only the first two terms of $\bar{V}$, and using operator equivalents we find

$$\bar{V} = \frac{q|e|^2}{a}N\left\{6 + \frac{<r^2>\alpha}{4a^2}(L(L+1) - 3L_z^2) + \ldots\right\} \tag{3.54}$$

where the average of r^2 for 3d^6 Fe^{2+} is $<r^2> = 2.026a_0^2$, a_0 being the radius of the first Bohr orbit of hydrogen. The constant $\alpha = -2/21$ and L is the total orbital angular momentum of Fe^{2+} for its ground term which by Hund's Rules is L = 2. We drop the first constant term in eq. (3.54) as this merely shifts all levels up or down in energy depending on its sign. In order to estimate the range of acceptable values for $\bar{V}$ in (eq. 3.54) we rewrite this expression as

$$\bar{V} = \frac{\Delta E q N}{x^3}(-4800)cm^{-1} \tag{3.55}$$

where we have used the facts that $e^2/a_0 \simeq 10^5 cm^{-1}$ and $1eV \simeq 8000cm^{-1}$. E is the energy separation in dimensionless units from $L(L+1) - 3L_z^2$ when the component of orbital angular momentum takes the values $\mp 2, \mp 1$ and 0. Thus positive values of ΔE can be 9, 12 and 3. The parameter q represents the valence of the carbon atoms which can lie between 2 and 4 and N is the total number of carbon atoms in the ring which we estimate can range between 30 and 36. Here, x represents the multiple of a_0 which makes up the radius of the circle, i.e., $a = xa_0$. We believe x can vary approximately between $x = 10$ and $x = 20$. The largest positive value of $\bar{V}$ is, therefore,

$$\bar{V} \simeq \frac{12 \times 4 \times 36}{1000} \times 4800 = 8294.4cm^{-1} = 1.04eV \tag{3.56}$$

The energy corresponds to a wavelength which is almost 1200 nm and lies approximately in the middle of the range shown in Figure 3.42 stretching from

600 to 1700 nm. The values above 1200 nm correspond to lower energies and are easy to account for due to the possibility of a larger ring size, smaller valence or smaller values of ΔE. We envisage, as we see later, a proton in close proximity to the ring which will then become distorted preferentially on one side only. Thus the effective radius, a, of the ring for a percentage of the carbon atoms on one side of it will become less. As the quantity a appears as a^{-3} in expression (eq. 3.54) or (eq. 3.55), this distortion will have the effect of increasing the magnitude of $\bar{V}$ in (eq. 3.55). This increase will clearly depend on the number of carbon atoms affected and the average fractional decrease in a, i.e., $-3\delta a/a$. We estimate that for $|\delta a/a|$ 5/100 then the value of $\bar{V}$ in eq. (3.56) will be increased, for half the carbon atoms affected, by approximately 6%. This is quite sufficient to bring $\bar{V}$ close to $1.10eV$ in agreement with experiment. In order to explain the peak below 1200 nm one needs to revisit the structure of the ring. The trivalent nitrogen tends to be closer to the centre of the ring and could also easily increase the energy of the transition.

3.9.4 A Proposed Mechanism

The question of electronic [171-232], protonic [73-259] and ionic [349, 41] conduction in biological systems, especially proteins and enzymes, has received considerable attention in the last two decades. Although dry state activation energies for biopolymers have been reported in the range 2.2 to 3.7 eV [95] this potential barrier can be significantly reduced to the values observed for semiconductors which are 1.1 eV or less [28] as result of hydration which, of course, is the natural state in biological systems. A particular mechanism of electron and proton mobility involves transfer of electrons from donor to acceptor sites via a ferrying process facilitated by protons. We believe that this is indeed the case with a protonic passage through channels in mitochondrial walls.

Gilmanshin and Lazarev [122] claim that the transfer of electrons along a protein is accomplished by a series of redox centres incorporated into the protein structure. This allows for the directionality of electron transfer over distances of between 0.3 and 3.0 nm. The protein-electron carrier system consists of protein, possibly localised within a mitochondrial wall, which has a single redox centre. The redox centres are usually prosthetic groups containing molecules of non-proteinous origin which have conjugated orbital systems and often incorporate metal ions. The function of the prosthetic group is several fold. It ensures the fixation of charges and dipoles in its microsurroundings and catalyses electron transfer. Secondly, the orientation of the prosthetic group relative to other proteins may enhance recognition of the redox centre. Thirdly, the prosthetic group may act to influence the electronic state of associated proteins or vice versa, and result in a degree of isolation from the polar solvent. Finally, it may control electron transport through its oxidation or reaction which alters the carrier concentration. Ichinose et al. [154] are in general agreement with the statement that electron transfer in biological systems occurs via a redox scheme but are critical of electron tunnelling over distances of 3.0-7.0 nm as some have reported. Instead, Ichinose et al. propose that protein is an insulator at physiological temperature and that electron transport is mediated by protons.

In terms of the energies and individual mechanisms for the electronic transitions in the cytochrome oxidase we propose that the first step in the process involves the excitation of the ferrous ion. This is most likely caused by a kinetic energy transfer from the decelerating proton approaching a positively charged heme group. It would only require a slow-down of the proton from a velocity seven times greater than the rms value to the thermal average. Once the ferrous ion is sufficiently excited an electron can tunnel through a potential barrier to become loosely bound to the proton as illustrated in Fig. 3.42. The virtual weakly bound hydrogenic system formed may then relax

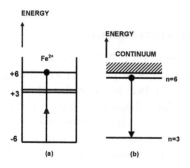

FIGURE 3.42: Excitation of Fe^{+2}, transfer of an electron to a proton and subsequent descent in energy from the n=6 to n=3 state of hydrogen. (A) Energy levels of Fe^{2+} labelled with the component of angular momentum. (B) Two energy levels of hydrogen labelled with the principal quantum numbers n = 6 and n = 3.

from an n = 6 state to an n = 3 state to emit IR radiation of 1.13 eV. The iron remaining in the heme group will be a ferric ion, in an S-state ground state, which will not couple strongly to its immediate ring of carbon atoms. The loosely bound hydrogen electron, when the proton moves past the heme group, would then be transferred to a local acceptor site, so that when donor and acceptor relax, the iron is restored to its Fe^{2+} valence state with six electrons and the process begins again as the proton moves further through the mitochondrial membrane (see Fig. 3.43).

All living systems spontaneously emit biophotons [126] in a broad range from thermal radiation in the infra-red through visible up to ultra-violet. The intensity of biophotons has been detected to range from a few photons/second/

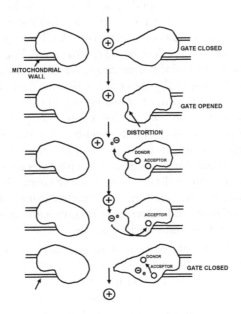

FIGURE 3.43: The passage of a proton through an opening in the mito-chondrial wall ferrying an electron between donor and acceptor sites. The proton "opens the gate" by distortion of the heme group, accepts a loosely bound electron, freely diffuses to the acceptor site where it then donates the electron. Donor and acceptor relax to their initial state and the "gate closes".

cm^2 up to several hundred photons/second/ cm^2. In most cases the spectral distribution is broad and rather flat except in special situations such as cell division. The origin of biophoton research can be traced back to the pioneering work of A.G. Gurwitsch [127] and can be linked to the electromagnetic theory of life proposed by Burr and Northrop [51]. In 1955 Colli [71] succeeded in proving the existence of biophoton emission in the visible between 390 and 650 nm.

Today, bioluminescence is a well documented phenomenon which can be traced to bacteria whose light emissions can be visible to the naked eye. Various higher organisms such as squid, jelly fish and fireflies emit light using the enzyme luciferase. We note that for fireflies and fishes this luminescence serves interorganism communication purposes. Colli's experiment were later supported by the results of Veselovskii [337], Ruth and Popp [276], Tarasov et al. [311], all of whom studied different organisms. Significantly, for the purpose Vladimirov and L'vova [339] detected emissions from isolated mitochondria of rat liver that occurred most intensely when conditions were optimum for oxidative phosphorylation. Zhuravlev et al. [355] worked with isolated rat liver mitochondria, finding that mitochondrial luminescence requires ADP and oxygen, and that uncoupled electron transport (non-phosphorylating) can contribute to the observed luminescence. Since photons are emitted when electrons jump from molecular excited states to lower energy levels, it should not be too surprising to find that photons are emitted from mitochondria, where oxidative phosphorylation and electron transport can provide the energetic pathways to boost electrons into excited states.

For much more detailed information about biophotons the reader is referred to *http://www.datadiwan.de/iib/ib_oooe_.htm* (the Web site of the International Institute of Biophysics), and on bioluminescence we refer to the reader to consult *http://www.dcn.davis.ca.us/go/karl* (the Web site of K.Simanonok). Interestingly, the latter source comments on the source of bioluminescence in eukaryotic cells in the following way: "Unlike bacteria and other organisms displaying macroscopic bioluminescence however, precise mechanisms for weak eukaryotic photon emissions remain to be discovered". It was our intention in this section to provide a small insight into this question.

3.10   Membranes and vesicles: phase transitions in lipid bilayers and shape selection

Amphipathic molecules adsorb themselves to the air-water or oil-water interfaces such that their head groups are facing the water environment. They aggregate to form either spherical micelles or liquid crystalline structures. In general, amphipathic molecules can be anionic, cationic, non-ionic or zwitterionic. The relative concentrations of these surfactants in an aqueous solution affect its physical and chemical properties. At a specific value, called the critical micelle concentration (CMC), micelles containing 20-100 molecules are formed spontaneously in the solution with the hydrophilic head groups exposed and the hydrophobic tails hidden inside the micelle. The principal driving force for micelle formation is entropic due to a negative free energy change accompanying the liberation of water molecules from clathrates. When phospholipids are mixed in water, they form double layered structures since their hydrophilic ends are in contact with water while the hydrophobic ends face inwards touching each other.

3.10.0.1 Membrane and Membrane Proteins

Biomembranes compartmentalize areas of different metabolic activity in the cell. They also regulate the flow into and out of cells and cell compartments. Membranes are also sites of key biochemical reactions. As previously mentioned, the cell membrane forms a thin, nearly invisible continuous boundary region that completely surrounds the cytoplasm of the cell. It also surrounds the endoplasmic reticulum and the nuclear membrane, and connects the two together. Membranes are composed of phospholipids, glycolipids, sterols, fatty acid salts and proteins. A simple representation of a phospholipid with the globular head group structure representing the hydrophilic (or water loving) section of the phospholipid is shown in Fig. 3.44. The tails that come off of the sphere represent the hydrophobic (or water fearing) end of the phospholipid. The two long chains coming off of the bottom of this molecule are made up of carbon and hydrogen. Because both of these elements share their electrons evenly, these chains have no charge. These molecules are not attracted to water; as a result water molecules tend to push them out of the way as they are attracted to each other. At the other end of the phospholipid are a phosphate group and several double-bonded oxygens. The atoms at this end of the molecule are not shared equally. This end of the molecule has a charge and is attracted to water. In Fig. 3.45 we show a simplified representation of a cellular membrane.

The functions of a cell membrane can be listed as follows:

- it is a selectively permeable barrier between two predominantly aqueous compartments,

- it allows compartmentalization of the various structures in the cell,

- it enables the formation of a stable and fluid medium for reactions that are catalyzed,

- it provides a flexible boundary between the cell or an organelle and its surrounding medium.

- it maintains an electric potential difference, participates in signal transmission to the actin cycloskeleton (via integrins) and provides adhesion

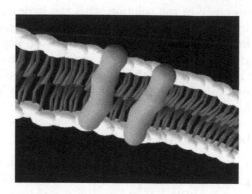

FIGURE 3.44: A simple representation of a phospholipid with the globular head group structure representing the hydrophilic (or water loving) section of the phospholipid.

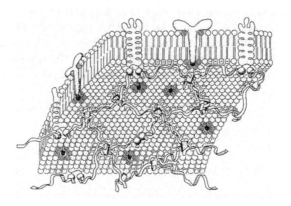

FIGURE 3.45: Schematic of a typical cellular membrane.

forces for the cells to their substrates (controlled by membrane elasticity).

- it enables mass transport (via ion channels)

The fluid mosaic model of Singer and Nicolson (1972) views the membrane as a fluid bilayer of amphipathic complex lipids with proteins embedded in it and spanning it as shown in Fig. 3.45. The relative abundance of proteins in a membrane varies from species to species and it correlates with the metabolic activity. For example, the mitochondrial wall contains large amounts of protein (52-76%) and smaller amounts of lipids (24-48%) facilitating its high metabolic activity. Conversely, the inactive membrane of the myelin sheath in neurons contains only 18% of proteins and 79% of lipids.

A double layer of phospholipid molecules with a variety of imbedded proteins makes up the plasma membrane or the outer surface of a cell. This plasma membrane does not resemble the surface of a fluid or even the interface between two fluids. The reason for this is that the plasma membrane has an essentially fixed surface area, i.e., there are only a fixed number of phospholipid molecules and proteins which, when packed together, make up the membrane. Each lipid molecule or protein has a preferred surface area so, unlike the surface of a fluid, the plasma membrane is for practical purposes inextensible.

The various components of membranes are subject to rapid movements. Rapid lateral movement of lipids has a diffusion constant of approximately 10^{-8} cm^2/s while proteins between 10^{-10} and 10^{-12} cm/s. On the other hand, flip-flop movements across the membrane are slow, on the order of 10^5 s. Indeed, the phospholipids of the membrane may undergo a phase transition from a gel phase to a liquid crystal phase. This may take place as a result of changing the temperature, external pressure of even membrane composition (see Fig. 3.46).

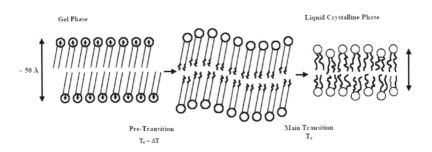

FIGURE 3.46: Membrane phase transitions.

Depending on the prevailing conditions, membrane dynamics has been described either using continuous Landau-Ginzburg models developed by Owicki et al. [250] and de Gennes [79]. Alternatively, for the development of phospholipid tails' twists discrete variable models have been proposed making use of a quasi-spin representing the state of each individual hydrocarbon chain. An Ising Model approach has been advocated by Zuckermann et al. [356] and Pink [254] where the Hamiltonian for the membrane is postulated in the form

$$H_T = -\frac{1}{2} \sum_n \sum_m J_{nm}(D_{nm}) L_{ni} L_{nj} + \sum_n (\pi A_n + E_n) L_{ni} \qquad (3.57)$$

where i labels chain sites, n enumerates chains, L_n is the chain length (depends on the number of twists), π is the lateral pressure and E_n is the internal energy of the n-th chain. Typical of 2D Ising models is the emergence of an order-disorder phase transition at a finite temperature that depends on the number of nearest neighbors, the strength of their interactions and the applied external field (pressure).

Suppose we increase the osmotic pressure of a cell. The cell will try to swell but this is prevented because the surface area of the plasma membrane is nearly fixed, i.e., an elastic stress will be built up inside the membrane and if this is too great the cell will burst - a condition called lysis. To find the stress at which the membrane will burst suppose we cut the membrane along some line of length ℓ. To prevent the two sides of the cut from separating a force, F, must be present which is proportional to ℓ. Writing $F = \gamma \ell$, then the proportionality constant, γ, is called the elastic tension in the wall. This tension is not the surface tension, T, of a fluid but it plays a similar role. The excess pressure inside a bubble of radius R, over and above atmospheric pressure, is $2T/R$. It turns out that a similar relation to this may be used to compute the excess pressure, ΔP, inside a spherical pressure vessel if we replace the surface tension, T, by the elastic tension γ. Thus we have

$$\Delta P = \frac{2\gamma}{R} \qquad (3.58)$$

The elastic stress, σ, inside the vessel wall is related to the elastic tension by

$$\gamma = D\sigma \qquad (3.59)$$

where D is the wall thickness. The reason for this is that the surface area of the cut is ℓD and hence the force per unit area on the surface of the cut is $F/\ell D = \gamma/D$ and this is the stress of the vessel wall. From the two equations above, the elastic stress, σ, is given by [25]

$$\sigma = \frac{R\Delta P}{2D} \qquad (3.60)$$

so that the vessel will burst when this stress exceeds the fracture stress of the material from which the vessel was made.

3.10.0.2 Mass Diffusion Across a Membrane

A large amount of diffusion in biological organisms takes place through membranes. These membranes are very thin, from 65×10^{-10} to 100×10^{-10} m across. Most membranes are selectively permeable; that is, they allow only certain substances to cross them because there are pores through which substances diffuse. These pores are so small (from 7×10^{-10} to 10×10^{-10} m) that only small molecules get through. Other factors contributing to the semi-permeable nature of membranes have to do with the chemistry of the membrane, cohesive and adhesive forces, charges on the ions involved and the existence of carrier molecules. Diffusion in liquids and through membranes is a slow process.

We can apply Fick's Law to the transport of molecules across a membrane which can be a biomembrane, e.g., the phospholipid layers surrounding cells, or an artificial membrane such as that used in a dialysis machine. Consider as an example a container of sugar water that is divided by a membrane of thickness Δx. Assume also that the concentration of glucose on the left side is c_L and that on the right side is c_R. The glucose diffusion current across the membrane, according to Fick's Law, is given by [32]

$$I = k \frac{D}{\Delta x} A \Delta c \tag{3.61}$$

where $c = c_R - c_L$, k is the diffusion constant and A is the cross sectional area of the membrane.

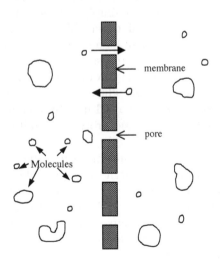

FIGURE 3.47: Schematic of a semi-permeable membrane.

3.10.0.3 Membrane Proteins

The cell membrane's function, in general, revolves around its membrane proteins. Floating around in the cell membrane are different kinds of proteins. These are generally globular proteins. They are not held in any fixed pattern but instead move about in the phospholipid layer. Generally these proteins structurally fall into three categories:

(a) carrier proteins that regulate transport and diffusion,

(b) marker proteins that identify the cell to other cells,

(c) receptor proteins that allow the cell to receive instructions, communicate, transport proteins regulate what enters or leaves the cell.

Membrane proteins are either (a) peripherical or (b) integral. Peripherical proteins are bound electrostatically to the exterior parts of head groups and hence can be easily extracted. Integral proteins are tightly bound to lipid tails and are insoluble in water. Steroids are sometimes a component of cell membranes in the form of cholesterol. When it is present, it reduces the fluidity of the membrane. However, not all membranes contain cholesterol. Transport proteins come in two forms. Carrier proteins are peripherical proteins that do not extend all the way through the membrane. They bond and drag specific molecules through the lipid bilayer one at a time and release them on the opposite side. Channel proteins extend through the lipid bilayer. They form a pore through the membrane that can move molecules in several ways. In some cases the channel proteins simply act as a passive pore. Molecules randomly move through the opening via diffusion. This requires no energy and molecules move from an area of high concentration to an area of low concentration. Symports also use the process of diffusion. In this case a molecule that is moving naturally into the cell through diffusion is used to drag another molecule into the cell. For example, glucose hitches a ride with sodium.

Marker proteins extend across the cell membrane and serve to identify the cell. The immune system uses these proteins to tell own cells from foreign invaders.

The cell membrane can also engulf structures that are much too large to fit through the pores in the membrane proteins. This process is known as endocytosis and in it the membrane wraps itself around the particle and pinches off a vesicle inside the cell. The opposite of endocytosis is exocytosis. Large molecules that are manufactured in the cell are released through the cell membrane. A prominent example of this process is the exocytosis of neurotransmitter molecules into the synapse region of a nerve cell.

3.10.0.4 Electrical Potentials of Cellular Membranes

First measurements of the electrical properties of cell membranes were made on red blood cells by H. Fricke and on sea urchin cells by K.S. Cole [70] and it was found that membranes act as capacitors maintaining a potential difference between oppositely charged surfaces composed mainly of phospholipids with proteins embedded in them. A typical value of the capacitance per unit area C/A is about $1\mu F/cm^2$ for cell membranes. This relates to the membrane's dielectric constant κ via

$$\frac{C}{A} = \frac{\varepsilon\varepsilon_0}{d} \tag{3.62}$$

where $\varepsilon_0 = 8.85 \times 10^{-12}C^2/Nm^2$ giving a value of $\varepsilon \simeq 10$ which is greater than $\varepsilon \simeq 3$ for phospholipids above resulting from the active presence of proteins. Note that potential differences across cell membranes are created and maintained at an energy cost through the action of sodium/potassium pumps. In the normal resting state, the cellular membrane is much more permeable to potassium ions than to sodium ions which results in an outward flow of potassium ions and the voltage inside the cell is $-85mV$. This voltage is called the resting potential of the cell. If the cell is stimulated by mechanical, chemical or electrical means sodium ions diffuse more readily into the cell since the stimulus changes the permeability of the cellular membrane. The inward diffusion of a small amount of sodium ions increases the interior voltage to $+60mV$ which is known as the action potential of the cell. The membrane again changes its permeability once the cell has achieved its action potential and potassium ions then readily diffuse outward so the cell returns to its resting potential. Depending on the state of the cell, the interior voltage can therefore vary from its resting potential of $-85mV$ to its action potential of $+60mV$. This results in a net voltage change of $145mV$ in the cell interior. The voltage difference between the two sides of the membrane is fixed by the concentration difference. Having a salt concentration difference across a membrane and allowing only one kind of ion to pass the membrane produces a voltage difference given by the formula

$$V_L - V_R = \frac{k_B T}{e} \ln\left(\frac{c_R}{c_L}\right) \tag{3.63}$$

and called the Nernst potential. This is the basic mechanism whereby electrical potential differences are generated inside organisms. Note that the Nernst potential difference only depends on the concentration ratio. In Table 3.10 are some concentration ratios for cellular membranes and the corresponding value of the Nernst potential computed from the equation above.

According to Froehlich [112] the phospholipid head groups in membranes play an important role in cell-cell interactions due to their oscillating dipole moments as shown schematically in Fig. 3.49. He envisaged that their mutual interactions would lead to microwave frequency coherent oscillations that would be cell-specific and hence resonantly couple to other oscillations of same type cells in the vicinity. This hypothesis still awaits experimental confirmation.

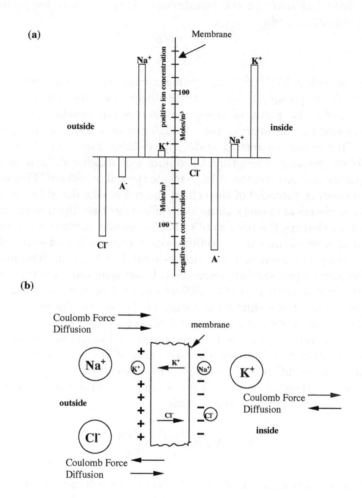

FIGURE 3.48: a) Ion concentrations outside and inside a cell. Positive ion concentrations are graphed above the line and negative ion concentrations below. b) The membrane is permeable to K^+ and Cl^- ions, which diffuse in opposite directions as shown. Small arrows indicate that the Coulomb force resists the continued diffusion of both K^+ and Cl^- ions.

TABLE 3.10: Concentration Ratios for Cellular Membranes and the Corresponding Nernst Potential in Equation

Ion	c(r)/c(L)	Valency	ΔV (millivolt)
Na+	10	1	62
K+	0.03	1	-88
Cl-	14	-1	-70

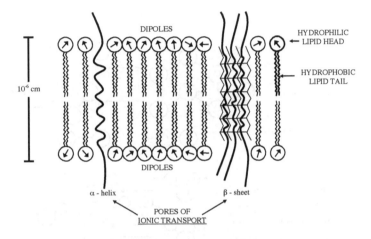

FIGURE 3.49: Membrane dipole dynamics.

However, as shown in Fig. 3.50 the dipole dynamics of membrane's phospholipids is very complicated with many different modes of vibration assigned to different motions of their molecular groups.

3.11 Motor Proteins and Their Role in Cellular Processes: Dynein, Kinesin, Dynactin, Myosin, Force Generation

One important method of generating force and movement within the cell is via motor proteins. The assembly and disassembly of the cytoskeleton in conjunction with force generation by motor proteins is thought to be the main mechanism for mitosis, as well as for organelle transport within the cell [187]. Approaching this problem from a biophysical viewpoint involves examining the interaction between the motors and the cytoskeletal filaments as well as interactions between the motors and the cytoplasm. Advances have

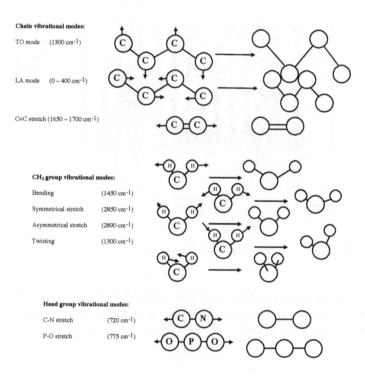

Chain vibrational modes:

TO mode (1300 cm^{-1})

LA mode $(0-400 \text{ cm}^{-1})$

C=C stretch $(1650-1700 \text{ cm}^{-1})$

CH₂ group vibrational modes:

Bending (1450 cm^{-1})

Symmetrical stretch (2850 cm^{-1})

Asymmetrical stretch (2890 cm^{-1})

Twisting (1300 cm^{-1})

Head group vibrational modes:

C-N stretch (720 cm^{-1})

P-O stretch (775 cm^{-1})

FIGURE 3.50: Some of the key vibrational modes of the membranes molecular groups.

recently been made in our understanding how myosin and kinesin work, motor proteins that interact with actin and with microtubules, respectively [16]. Motor proteins share two essential functional components: a motor domain and a cargo binding domain. Most also contain a central linking domain that can be quite extended.

There are three large families of naturally-occurring motor molecules: the myosins, the kinesins and the dyneins. All function by undergoing shape changes, utilizing energy from the biological fuel adenosine triphosphate (ATP). Each family has members that transport vesicles through the cell's cytoplasm along linear assemblies of molecules-actin in the case of the myosins and tubulin for both of the other families. The kinesins and dyneins move or 'walk' along microtubules carrying their cargo. We describe these and remaining major motors proteins in this section.

The motors are ATPases that move objects on the filament surface. The axonemal dyneins drive MT sliding in cilia and flagella; cytoplasmic dynein and kinesins move many objects in opposite directions in the cytoplasm. In general, kinesins move toward the plus ends of MTs (outward) and dyneins move toward the minus ends (inward). The vesicles are attached to MAP's such as kinesin moving along the microtubule functioning like a conveyer belt (see Fig. 3.51). In neurons, as the microtubules grow from the cell body, the plus end is more peripheral. These proteins have head regions that bind to microtubules and also bind ATP that is the source of biochemical energy. The head domains are thus ATPase motors. The tail domain binds to the organelle or vesicle to be moved. It is not known at a molecular level how the energy from ATP breakdown is converted into vectorial transport. There exist a number of mathematical models, however, that describe motor protein motion quite accurately.

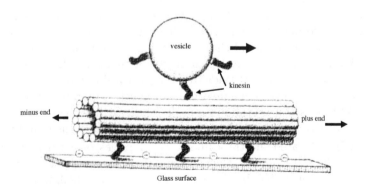

FIGURE 3.51: Vesicles are shown schematically to be attached to MAPs that move along the microtubule conveyer belt by a kinesin motor.

Below, we discuss each major class of motor proteins separately.

3.11.1 Myosin

Myosins are a family of motor proteins that move along actin filaments, while hydrolyzing ATP. Myosin is, therefore, an ATPase enzyme that converts the chemical energy stored in ATP molecules into mechanical work. The motive power for muscle activity is provided by myosin motors, organized as thick filaments which interact with an array of thin actin filaments to cause the shortening of elements within each myofibril. This shortening is achieved by relative sliding of the myosin and actin filament.

Myosin I has only one heavy chain, with a relatively short tail, plus two or more light chains in the neck region. While it binds to the plasma membrane, it may pull the membrane forward as it walks along a growing actin filament toward the plus end.

Myosins I and V, both of which bind to membranes, are postulated to have roles in the movement of organelles along actin filaments as well as movements of plasma membrane relative to actin filaments. They have been found to associate with Golgi membranes, which give rise to secretory vesicles, including synaptic vesicles.

Myosin II, the form found in skeletal muscle, is sometimes referred to as conventional myosin. It includes two heavy chains, each with a globular motor domain [275] that includes a binding site for ATP and a domain that interacts with actin. Tail domains of the heavy chains associate in a rod-like helical coiled coil. Light chains associate with each heavy chain in the neck region (a total of four light chains for myosin II). Light chains of different myosin types are calmodulin or calmodulin-like regulatory proteins. They may provide stiffening in neck domains.

Myosin II heads interact with actin filaments in a reaction cycle that may be summarized as follows. ATP binding causes a conformational change that causes myosin to detach from actin. The active site closes and ATP is hydrolyzed, as a conformational change (cocking of the head) results in myosin weakly binding actin at a different place on the filament. Pi release results in a conformational change that leads to stronger myosin binding, and the power stroke. ADP release leaves the myosin head tightly bound to actin. In the absence of ATP, this state results in muscle rigidity called rigor. In non-muscle cells, myosin II is often found associated with actin filament bundles.

Myosin II is located in stress fibers, i.e., bundles of actin filaments that extend into the cell from the plasma membrane. These contract under some conditions. Myosin II has been shown by fluorescent labelling to be located predominantly at the rear end of a moving cell, or in regions being retracted. Contraction at the rear of the cell probably involves sliding of actin filaments driven by bipolar myosin assemblies. The coiled coil tail domains of myosin II molecules may interact to form anti-parallel bipolar complexes. These may contain many myosin molecules, as in the thick filaments of skeletal muscle, or they may consist of as few as two myosin molecules.

Myosin crossbridge acts as a rigid rod combined with an elastic element. The following is a scheme following Huxley [150] that relates conformational changes at several stages where the largest movements are associated with the binding of ATP and the subsequent release of ADP (see Fig. 3.52):

(a) The head binds to actin;

(b) The crossbridge movement during the power stroke is initially taken up by stretching the elastic element.

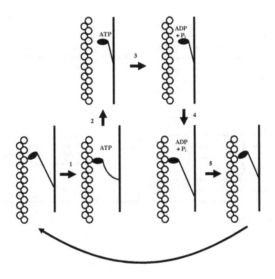

FIGURE 3.52: Scheme relating conformational changes at several stages; the largest movements are associated with the binding of ATP and the subsequent release of ADP.

Myosin V has two heavy chains like myosin II, but has a shorter coiled coil region followed by a globular domain at the end of each heavy chain tail. There are calmodulin light chains in the neck region. Movement of myosin V along actin is processive, consistent with the role of myosin V in transporting organelles along actin filaments. Each of the two myosin V head domains dissociates from actin only when the other head domain binds to the next actin filament with the correct orientation, about 13 subunits further along the helical actin filament.

The ATP-dependent movement of myosin heads along actin filaments is accompanied by ATP hydrolysis. ATP binds to the myosin head adjacent to a 7-stranded β-sheet. Loops extending from β-strands interact with the adenine nucleotide. The nucleotide-binding pocket of myosin is opposite a deep cleft that bisects the actin-binding domain. Opening and closing of the cleft is proposed to cause the head to pivot about the neck region, as occupancy of the nucleotide-binding site changes and as myosin interacts with and dissociates from actin. Actin filaments may move relative to one another, as heads at the opposite ends of bipolar myosin complexes walk toward the plus ends of adjacent antiparallel actin filaments.

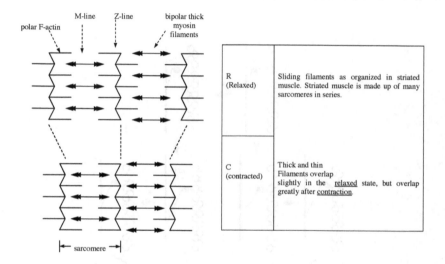

FIGURE 3.53:   Actin-myosin complex in muscles.

Consistent with the predicted conformational cycle, different conformations of the myosin head and neck have been found in crystal structures.

Ca^{++} regulates actin-myosin interaction in various ways, in different tissues and different organisms. Some myosins are regulated by Ca^{++} binding to calmodulin-like light chains, in the neck region. Others are regulated by phosphorylation of myosin light chains, catalyzed by a Ca^{++}-dependent kinase or by a kinase that is activated by a small GTP-binding protein. Meanwhile, a complex of tropomyosin and troponin (which includes a calmodulin-like protein) regulates actin-myosin interaction in skeletal muscle. Caldesmon, a protein regulated by phosphorylation and by Ca^{++}, controls actin-myosin interaction in smooth muscle.

3.11.2 Kinesin Family

Kinesin is abundant in virtually all cell types, at all stages of development, and in all multicellular organisms. While a majority of kinesin appears to be free in the cytoplasm, some is associated with various membrane-bounded organelles, including small vesicles, the endoplasmic reticulum and membranes that lie between the ER and Golgi. It is accepted now that kinesin transports membrane-bounded organelles and perhaps macromolecular protein assemblies along microtubules. Kinesin is the motor used for fast axonal transport, and is also employed in mitosis and meiosis. It is a force-generating motor protein, which converts the free energy of the gamma phosphate bond of ATP into mechanical work. This work is used to power the transport of intracellular organelles along microtubules. These microtubules always glide such that kinesin moves toward the plus, or fast-growing, end of the microtubule. This polarity and the orientation of microtubules within cells show that kinesin is an anterograde motor [59]. Interestingly, other kinesin proteins, whose motor domains have high sequence homology and nearly identical structure to kinesin, move in the opposite direction (namely, Ncd proteins; see below).

By decreasing the density of kinesin on the surface of a microtubule, it has been shown that a single kinesin molecule suffices to move a microtubule through distances of several microns (see Fig. 3.54). Interestingly, a single motor can move a microtubule as quickly - at a rate of about 1 micron per second - as ten or one hundred motors.

Virtually all members of the kinesin family contain an extended region of predicted alpha-helical coiled coil in the main chain that likely produces dimerization. Kinesin itself is a heterotetramer of two heavy and two light chains. Dimers of isolated head domains of kinesin possess the required ATPase properties. A large body of structural, biochemical and biophysical evidence shows that kinesin has just one binding site per tubulin dimer [306], and that the motor takes 8-nm steps from one tubulin dimer to the adjacent one in a direction parallel to the protofilaments [145]. Since the isolated motor domain of kinesin can hydrolyze up to 100 ATP molecules per second, it is likely that each step corresponds to one cycle of the ATPase reaction. The current model for how motor proteins generate force is that the motor contains an elastic element, a spring, that becomes strained as a result of one of the transitions between chemical states: this strain is the force that the motor puts out, and the relief of this internal strain is the driving force for the forward movement [82].

There are 40-50 proteins in the kinesin superfamily found to date. Kinesins have a common motor domain and a variable tail domain. The structure of kinesin reveals the existence of a heavy (KHC) and a light chain (KLC) of 110-130Kda and 60-80Kda, respectively.

The motor domain [178, 181] shown in Fig. 3.55 defines the direction and rotational freedom of kinesin while the tails are interpreted as acting as a spring. While the exact nature of the movement of the two headed kinesin

FIGURE 3.54: A kinesin molecule moves a microtubule along a glass surface.

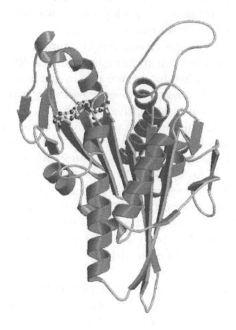

FIGURE 3.55: Crystal structure of the motor domain.

has not been explained so far, the current opinion among biologists and bio-chemists leans towards the so-called hand-over-hand model of kinesin motion.

Electron microscopy of head domains bound to microtubules indicates that the orientation of kinesin and Ncd heads is similar in spite of their different directions of movement [102]. In addition, part of the kinesin head domain appears to swing in a nucleotide-dependent manner that may be related to the power stroke generation of movement. The kinesin and Ncd motor domains resemble asymmetric arrowheads with dimensions of approximately $75 \times 45 \times 45 \text{Å}$. The bound MgADP lies in an exposed surface cleft. The core of the motor domain is an eight-stranded beta-sheet, flanked by three alpha-helices on each side. Although Ncd and kinesin move in opposite directions along microtubules, their 3-D structures are remarkably similar. The eight-stranded beta-sheet and six alpha-helices of the core, and the three anti-parallel beta-strands that form the small lobe are essentially identical in length, position and orientation in both motors. The position of the MgADP is virtually identical in both motors. The largest differences between kinesin and Ncd are in the surface loops near the nucleotide-binding pocket. The N- and C-termini of the Ncd and kinesin motor domains are positioned similarly in space and located 9Å from one another on the side of the motor opposite to the nucleotide.

3.11.3   Ncd Dimer Structure

The X-ray crystal structure of a dimeric form of the minus-end-directed Ncd motor has been determined to 2.5 Å resolution. The model shows that the Ncd neck, the region that links the motor catalytic core to the coiled coil stalk, is helical, rather than an interrupted beta-strand as in kinesin. Random mutagenesis of the Ncd neck results in a plus-end-directed motor. Ncd has now been made to move like kinesin, towards plus ends, by fusing regions from outside the kinesin motor domain to the Ncd motor. That is, the directionality of the Ncd motor is reversed when the motor is joined to the N terminus of KHC, replacing the KHC motor domain. This suggests that direction of movement can be determined in part by a subdomain that is transferable between proteins.

Conversely, kinesin, a plus-end-directed motor, has been made to move towards microtubule minus ends by fusing the stalk and neck of minus-end-directed Ncd to the conserved KHC motor core. Mutation of the Ncd neck in the NcdKHC chimeric motor reverted the motor to slow plus-end movement, showing that minus-end determinants of motor directionality are present in the Ncd neck and that plus-end determinants also exist within the KHC motor core.

Existing models of motor protein movement can be divided into those with diffusion and those with a power stroke. There have been suggestions that motor proteins may be directed by electrostatic interactions with the protein filaments on which they walk; in particular the binding of kinesin to MTs has been shown to be primarily electrostatic. The motion of the motor protein in either model is fuelled through the hydrolysis of ATP since each step corresponds to one cycle of the ATPase reaction [21]. The diffusion models require an oscillating potential that is assumed to be driven by a conformation change of the motor. In the power stroke models by contrast, it is the motor protein whose structure changes. Ratchet potential models can be based on the Langevin-equation with a fluctuating force [84] and/or with a Gaussian noise term or they can be developed through the Fokker-Planck formalism [267] for the probability distribution function $P(x, t)$ that describes the motion of the motor in a statistical manner [165]. While such models have been fairly successful in representing the gross features of the experimental data, the use of a hypothetical potential and a number of arbitrary parameters casts doubt on their correctness. The use of a point mass approximation for kinesin is also highly questionable since it is even larger than the tubulin dimer. Meanwhile, the emphasis placed on stochasticity may also be unwarranted for two-headed motors whose motion appears to be very deterministic. Therefore, it is a chemo-mechanical cycle that determines the key properties of the KN walk.

3.11.4 Dynein

Dyneins move towards the minus end of the microtubule, which tends to be anchored in the centrosome of the cell. Dynein motors also cause sliding between microtubules that form the skeleton of cilia and flagella. Other ciliary structures resist this sliding with the result that bends form along the length of the cilium or flagellum and propagate from base to tip or from tip to base. The propagation of these bends requires co-ordinated action of the several types of dynein motors present in the cilium or flagellum and the structures providing the resistance to sliding. Dynein is also found in the cytoplasm (dynactin).

It is a minus-end directed motor presumed to drive fast retrograde vesicle transport in axons and other cells toward the centrosome. Cytoplasmic dynein is a rather large multi-subunit protein complex composed of two identical heavy chains of about 530 kDa each, two 74 kDa intermediate chains and about four 53-59 kDa intermediate chains. In addition, there are several less well-known light chains. The heavy chains each contain four ATP-binding sites, including the ATP-hydrolytic site that provides the energy for its movement along microtubules, and a microtubule binding site. The 74 kDa intermediate chains are thought to bind the dynein to its cargo, whether that cargo is a membrane-bounded vesicle in a neuron, a Golgi vesicle, a kinetochore or a mitotic spindle astral microtubule. The dynein then provides the force to move this cargo along a microtubule toward its minus end. In cilia and flagella, axonemal dynein motor molecules are attached to nine microtubule doublets arranged cylindrically around a pair of single microtubules. The dynein motors undergo a cycle of activity, during which they form a transient attachment to the doublet, and push it towards the tip of the cilium or flagellum. At a particular ATP concentration, microtubule gliding velocities are found to increase with microtubule length.

3.11.5   Myosin V

Myosin V is the best understood out of the myosin family. It walks in a processive manner along actin and was the first processive myosin found, in 1999 by Mehta et al. [218]. Each head has a duty ratio (occupancy ratio of strong to strong + weak + dissociated states) that approaches 1 for high actin concentrations. This means that when one head is dissociated, the other is most likely strongly bound long enough for the dissociated head to swing forward by a lever arm movement (the working stroke of the attached head is $\sim$ 21nm) and move in a biased brownian diffusion of 15nm towards the next preferred attachment site 36nm further along the actin [334]. This is supported by a lot of evidence of "hand-over-hand" walking along actin filaments, with its light chain binding domains tilting as it does. Each head has a pace of 72±10nm and has been observed by both fluorophore and differential quantum dot imaging [351, 299, 345], and may take "sub-steps" of either 52-23nm or 42-33nm (lever plus diffusion).

During the actomyosin ATPase cycle, there is an alternation between strongly binding with actin during ADP and nucleotide-free (rigour) states, and weakly binding during ATP and ADP-P_i states. When myosin binds ATP, it must undergo a conformational change before it can hydrolyse ATP (M^* to M^{**}). $M^{**} \cdot ADP \cdot P_i$ state does not rapidly release P_i, and a new state that encourages P_i release must occur when rebinding to actin. After P_i is released, a state is formed with strong binding of both ADP and actin ($A \cdot M \cdot D^S$), followed by a weak ADP binding, strong actin binding state ($A \cdot M \cdot D^W$). The ADP is then released, forming the rigour complex. ATP attaches to the myosin head and causes dissociation from the actin, forming the post-rigour complex, also known as near rigour, or the open state, because of the position of switch II. After hydrolysis, the pre-powerstroke state is formed, also known as the transition state or the closed state [307].

There is a direct linear correlation between lever arm length and stroke size and step size of myosin V, but it is not known whether force production is from the lever arm movement [261, 278, 229]. However, there is conflicting evidence that shorter lever arms (1/6 length) do not seem to affect the step size of myosin V [309], which may be due to diffusion, the motor domain, or something else. Also, Tanaka noticed a 10:1 ratio of forward to backwards steps.

Image averaging of myo5 on actin filaments is interpreted as the lead head being in pre-powerstroke, and the rear head separated by 36nm in post-powerstroke [344, 50]. Recent kinetic experiments show that both heads should have lost their P_i; therefore, the lead head must be in a post-prepowerstroke state [273].

The reason myo5 has such a high duty ratio without strain is due to its

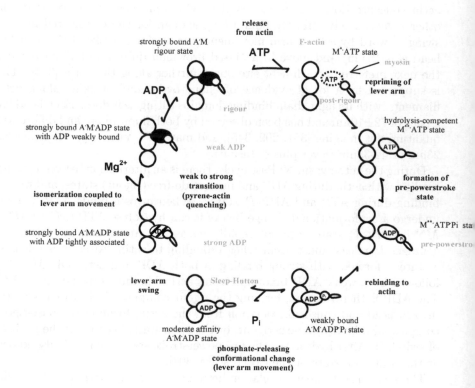

FIGURE 3.56:   The actomyosin ATPase cycle.

TABLE 3.11: A Probable Cycle of Actomyosin Motion

Head 1	Head 2	rate (s^{-1})	temp $(^\circ C)$	Notes
$M \cdot ADP \cdot P_i$ Rear, detached prepowerstroke	$M \cdot ADP \cdot P_i$ Front, detached prepowerstroke			**Both:** First act in encounter. Both heads attach to act in simultaneously, possibly 8nm or 2 monomers apart .
$A \cdot M \cdot ADP \cdot P_i$ Rear, attached weakly bound	$A \cdot M \cdot ADP \cdot P_i$ Front, attached weakly bound	200	20	**Both:** Release P_i. Conformational change.
$A \cdot M \cdot ADP$ Rear, attached Sleep-Hutton	$A \cdot M \cdot ADP$ Front, attached Sleep-Hutton			**Both:** Lever arm swing. No direct measurement of Sleep's affinity.
$A \cdot M \cdot ADP$ strongly bound actin ADP tightly associated	$A \cdot M \cdot ADP$ strongly bound actin ADP tightly associated	44	20	**Both:** Simultaneous accelerated "weak to strong" transition.
$A \cdot M \cdot ADP$ strongly bound actin weakly bound ADP	$A \cdot M \cdot ADP$ strongly bound actin weakly bound ADP	30		**Rear:** Accelerated ADP release from rear head during first step. (Geometry) **Front:** ADP release from front is gated. If rear detachment prevented due to high [ADP]/[ATP], then ADP dissociates from front at 0.3 s^{-1} . Involves release of Mg^{2+} (probably).
$A \cdot M$ strongly bound rigour state	$A \cdot M \cdot ADP$ strongly bound actin weakly bound ADP			**Rear:** Fast ATP hydrolysis. Myocin head is released from actin.
$M^* \cdot ATP$ post-rigour	$A \cdot M \cdot ADP$ strongly bound actin weakly bound ADP			**Rear:** Lever arm of head 1 still bent. M^* and M^{**} are two different fluorescent states. Arm repriming.
$M^{**} \cdot ATP$ hydrolysis-competent	$A \cdot M \cdot ADP$ strongly bound actin weakly bound ADP			**Rear(Front?):** Needed conformational change so myosin can hydrolyze. Format ion of the prepowerstroke.
$M^{**} \cdot ADP \cdot P_i$ **(Front)** prepowerstroke	**(Rear)** strongly bound actin weakly bound ADP			**Front:** in front now (?) and diffusing to find area of strong binding along actin ($\sim$ 36nm), which is further than the stroke. Bind to actin.

Head 1	Head 2	rate (s^{-1})	temp $(°C)$	Notes
$A \cdot M \cdot ADP \cdot P_i$ **(Front)** Front, attached weakly bound	$A \cdot M \cdot ADP$ **(Rear)** strongly bound actin weakly bound ADP	200	20	**Front:** Release P_i. Conformational change.
$A \cdot M \cdot ADP$ **(Front)** Front, attached Sleep-Hutton	$A \cdot M \cdot ADP$ **(Rear)** strongly bound actin weakly bound ADP			**Front:** Lever arm swings forward during transition to strong-strong state.
$A \cdot M \cdot ADP$ strongly bound actin ADP tightly associated	$A \cdot M \cdot ADP$ **(Rear)** strongly bound actin weakly bound ADP	25	20 under no strain	**Front:** Weak-to-strong transition. May have a strong temperature dependence
$A \cdot M \cdot ADP$ strongly bound act in weakly bound ADP	$A \cdot M \cdot ADP$ **(Rear)** strongly bound actin weakly bound ADP	12-16	25	**Rear:** ADP is released at 13/s at the same time as front undergoes weak-to-strong at 25/s. Very close to 12-15s^{-1} maximal rate steady state of actin-activated ATPase. Predominant steady state intermediate. Rate-limiting step.
$A \cdot M \cdot ADP$ strongly bound actin weakly bound	ADP $A \cdot M$ strongly bound actin rigour state			Wash. Rinse. Repeat.

slow ADP release and its fast ATP hydrolysis and P_i release [307]. Because there is lever arm motion correlated with ADP release, this can be how the rear head releases its ADP before the front head, with a stress or kink on the front head preventing its arm from moving forward until the rear head detaches with an ATP binding to it, and/or possible acceleration of release of the rear head from lever arm stress [299].

Forkey et al. [107] discusses 3D geometry of myo5, specifically saying that there is a lever arm rotation and the single light chains tilt between two well-defined angles. Concentration of ATP affects these angles. Dwell time has ATP dependent and independent processes. The axial angle goes from 28° to 75° where the azimuthal angle changes from 45 to 30°.

Ali et al. [10] found that myosin rotates in a left-handed screw fashion along a right-hand screw actin when unconstrained. Where others' experiments show a step size of 36nm, they found that if the actin was off of surfaces (like a highwire), it has steps of 34.8nm, with an average rotation of 6° to the left. This equates to a head moving 69.6nm per step. Oddly enough Ali [11] shows myo6 to have a longer step than myo5.

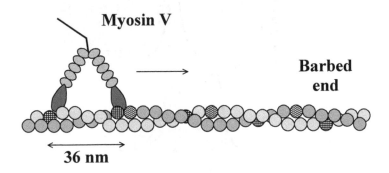

FIGURE 3.57: Interpretation of the left-handed rotation. Stepping on every 11th actin subunit (cross-hatched) would result in left-handed spiral movement, whereas linear movement is expected on the 13th subunits (wavy lines). The estimated step size of 34.8 nm is between the 11th and 13th subunits. (From [10].)

Nguyen & Higuchi [240] look at the regulation of Myo5 motility by calmodulin, whereas Rosenfeld et al. [274] says magnesium regulates ADP dissociation, which seems to be the limiting factor (according to Sweeney et al. [307]).

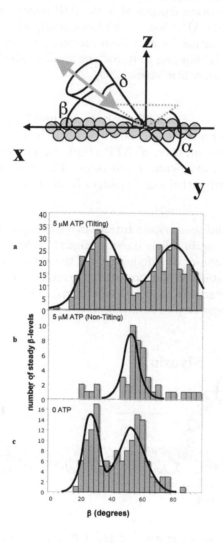

FIGURE 3.58: (From [107]). Distributions of average β-values in the presence and absence of ATP. (a) Myosin V molecules translocating and showing repeated tilting events at 5 μM ATP. (b) Molecules translocating in the presence of 5 μM ATP but not showing angle changes. (c) Myosin V co-localized with actin filaments in the absence of ATP. Solid lines indicate the best fits to single or double Gaussian-shaped profiles. Parameters of the fits are given in Table 3.12.

TABLE 3.12: Fits of Single or Double Gaussian Profiles to the Histograms in Fig. 3.58 (5 μM tilting/non-tilting and 0 ATP)

ATP concentration	5 μM tilting	40 μM tilting	5 μM non-tilting	40 μM non-tilting	0 μM
η	86	100	48	101	140
Fraction in First Peak	0.48± 0.04	0.46±0.03	–	–	0.42±0.04
$\bar{\beta}_1$	30.0±0.9	27.1±0.6	–	–	26.6±0.6
σ_{β_1}	10.0 ± 0.9	8.92±0.6	–	–	4.78±0.6
δ_1	34.4	31.6	–	–	36.1
σ_{δ_1}	13.6	11.4	–	–	7.6
$\bar{\beta}_2$	71.7±1.1	77.8±1.3	52.0±0.5	54.2±0.7	51.3±0.9
σ_{β_2}	12.4±1.5	13.1±1.5	4.5±0.5	6.2±0.7	7.8±1.0
δ_2	37.3	30.4	44.9	41.9	46.4
σ_{δ_2}	11.2	12.1	9.0	6.3	6.0

3.11.6 Myosin VI

Myosin VI is unique to most other myosins in that a dimeric myo6 travels towards the minus end of its actin filament, which is believed to be towards the centre of the cell [77]. Unlike myosin V, which is believed to operate on a simple lever-arm of a converter region and neck domain, myosin VI is believed to have additional features integrated, although it is still believed to be hand-over-hand [352]. It is possible that myo6 may operate within the cell as both a processive dimeric motor, as well as a monomeric non-processive motor, with each monomer having a large working stroke despite having a short lever arm. However, the properties and functions of myo6 are poorly understood compared to myo5. For example, it is not yet known if myosin VI exists as a stable dimer within the cell [193].

Myo6 has a standard organization, with an N-terminal motor domain, a neck region that consists of one IQ motif and a C-terminal domain. The motor domain contains a unique 22 amino acid insert near the ATPase site which may be involved in controlling the release of nucleotide from the nucleotide binding pocket, and in between the motor domain and the single IQ motif there is a 53 amino acid insert that is predicted to be the reverse gear that allows myosin VI to move opposite to its relatives [348]. Tsiavaliaris et al. engineered a backwards walking myosin in 2004 by adding an extra domain into the neck region, although it is not known if this is how myo6 works.

To perform the duties it is believed to do within the cell, myo6 must have a high duty ratio. Because it was originally believed to be a dimer with two heads due to a tail region that was predicted to be a coiled coil (more detailed COILS analysis indicates loops may occur, and the central part is more likely a salt-bridged stabilized helix [52]) most *in vitro* motility studies have used artificial myosin VI dimers that use a C-terminal leucine zipper or smooth muscle myo2 rod domains, but are also missing the important cargo-carrying tail. These constructs have a duty ratio of 90% [80], and move processively with a movement of 30nm per step [270], [244] which, with the lever arm of one IQ motif and the 53 residue insert giving a lever arm movement of 8- 10nm maximum, cannot be explained by a simple lever arm theory ... or can it? Refer to Balci et al. [29] for a model that may describe their interhead distances of near-rigour myosinVI dimers. Ali et al. [11] found that the unconstrained motion of a myo6 dimer to actually be $> 36nm$, the pseudo-repeat of an actin filament, as was observed with a right hand spiral of the myo6, which is opposite to the direction of the actin "spiral".

Rock's [270] model, which has been refined by Altman [3], has a large diffusive search of 20nm by the lead head for the next preferred binding site after the working stroke of 10nm. Nishikawa's [244] model has the first binding of a myo6 head causes a conformational change to the actin filament, allowing the second head to bind 30nm away. All of the models at this point are speculative, as the structure of myo6's tail domains needs to be determined.

Calcium binding to the calmodulin in the neck region causes a loss of coordination between heads, and velocity is reduced by 3x as processivity is lost [231]. This may have a regulatory role, along with different myo6 isoforms (4 different combinations in its tail domain of two unique inserts, each with different tissue distributions), phosphorylation at T^{406} (conflicting data between Yoshimura [353] and Morris [231]) and conformational changes (possible dimerization of monomers, though there is a lack of evidence for or against this).

Experiments with optical tweezers showed that single actomyosin (monomer) attachment events produced a working stroke of 18nm [193], in two steps: 17nm associated with Pi release, and 1nm associated with ADP release, when it goes into the rigour state [334].

3.12 Directed Binding as a Model of Kinesin Walk - from [43]

Different kinesins (KNs) belong to a superfamily with more than 250 members that are ubiquitous in eukaryotic cells and play an important role in force generation as recently reviewed by, e.g., Schliwa [291] and Howard [145]. These KN molecules share a common motor domain of approximately 330-350 amino acids. KNs force-generation originates from the conversion of the free energy of nucleoside triphosphates (typically ATP) enabling mechanical work that powers the transport of intracellular organelles along microtubules (MTs). Generally, the ATPase is strongly promoted by MTs. Many KNs move processively towards the + end of a MT, while some move to the − end.

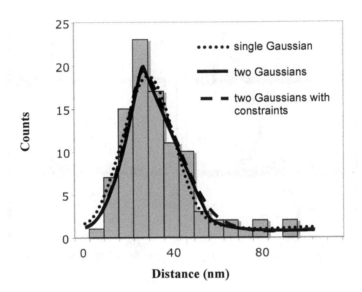

FIGURE 3.59:   Histogram of measured separations between the myosin VI heads and three Gaussian fits to the data. The dotted line is a single Gaussian fit representing the hand-over-hand model, the solid line is an unconstraint fit of two Gaussians and the dashed line is a fit of two Gaussians with constraints [29].

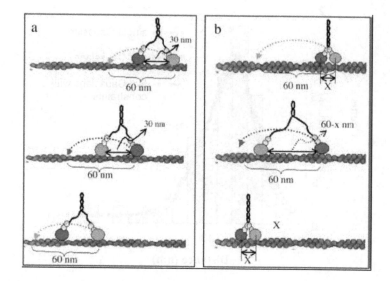

FIGURE 3.60:   (a) Hand-over- hand mechanism of walking. (b) An alternative model of motility in which the separation between the myosin heads alternates between x and 60 x nm, whereas each head takes 60 nm steps sequentially. Cryoelectron microscopy images suggest that x could be much less than 30 nm [29].

KNs are comprised usually of globular head regions and extended coiled-coil tail sections as shown schematically in Fig. 3.61.

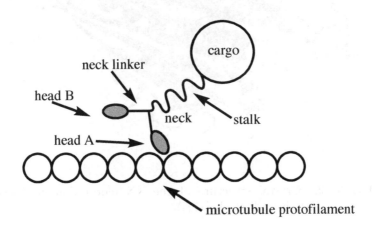

FIGURE 3.61:   A sketch of a KN molecule with cargo moving along a MT.

Conventional KN is a heterotetramer consisting of two heavy (between 120 and 130 kDa) and two light chains (60 to 70 kDa) [178] resulting in an overall molecular mass of about 400 kDa [2, 183, 293]. Each heavy chain contains a globular motor domain (with dimensions of approximately 10nm × 9 nm; [140]) with both an MT-binding site and an ATPase-active center. A flexible domain attached to the neck region allows a significant spatial extension between the two heads [15]. A stalk region involves the cargo binding site. The light chains seem to have regulatory functions [336].

Biochemical evidence shows that there exists one binding site on the KN head per tubulin dimer and that the KN head interacts with the β-tubulin monomer and the flanking monomers [300]; hence it overlaps with both α and β tubulin monomers [130]. Significantly, KN's motor domain has been crystallized and structurally analyzed (see Fig. 3.62).

It is known that two-headed conventional KN molecules move approximately a hundred steps along a MT before they completely dissociate from an MT into the solution [37, 78, 131, 146]. This so-called processivity of a KN molecule is defined by moving continuously along the MT surface without dissociating completely. Note, however, that another possibility exists for processive motion. Namely, both heads can be in an intermediate (transition) state that keeps the molecules in the vicinity of the filament due to an attractive force acting on KN. The existence of such attractive potentials has been explicitly included in various theoretical models (see below). In other words, this quasi-bound state corresponds to a process of binding while the

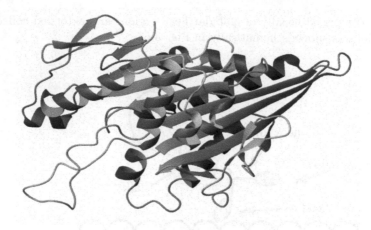

FIGURE 3.62: Crystal structure of the KN motor domain (based on Kikkawa et al. [170], Koradi et al. [177]).

bound state corresponds to its completion. Which of these two possibilities is utilized by KN still requires careful experimental determination.

Both heads translocate by 16-nm steps [211] and it was experimentally demonstrated that an 8-nm KN center-of-mass movement is correlated with an ATP hydrolysis event. However, the rate of hydrolysis is dependent on the concentration of ATP and magnesium ions. The maximum propagation velocity of conventional KN is found between 0.6 μm/s and 2.0 μm/s [40,328, 342] and it depends on various chemical and physical factors, in particular on the concentration of ATP.

For the two-headed processive KN motors, the essential structural elements for force generation are located within each of the two globular heads [69, 128, 129]. KN moves along protofilaments that are parallel to the long axis of the MT [76, 263] in the case of the 13 protofilament MTs, helically, however, in case of the 12 or 14 protofilament MTs. Young et al. [354] showed that one-headed KN is often non-processive with a different mechanism of motion compared to the two-headed KN. Inoue et al. [159] showed that the average travel distance and speed for the one-headed KN are reduced by a factor of three compared to the two-headed variety and that it moves stochastically backwards and forwards with a directional tendency towards one end of a MT. One-headed KNs may interact, however, also processively due to dimerization [315] or by reversible binding to the highly negatively charged C-terminal region of tubulin [248].

Motor protein motion has been theoretically described using both mechanical and chemical approaches. Some of these models can be suitable for one-headed motor proteins but are not as suitable for the description of processive two-headed KN. Especially lacking is the inclusion of spatial dimensionality to show hand-over-hand motion of the two-headed KN. Furthermore, each KN head is an extended object comparable in size to the tubulin dimer. The assumption of hand-over-hand motion cannot be captured by one-dimensional mechanical (ratchet) models. In fact, most of these latter models use an asymmetric MT potential [207, 208] to explain directionality of motion. Such asymmetric potentials acting on a point mass (instead of an extended object) as envisaged by ratchet models have not been directly shown from either ab initio theoretical studies or any direct experiment. Moreover, there exists a KN analog, called Ncd, that has a very similar motor domain as conventional KN but it moves in the opposite direction contradicting the role of asymmetric potentials. Chemical kinetics, on the other hand, does not even directly deal with motion per se and hence it offers no insight into the directionality of motion.

In the following we broadly classify the proposed models into (a) chemically-based and (b) mechanically-based ones. Due to temperature, chemical reactions as well as mechanical motion in a heat bath are stochastic in nature. Thus, stochasticity is not a distinguishing feature among theoretical models.

Let us first discuss the relationship between mechanical and chemical steps. It is clear that the mechanical steps are correlated with chemical reactions, but how this happens in detail is still an open issue. The following assumptions have been partly implemented in a number of existing theoretical models:

Independence of the biochemical cycle and mechanical movement: Tight-binding models assume that the biochemical cycle is independent of the mechanical movement that simply follows the changing protein conformations [151, 187]. This is often realized by a static asymmetric potential (ratchet) together with an oscillating force mimicking the chemical reactions.

Alternating chemical and mechanical transformations: Here, it is assumed that the first chemical reaction will end in a new chemical state without a change in the mechanical state (position coordinate). Next, the position coordinate will change until the equilibrium position is reached. Then the next chemical reaction will follow, again followed by mechanical movement. This is repeated through the whole mechano-chemical cycle. The mechanical aspect of this process should lead to the walking mode of KN. It is also assumed that the chemical reaction can only start at a specific spatial position. On the other hand, the chemical reaction should not affect the spatial position.

Combined chemical and mechanical transformations: Here, we have a combination of a number of possibilities. For example, the new chemical reaction starts before the mechanical step has ended or the mechanical step starts before the chemical reaction has ended. Another possibility is to combine chemical reactions with mechanical steps. This combined picture is, in our opinion, the most realistic description, but it is not easy to implement in a quantitative model.

3.12.1   Chemical Reaction-based Models

These models are very useful from the point of view of direct comparison with experimental data. They are defined by chemical reaction cycles. Commonly, chemical reactions are combined with conformational changes of the entire motor that correspond to a mechanical step. Such conformational changes in our interpretation represent a transition to a new relaxed state in which the mechanical step will end. Therefore, tensions caused by such chemical reactions are relieved as driving forces for the motor.

Assuming that a load force acts on the motor, every reaction, which is combined with a mechanical step, must have a force-dependent reaction rate. These force-dependent reaction rates define virtually everything that can be calculated within such models. Solving the stationary reaction equations (which are algebraic equations) directly leads to the Michaelis-Menten form for the velocity usually written as

$$v = \delta k_{ATPase} = \delta k_{cat} \frac{[ATP]}{K_M + [ATP]} \tag{3.64}$$

where δ is the step size, [ATP] represents the ATP concentration, K_M is the Michaelis-Menten constant and the $k's$ denote the corresponding reaction rates.

For example, Schnitzer et al. [292] proposed a KN work cycle involving chemical energy conversion due to nucleotide hydrolysis and assumed that chemical reaction rates depend on the load force F via an Arrhenius relation

$$k = k_0 exp(-Fd/kT) \tag{3.65}$$

where d is the path distance traveled in the reaction. In the reaction cycle, the swing stage with an ADP-bound free head was assumed to have a non-zero probability of dissociation of the two-headed KN from the proto lament. Experiments produced plots of velocity as a function of ATP concentration reaching saturation velocity of 1000 nm/s at 1000 μ M of ATP. Furthermore, the velocity as a function of the load was seen dropping monotonically from 50 nm/s at 5 μM of ATP down to 5 nm/s at 5-6 pN of the load force.

According to Howard [145], the hand-over-hand model for dimeric KN is based on the scheme:

$$K \underset{k_{-1}}{\overset{k_1}{\rightleftharpoons}} KT \overset{k_2}{\rightarrow} KP \overset{k_3}{\rightarrow} K \tag{3.66}$$

with reaction rates denoted by $k_1[ATP]$, k_{-1}, k_2 and k_3, respectively. An Arrhenius assumption for the reaction rates was also made in his work. The solution for the above models in terms of propagation velocity is given by Eq. 3.64.

Peskin and Oster [252] assumed that a hydrolysis of ATP must be coordinated with relative positions of the two heads to provide an explanation of the behavior of KN. Recently, Mogilner et al. [228] developed a chemical kinetics model that accounts for structural changes in the neck linker region as affecting the force dependence of the motion using a bistable nucleotide-dependent behavior of the neck linker. The latter model reproduces the shape for the velocity versus load force dependence $v(F)$ very well. Block [36] stated that KN binds to the β-tubulin unit, possibly moving along the seam of the B lattice of MT. He suggested abandoning the notion of a power stroke, maintaining instead that KN motion is dominated by biased diffusion.

In the power stroke models it is the motor protein whose structure changes. The motor protein has two or more distinct states where at least one conformational change occurs and is driven by ATP hydrolysis. In conclusion, the reaction cycles proposed differ somewhat from paper to paper (see also Hancock and Howard [132] for a discussion of models).

The motion of the motor protein in any model is fueled by ATP hydrolysis. Note that all chemical reaction-based models propose a KN work cycle involving chemical energy conversion. Some of them do not calculate the velocity-force relationship but give important experimental results regarding the structure of KN. Other models calculate also the force dependence of the velocity mostly using the Arrhenius assumption.

Comments regarding the Arrhenius assumption Consider the basic idea in the well-known Arrhenius relation for k_+^0,

$$k_+^0 = A exp(-E_b^+/kT) \tag{3.67}$$

which is the forward rate from the initial state to the nal state in a chemical reaction. The common explanation is to assume a reaction potential, with two minima at the initial state and the nal state respectively, and a maximum at the so-called reaction barrier (see Fig. 3.65). In Eq. 3.67, E_b^+ is the energy difference between the barrier energy and the initial state energy. Assuming the presence of an additional load potential Fx, we obtain instead

$$k_+(F) = k_+^0 exp(-Fd_+/kT) \tag{3.68}$$

Here, d_+ is the reaction coordinate distance from the initial state to the barrier. Considering the load dependent backwards reaction $k_-(F)$ that starts from the final state to the initial state, we have instead

$$k_-(F) = k_-^0 exp(-Fd_-/kT) \tag{3.69}$$

Here, d_- is the reaction coordinate distance from the nal state to the barrier. We see from Fig. 3.65 that the sum

$$d = d_+ + d_- \tag{3.70}$$

and *not* d_+ or d_- alone is the distance traveled in a given step.

Furthermore, the calculation of $k_+(F)$ and $k_-(F)$ uses the assumption that the prefactor A in Eq. 3.67 is independent of the force. This is not experimentally justified; see Keller and Bustamante [167]. Indeed, model calculations using a given reaction potential show a somewhat different behavior. The Arrhenius assumption should therefore use Eq. 3.70. For steps which are mostly mechanical, other methods, if possible, should be preferred.

3.12.2   Mechanically Based Models

In these models, KN is considered to be one or two mass points moving in a potential, which has a static part and sometimes a part which can change in time. However, KN is moving in a fluid at a finite temperature; it experiences overdamped motion. Two types of description are therefore possible:

1. The Langevin equation is an equation of motion with a randomly fluctuating force which accounts for the effects of finite temperature. Consequently, the calculated co-ordinates also exhibit a fluctuating behavior.

2. The Fokker-Planck equation deals directly with the probability of finding the mass point at a particular position in space at time t.

Both methods are only different mathematical representations of the same physical problem, i.e., the biased Brownian motion in the presence of external potentials [22]. Since the movement of the particle is overdamped, inertial terms (i.e., the second time derivatives of position) are conveniently neglected.

For example, for a single mass point the Langevin equation in one dimension reads [84]:

$$\xi \frac{dx}{dt} = -\partial_x V(x) + F(t) \tag{3.71}$$

Here, x is the spatial co-ordinate along the MT long axis, $V(x)$ is the effective potential, ξ is the friction coeffcient and $F(t)$ the uncorrelated random force [21, 260], with

$$< F(t) >= 0, < F(t)F(t') >= 2\xi kT\delta(t - t') \tag{3.72}$$

according to the so-called fluctuation-dissipation theorem.

The Fokker-Planck equation, on the other hand, calculates the probability density $P(x,t)$, to find the particle at the position x at the time t, and for the same problem it reads [34]

$$\frac{\partial P(x,t)}{\partial t} = \left(\frac{\partial}{\partial x} V(x) + \frac{\partial^2}{\partial x^2} D \right) P(x,t) \tag{3.73}$$

where $V(x)$ is the substrate potential and D is the diffusion coefficient $D = k_B T \xi$. The Fokker-Planck equation, if it can be solved, is more powerful than the Langevin approach, since it directly gives quantities that can be averaged and compared with experiment.

We can divide mechanically based models in the following way: a) models with oscillating potentials (tight binding models), and b) models that use alternating chemical and mechanical transformations. The idea of tight binding implies that the chemical cycle produces an oscillating potential. This potential is assumed to be relatively flat and asymmetric. Diffusion results in a forward propagation of the motor protein due to the asymmetry of the potential [165]. For example, Abad and Milke [1] described Brownian motion in fluctuating periodic potentials resulting in a rectified motion of the motor.

Gibbons et al. [120] developed a dynamical model for KN-MT motility assays that is a 2D stochastic model for the dynamics of KN coated MTs. In it, viscosity is used and MT-KN interactions are represented by a potential while KN is modeled as a spring with a mean attachment lifetime to the MT. However, the existence of the spring in this model is only required for load application and directionality. Furthermore, no account of chemical reactions taking place in the system is given.

3.12.3   Models with Alternating Chemical and Mechanical Transformations

The development of mixed mechano-chemical models for KN walks can be traced back to Fisher and Kolomeisky [101]. In these models the two KN heads are represented by moving mass points, which are described using either a Langevin or a Fokker-Planck equation. Switching between different states represents transitions due to (implicit) chemical reactions that result in different effective potentials the masses experience. To achieve directionality, an asymmetric potential experienced by the molecule due to the MT has to be assumed. However, to date there is no concrete empirical evidence showing this asymmetry. Since KN and other motors like Ncd have virtually the same motor domains, the differences in the walking direction between KN and Ncd cannot be explained within this formalism.

An example of this approach applied to the two-headed KN was given by Derényi and Vicsek [82] who employed two coupled Langevin equations each describing one head as a single point mass, subjected to a load force and a stochastic noise term due to fluctuations. Each head was assumed to interact with a periodic asymmetric potential representing the MT's attraction. The interaction between the heads was provided by an elastic spring. Note that a more realistic model should incorporate a two- or even three-dimensional energy landscape [338]. The so-called uniform ratchets refer to models in which explicit inclusion is made of the mechanical degrees of freedom and the enzymatic activity of the motor is independent of the position along the filament.

The non-uniform ratchet models include the spatial co-ordinate and assume an enzymatic activity described by a combination of master and Fokker-Planck equations. For example, Lipowsky and Harms [192] proposed a hand-over-hand model using non-uniform ratchets corresponding to several internal states of KN. An example was solved which involved a doubly bound state, two excited states with one head unbound, and an unbound state. Transitions between conformational states in this model depend on the spatial coordinate x along the filament. The dimeric KN model was characterized by internal states of KN with different effective potentials signifying a chemical transformation. The calculated load-dependent velocity was in good agreement with experimental data [75]. The advantage of this method is that only stationary states need to be considered, which can be solved exactly in one dimension.

Keller and Bustamante [167] discussed mechano-chemical energy transduction of molecular motors. They analyzed a minimal family of motor models depending on the chemical reaction schemes: (a) mechano-chemical binding model, (b) mechano-chemical reaction model, (c) mechano-chemical release model and (d) mechano-chemical trigger model. The general characteristics that emerged are correct. However, subsequent applications require the knowledge of chemical reaction rates.

Other Models

Recently, Astumian and Derenyi [23] proposed a Brownian ratchet model for the twoheaded KN that depends on temperature exponentially and claimed that directionality of motion is due to the chemical mechanism of ATP hydrolysis (space directed) and not due to structural features. One-dimensional diffusion of the motor was claimed to suffice and the motor would work even if it were not to dissociate from the filament. In the case of two-headed motors, correlations between the two heads require the introduction of internal degrees of freedom such as springs and interactions between them [2, 90, 252]. Finally, while the vast majority of the mechanical models proposed involve either the Langevin or the Fokker-Planck approach, Arizmendi and Family [19] used a master equation approach to the modeling of molecular motor motion.

Advantages and Disadvantages of Existing Models

Many of the existing models concentrate on calculations of the propagation velocity of KN under load. Since very different assumptions and computational methods often lead to equally good results in this respect, other experimental data should be included and explained in addition to the velocity versus load dependence. Below, we give a short summary of the analysis:

As mentioned earlier, the walking direction for (conventional) KN is from the minus- to the plus-end of an MT. Most models explain it using an asymmetric potential between the KN heads (represented as point masses) and the MT. This type of potential does give a preferred direction. Hence KNs possessing virtually the same head structure should move in the same direction. However, there is experimental evidence that contradicts this assumption. As is now known, KN heads are highly conserved independent of the KN directionality. Interestingly, all the isolated natural KNs with C-terminal bound motor domains propagate in the direction of the MT minus ends [96]. It has been shown experimentally that KN's directionality is mainly determined by the neck region. This was done by constructing chimeric KNs and other genetically engineered ones, often derived from the plus-end motor conventional KN and the minus-end motor Ncd. Therefore, construction of KN-derived motor molecules working in both directions became possible. The neck linker region, in the case of conventional KN, is also essential for processivity and regulation of ATP hydrolysis [97]. As will be shown in our model, the internal structure of the KN molecule in its relaxed state, and not the asymmetry of the binding potential for the heads, determines the KN walking direction which is consistent with the experimental observations discussed above.

The two KN heads are extended objects which are oriented at roughly 120° in the ADP state with respect to each other and have identical structure and geometry [178]. However, the neck is not symmetric. The mode of walking is assumed here to be hand over hand. The difference is explained in the presence of torsion in the KN walk as is described below. Hence, a model describing this motion must be developed in two dimensions for every head viewed an extended object to show a defined direction. Such a movement cannot be described by models in one-dimensional space that use point masses to represent KN heads. Experimental results indicate that the free (unattached to MTs) heads for processive plus-end oriented KNs point in the walking direction [20, 141]. This orientation of the free KN head in relation to the movement direction has not been properly explained in existing models. Differences, however, are revealed for the positions of the heads in relation to the MT long axis which appear to be mirror symmetric.

In the following sections we concentrate on the most important aspects of the walking mode of KN and propose a model that provides both qualitative and quantitative agreement with the key experimental facts regarding directionality and processivity of KN.

Although some experimental findings seem to engender controversy, the following facts seem to be generally acceptable.

The binding of both KN heads is equally directed with respect to MT's polarity. This is obvious since the chemical and 3-D structure of a KN head is not symmetric. Recognition and binding of a head to the MT is only possible for a special position and direction of the head relative to the MT. This should apply not only if one of the heads binds, but also if two heads bind simultaneously.

X-ray diffraction for crystallized KN was interpreted as showing a rotationally symmetric form with a relative angle between the two heads of 120° for the "relaxed" state of the KN molecule, in the absence of binding or other forces acting on it [178]. This means that by reducing the angle between the heads from 120° to zero the two heads would become co-located and co-oriented. Note that this does not imply a mirror reflection symmetry of the two heads.

A simultaneous binding of both heads to the same protofilament of a MT results in a tense form for the molecule as defined by a strand between core heads and the dimerization domain that should act like a torsion spring. This latter portion of the molecule is called here the neck linker (see Fig. 3.61).

Arnal and Wade [20] as well as Mandelkow and Hoenger [212] showed by MT decoration experiments for a variety of KNs that the free head points in the walking direction when the other head is bound to the filament.

The Different Chemical States of the Heads and the KN-MT Complex

To develop a conceptual framework for the model we assume that each KN head exists in one of the following three chemical states:

- **Chemical State A**: A nucleotide-free (empty) state. A head in the empty state can bind to an MT.

- **Chemical State B**: An ATP state. In this state an ATP- magnesium complex is attached to the head. ATP can only attach to a KN head that is bound to an MT. An ATP head can either remain bound to the MT or detach from it. However, this is a metastable state due to ATP hydrolysis.

- **Chemical State C**: An ADP state. This state is a result of an ATP hydrolysis and a removal of the phosphate group Pi. In this state an ADP-molecule is attached to the head. A KN head in an ADP-state is either free, or in the process of detachment from a MT in which an intermediate weakly bound state can also appear.

During the mechano-chemical cycle each **KN-MT complex** can also be found in several different chemical states; three main ones being:

- **State α** : Head B is bound and head A points in the walking direction at an angle ϕ to the minus direction along the MT axis.

- **State β**: Head A is bound and head B points in the walking direction at an angle ϕ to the minus direction along the MT axis. The value of this angle has been determined in KN-MT decoration experiments. Note that the angle between the two heads in an unbound state of KN has been shown in x-ray experiments to be 120 degrees. This angle could be different than ϕ.

- **State γ** : Head A and head B are both bound, resulting in a tensed KN form.

Note that the α and β states are mirror-symmetric with respect to the MT axis. We will show below, that KN must contain at least two different springs and that state γ must have two different realizations with the springs, that are alternatively loaded.

The model we propose employs a two-dimensional potential which is periodic along the MT axis. No need for it to be asymmetric arises in our model. However, at least two space dimensions are needed, in order to describe different directions in space. The KN motor is represented here as consisting of two heads and a neck/stalk, each of which is considered to be an extended object. The KN molecule is processive with directed binding to the MT filament. Both heads are not only identical but also identically oriented with respect to their own axis and the MT filament. They are not mirror images of each other with respect to either the stalk axis or the MT axis. This is analogous to a hypothetical human with two right hands (see Fig. 3.63 below). Consequently, the binding of both KN heads to the MT filament must be identical in terms of both geometry and energy.

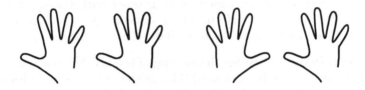

FIGURE 3.63: A hypothetical situation of a human being with two identical (right) hands corresponds to KN while the actual pair of a left and right hand does not.

Moreover, we show below, that KN must contain at least two springs, one located in the neck linker region which is what we use later.

The Two Springs

The individual states of the KN walking cycle are shown in Fig. 3.64, and are numbered from 0 to 4. The four transitions between the states (steps) are described below as follows: Initially, KN is in a relaxed state which is unbound to MT (see the middle panel of Fig. 3.64).

- **State 0**: In the first step one of the two heads binds to the MT. For illustration purposes, we assume that head B is bound first, head A is free, pointing in the walking direction at an angle ϕ to the other head and KN ends up in state 1. The heads A and B are equivalent with respect to the model.

- **State 1**: This step involves the binding of the other head, ending up in a state where the angle between the heads is 0°. To do this, the KN molecule has to be twisted starting from a relaxed position. In other words, an internal spring located at head A is being loaded as head A turns in a clockwise direction (which is the shorter way due to the energy minimization principle) to bind to the next possible binding site. KN is then in state 2, with spring A loaded, spring B unloaded.

- **State 2**: In this step the hydrolysis energy is used to release one head. Releasing one head will result in mechanical movement, since the KN will move towards its equilibrium position which is in a relaxed state. Two possibilities now exist. If head A is released, KN again ends up in state 1 with the center of mass remaining almost stationary. However, if head B is released, the mechanical step results in a swinging movement of head B around the still bound head A. Here, the center of mass will move by the unit of length in the periodic lattice of binding sites which is 8 nm for the KN-MT system. This then is a real walking step taking KN to state 3. Note that geometrically seen, states 1 and 3 are not identical but are mirror images of each other with respect to the MT axis. Note that for $\phi \neq 180°$ in this process the spring located at head B is also being loaded when head A binds. This can be seen in Fig. 3.66.

- **State 3**: In this step we see the binding of head B. The binding energy is used to load spring B, since head B turns counter-clockwise with respect to the correct binding direction. Therefore, KN ends up in state 4, but now spring B is loaded and spring A is unloaded.

- **State 4**: If head A is released, the mechanical step results in a swinging movement of head A around the still bound head B.

- **State 5**: This step is identical to step 1 and it starts a new walking cycle. Note that the cycle can start with step 3 where binding begins with head A. The probability of starting this process from state 0 and going into state 3 is the same as going to state 1 since both heads are identical.

Note that step 4 is the same as step 2, except heads A and B are swapped around. Analogously, states 1 and 3 are similar but not the same being mirror-symmetric, with the two heads swapped around again.

It is of absolute importance that there must exist at least two springs, each connected to one of the two heads. It is not at present known if there is a further spring in existence. Note that to load the spring located at head A, with KN initially in state 1, head A turns in a clockwise direction to bind to the MT. To load the spring located at head B, starting with state 3, head B is turned in a counter-clockwise direction to bind to the MT.

In our model, proposed as a model with the minimal number of necessary assumptions, the walking directionality of KN is explained using the principle of directed binding of the two heads, and an inclusion of only these two springs.

The Complete Cycle of Chemical and Mechanical Steps

Several different chemical reaction cycles describing the KN walk can be found in the literature. The mechanical hand-over-hand cycle, on the other hand, must be of the form: $1 - 2 - 3 - 4 - 1 \ldots$ (using the symbols introduced above). Note that between states 1 and 3 in the cycle the stalk turns by

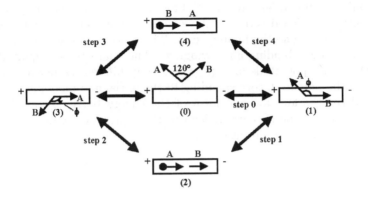

FIGURE 3.64: The mechano-chemical cycle for the plus-directed two-headed KN. The rectangle represents a MT. The plus and minus signs in the rectangles indicate the corresponding MT ends.

an angle which is approximately equal to the angle between the heads in a relaxed state and a tensed state. For simplicity we choose $\phi = 120°$, the same angle between the heads as for the unbound state. Also, we predict that the unbound head turns in such a way as to make the shortest move in the binding process.

There is also the possibility that KN in state 1 or 3 ends up in state 0 (shown in the center of Fig. 3.64) with both heads unbound. However, the whole cycle could then start again via either state 1 or 3. An example of this situation could be the sequence: $0 - 1 - 2 - 3 - 0 - 3 - 4 - 1 \dots$. Hence, absolute processivity is not the only possible outcome of this model.

Finally, in order for the mechanical cycle to proceed in the right direction, the ATP hydrolysis energy must exceed the binding energy, which itself must exceed the elastic spring energy (which is the energy of the loaded spring).

The Mechanical Steps seen as Chemical Reactions

We postulate the introduction of reaction barriers and energy differences as the essential parameters for the potentials, which are used for the calculation of the stochastic movement. Our model requires the use of at least eight degrees of freedom (8 position co-ordinates). The two heads have to be described by at least two mass points in order to account for directed binding. We use two spatial dimensions to see the hand-over-hand swinging motion of the two heads. Hence, we have to calculate the stochastic, two-dimensional movement of four mass points. The potentials, acting on these four mass points, contain a large number of parameters, which are used to fit to experimental data.

It must be kept in mind that most of the potentials are necessary to maintain the shape of the KN molecule, for example to determine the diameter of one head. Only two potentials are essential for the mechanical walking steps: one for the spring, which determines the swinging step and one for the binding potential to determine the binding step. These potentials specify the rate of the two reactions: binding and swinging. The characteristic quantities for such "chemical" potentials are: (a) the energy differences between the two minima of the potential, corresponding to the initial and final states (see Fig. 3.65) and (b) the location and form of the energy barrier separating these two states. These energy barriers lead to the typical Arrhenius temperature dependence of the reaction rates.

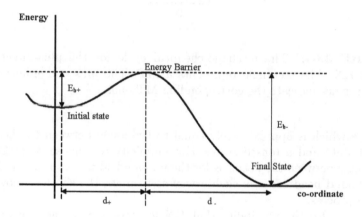

FIGURE 3.65:   A sketch of the potential energy barrier assumed in the model.

For a detailed mathematical description of the model, the reader is referred to Bolterauer, Unger, and Tuszyński [42].

Below we summarize the main differences between our model and the existing models:

- Our model is a two-dimensional model showing the hand-over-hand movement with respect to the surface of the microtubule.

- The heads of KN are extended objects. Hence, in two dimensions we require at least two mass points to define both position and orientation of each KN head.

- We use the fact that the binding of the heads onto the MT requires a proper orientational fit, i.e., has to be directed.

- We use the experimental result that the arrangement of the heads in the relaxed state of the kinesin molecule is symmetric.

- From 3 and 4 above we have concluded that when binding both heads simultaneously, the KN molecule generates internal stress.

- We show that there must exist at least two (torsional) springs, which are alternately loaded during the KN walk process (see Fig. 3.66).

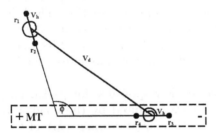

FIGURE 3.66: The interaction inside the KN molecule: The straight lines describe stiff springs, enforcing the internal distances within the heads and a constant distance between the heads. One head is bound to the MT while the other head is free. Also shown are the two torsional springs responsible for the relative position of the two heads.

- Our model states that if only one head of a processive kinesin is bound, the other unbound head points in the walking direction. However, this is also an experimental result for plus-end processive kinesins.

- The internal structure of the kinesin molecule defines the walking direction. A change of the internal structure could reverse the walking direction. For example in our model it is sufficient to change the angle ϕ in the relaxed state, to change the direction of walking and the stochasticity of the movement.

- We assume that the directionality-defining springs are located in the neck linker region connecting KN head to the dimerization domain.

- Analyzing the assumptions made in both the mechanically-based and chemical kinetics models, we see that both descriptions are included in our model.

3.13   Other Structures (ER, Golgi Apparatus)

Below we very briefly list and categorize some other important organelles in the cell. Spreading throughout the cytoplasm is the *endoplasmic reticulum* (E.R.). It is a folded system of membranes that loop back and forth giving it a very large surface area. This membrane provides a surface area for cell reactions. It is also the site of lipid production. The E.R. comes in two forms: (a) smooth E.R. which has no ribosomes associated with it, and (b) the rough E.R. that has ribosomes attached to it, giving it its textured appearance under light microscopy. These ribosomes manufacture proteins for the cell.

- The *Golgi apparatus* is responsible for packaging proteins for the cell. Once the proteins are produced by the rough E.R. they pass into the sack like cisternae that are the main part of the Golgi body. These proteins are then squeezed off into the little blebs that drift off into the cytoplasm.

- *Lysosomes* are called suicide sacks. They are produced by the Golgi body. They consist of a single membrane surrounding powerful digestive enzymes.

- Animal cells generally contain *centrioles* which have been presented above. Cilia and flagella are found in many different life forms.

- Plant cells generally contain *storage vacuoles, cell walls* and *plastids*. Vacuoles are large empty appearing areas found in the cytoplasm. They are usually found in plant cells where they store waste. As a plant cell ages, they get larger. In mature cells they occupy most of the cytoplasm.

- Cell walls are the rigid structure found surrounding plant cells. They provide support for the plant.

- Plastids are large organelles found on plants and some protists but not in animals or fungi. They can easily be seen through a light microscope. Chloroplasts represent one group of plastids called *chromoplasts* (colored plastids). The other class of plastid is called *leucoplasts* (colorless plastids); they usually store food molecules. Included in this group are amyloplasts or starch plastids present, for example, in a potato root cell.

- Finally, *contractile vacuoles* are organelles that are critical in enabling protozoa to combat the effects of osmosis. Protozoa must constantly excrete the water that enters through their membranes.

3.14   Large Polar Molecules

Large polar molecules, as embodied in inter- and intra-cellular biological materials, ferroelectric polymers and piezoelectric/ferroelectric liquid crystals, represent a class of electrically active materials that are crucial for many functions in living cells and either have had a tremendous impact on science and technology or have a great potential for making such impact. Especially important from the vantage point of this review is the fact that all of these materials are ideal prototypical systems for studying the science and technology of *"ferroelectricity at small length scales."*

This section examines two topics:

1. Bioferroelectricity and

2. Ferroelectric Polymers (Polyvinylidene Fluoride and its copolymers)

Ferroelectric liquid crystals (FLC's) have been one of the largest and most active branches of ferroelectricity for the past two decades. The interest in FLCs stems from the unusual and fascinating science of these materials as well as their technological applications in flat panel displays, optical image processing and other applications. Progress in both the science and technology of FLCs has been documented in books and in the proceedings of several large international meetings on the subject published in journals [258, 123, 35]. This progress in the past two decades has been impressive, and the field continues to be very active. A remarkable aspect of these materials is the ability of the organic chemist to synthesize a wide variety of chiral molecules that, in principle, allow tremendous tailorability of desired properties. Progress along these lines continues at a strong pace. Much of the work on FLC's is on thin films and small structures. Most impressive has been the discovery that FLC thin films exhibit long-range FE order at the single monolayer level, thus affording a unique class of materials to explore the fundamental issues of ferroelectricity at the very small length scale. Unfortunately, this large and specialized area that presents some unique and challenging issues at the nanoscale is beyond the scope of the present report. Consequently it will not be covered and the reader is referred to the vast literature on the subject for details.

3.14.1   Bioferroelectricity

The occurrence and role of piezoelectricity and pyroelectricity in biological systems have long been subjects of much interest as pertain to understanding certain functional processes in living organisms. Indeed, both phenomena are now known to be prevalent in biological materials. Specifically, it has been suggested that pyroelectricity is an essential and universal property of biomaterials [185].

This suggestion immediately raises the question about the possible occurrence of ferroelectricity in biological systems. This question has been around for years, and there is now a substantial body of evidence suggesting ferroelectric mechanisms for a number of biological functions. Examples include conducting membrane functions [188], brain memory [106] and the functioning of the postsynaptic membrane [157]. However, these suggestions remain controversial, and in no case can it be said that the occurrence of ferroelectricity has been definitively proven [188].

Two decades ago Kubisz et al. [180] investigated ferroelectric and piezoelectric properties of collagen. Collagen has since been shown to have both piezoelectric and ferroelectric properties [113]. Dielectric hysteresis loop and differential thermal analysis (DTA) studies of collagen were performed for temperatures ranging from 293 to 393 K. The DTA curve for collagen with 10% of water showed a sharp minimum at 353 K. A hypothetical dielectric hysteresis loop showed a maximum shift along the direction of the electric field applied in the vicinity of 375 K. Very large dielectric losses at low frequencies made it difficult to interpret the collagen results in terms of ferroelectricity. It is very likely that the well-documented ferroelectricity of bone material and tendon, going back several decades, is largely due to the presence of collagen in these structures [24].

In a special issue of Ferroelectrics, guest edited by Leuchtag and Bystrov [188], a collection of papers on Ferroelectric and Related Models in Biological Systems (or *Bioferroelectricity*) was published in which the papers featured focused on two distinct types of biological structures suspected of having ferroelectric properties. One type is *microtubules* and the other type is *voltage-dependent ion channels* found widely in cell membranes and associated with signal conduction in nerve and muscle cells. Both types of structures are proteins involved with information processing at the cell level. These structures will be discussed briefly below. Other areas of bioferroelectricity covered in the papers include the following. The use of infrared techniques to study the polarizability and transfer of protons in the hydrogen bonds important in biological structures was reviewed indicating a bonding mechanism with metal ions substituting for protons. The use of nuclear magnetic resonance was argued for the study of transitions between ferroelectric and superionically conducting states in a protein. A ferroelectric model was presented to explain the conduction of an impulse and its stimulation by a magnetic field. A review of a large body of evidence for ferroelectric phase transitions in nerve fibers, cells, and synapses was also presented.

More recently, attention has shifted to the molecules of the cytoskeleton as possible ferroelectrics. The cytoskeleton is a major component of all living cells. It consists of thin rod-like filaments that span the cytoplasm. There are three major types of filaments: actin-based filaments (i.e., microfilaments), tubulin-based filaments (i.e., microtubules) and intermediate filaments (e.g., neurofilaments, keratin). There are at least three well-studied mechanical functions of the cytoskeleton in vivo: providing the mechanical strength of the cell, segregating the chromosomes and participating in the transport of macromolecules via motor proteins.

Actin-based filaments are the simplest filaments, with a diameter of approximately 4 nm and variable lengths. The strand is a left-handed helix of actin monomers. Actin is the most abundant protein in the cytoplasm of mammalian cells, accounting for 10 - 20% of the total cytoplasmic protein. Actin exists either as a globular monomer (i.e., G-actin) or as a filament (i.e., F-actin). In the 1960's Kobayashi et al. [176] demonstrated the existence of a spontaneous permanent dipole moment of actin monomer that is oriented roughly perpendicular to the filament axis in F-actin. As will be shown below, a similar situation exists for microtubules except for a much larger magnitude of their dipole moment, and a greater complexity of the structure.

Thus, while, as already noted, there is some evidence suggesting ferroelectric mechanisms for a number of biological functions, in no case can it be said that the occurrence of bioferroelectricity has been definitively proven. What is clear is that this area is exciting, and its important challenges will undoubtedly attract a good deal of attention in the future. In a genuine sense, bioferroelectricity embodies "ferroelectricity at small length scales," and thus this topic is a natural for this review.

Our assessment of this area will focus on two prominent types of biological structures believed to exhibit ferroelectric behavior. These are *microtubules* and *voltage-dependent ion channels*.

3.14.1.1 Microtubules

As discussed earlier in this book, microtubules (MTs) are fibrous structures that constitute part of the cytoskeleton of living cells. They are made up of piezoelectric polymer of a protein, tubulin, which is present in most animals, plant and protis cells. In addition to providing structural strength and flexibility to the cells, MTs guide and transport cellular material towards or away from the center of the cell, play a key role in cell division and are also believed to be involved in intra-cell signalling [47]. In the recent past, a number of authors have suggested that these functions can be understood on the hypothesis that MTs exhibit ferroelectric-like properties [47 280 318]. Brown and Tuszyński [47] have reviewed aspects of the ferroelectric (FE) model of MTs. In this section we provide a brief summary of the properties and presumed physics of MTs. The reader is referred to the cited references for more details.

A microtubule is a hollow cylindrical structure whose wall is made up of parallel chains, or protofilaments, of tubulin dimer molecules longitudinally shifted from each other so as to result in a helical structure as depicted in Fig. 3.67(a). Each dimer, 8 nm in length, is composed of two units, α- and β-monomers (Fig. 3.67(b)). Each MT is generally made up of 13 protofilaments (Fig. 3.67(a)) which results in a tube diameter of 25 nm, and the length varies from submicrons to centimeters [280].

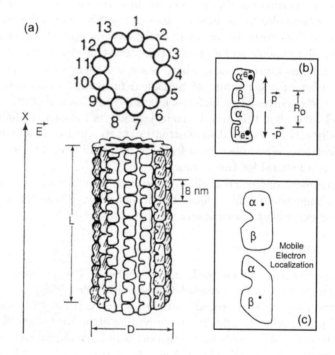

FIGURE 3.67: (a) Schematic representation of the structure of a micro-tubule; (b) two neighboring dimers with their opposite polarization; and (c) switching between the two conformational states. (After Sataric et al., Ref. [280].)

MTs form through the polymerization of $\alpha\beta$-tubulin dimers. This polymerization can be controlled by physical (temperature) and chemical (pH, concentration of protein and ions) means to produce closely or widely spaced MTs, centers, sheets, rings and other structures [85, 139] (Fig. 3.68), thus facilitating fabrication of nanowires, nodes and networks in the future. MTs have an electric and functional polarity. Table 3.13 summarizes the results of calculations [275] of tubulin's net charge and the dipole moment using the available data from the file 1TUB in the PDB (Protein Data Bank), i.e., excluding the C-termini that are known to account for approximately 40% of the total charge of the protein. Note that the x-direction in Table 3.13 coincides with the microtubule protofilament axis. The α-monomer is in the direction of increasing x values relative to the β-monomer. The y-axis is oriented radially toward the MT centre and the z-axis is tangential to the MT surface.

TABLE 3.13: Key Electrostatic Properties of Tubulin

Tubulin Properties	Dimer	α-monomer
charge (electrons)	-10	-5
dipole (Debye) overall magnitude	1714	556
x-component	337	115
y-component	-1669	554
z-component	198	-6

Magnesium ions (Mg^{2+}) are essential for microtubule assembly to occur. They bind in a pocket that appears to provide structural stability for the assembled tubulin dimers. Taking into account the bound water layer with electrons released due to the Mg^{2+} complex activity, it is highly plausible to envisage the MT structure as a ferroelectric system with the dipole moments of tubulin oriented as shown in Fig. 3.69 based on detailed molecular dynamics calculations [275]. The ends of a MT therefore possess a different amount of net charge, as was recently demonstrated through massive molecular dynamics computations [26]. Thus, it is concluded that the MT cylinder supports an intrinsic electric field, E.

Recently, Maniotis et al. [214] and others demonstrated that microtubules transduce mechanical forces over micron-long distances inside the cell even into the nucleus suggesting an efficient signaling phenomenon that can be related to microtubule piezoelectricity.

In proteins, virtually every peptide group in an a-helix possesses a considerable dipole moment on the order of $p_0 = 1.2 \times 10^{-29} Cm$ (= 3.5 Debye). All these dipoles are almost parallel to the helical axis giving rise to an overall dipole moment of this particular helix. It is generally accepted that this large

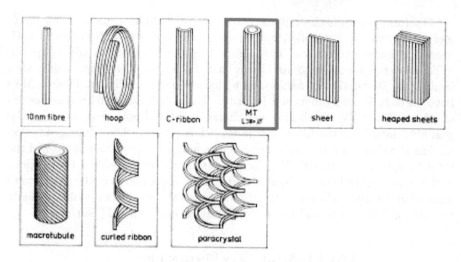

FIGURE 3.68: Various forms of tubulin assemblies. (After [303]).

dipole moment of an α-helix has an important biological role [143]. Since the tubulin dimer contains several α- helices which are not oriented randomly, it is not surprising that each tubulin dimer possesses a large net dipole moment, p. Tuszyński et al. [275, 323] have recently recalculated the values of the dipole moments and net charges using the sequences of the various homologous isotypes of tubulin biochemically characterized in the literature.

In order to experimentally determine the dipole moment of tubulin, Mershin et al. [220] and Schuessler et al. [283] performed a surface plasmon resonance study on tubulin dimers in solution. Analysis of their data yielded for tubulin a refractive index $n = 2.9$ and a dielectric constant $\epsilon(= n^2) = 8.4$ and $|p| \approx$ 1000 debye which agrees with the order of magnitude of the tubulin dipole moment calculated based on the atomic resolution structure. The differences between these two values (experimental and theoretical) can be attributed to the presence of C-termini with the surrounding counter-ions as well as the effects of water of hydration forming a layer around the exposed part of the protein.

Finally, there have been some preliminary experiments aimed at measuring the electric field around MTs [164] indicating that MTs could be ferroelectric. Stracke et al. [303] subjected MT's to moderate electric fields. The fields were applied to suspended microtubules and to microtubules gliding across a kinesin-coated glass surface. In suspension, microtubules without MAPs moved from the negative to the positive electrode at a pH of 6.8, indicating a negative net charge which has been effectively significantly diminished due to the counter ions in comparison to the bare charge of tubulin. An electrophoretic mobility of about $2.6 \times 10^{-4} cm^2/Vs$ was determined. Gliding microtubules migrated preferentially to the positive electrode, too. Recently, Dombeck et al. [87] used uniform polarity microtubule assemblies imaged in native brain tissue by second-harmonic generation microscopy which only worked for uniform polarity bundles but not for antiparallel arrangements indicating a non-linear polarization effect. Apart from the above observations, there exists little experimental evidence concerning tubulin's and MTs' electrical properties.

It is interesting to note that the idea that MT's are ferroelectric was already proposed some 30 years ago [9] on the basis of their piezoelectric properties. This was more recently argued by Nanopoulos et al. [237]. It is apparent that due to the strong curvature of an MT cylinder, the inner parts of the dimer structure are compressed in order to fit into a microtubule while the outer ones are stretched by a substantial amount of tension. This causes additional redistribution of excess negative charge that enhances the transverse component of the net dipole moment of every dimer comprising a MT. This effect is very analogous to that observed for the experimentally verified cylindrical hair cell properties [347]. This has also been corroborated by a detailed map of the electric charge distribution for the tubulin dimer [323].

Fig. 3.69 is a reconstruction of the electric polarization of a microtubule that takes into account the dipole moments of the individual dipoles and their orientation with respect to the microtubule axis [326]. It is interesting to note that while the net dipole moment of tubulin is very large, its symmetrical ordering around the MT axis gives rise to a net cancellation effect such that a minimal amount of torque is expected to arise from the application of an external electric field to a microtubule in solution.

The presence of oriented assemblies of dipoles in the wall of an MT is responsible for the observed piezoelectricity of MTs [47,280]. These properties suggest that it may be possible to orient arrays of MTs by the application of an electric field. Indeed, this has been demonstrated, and it has been further speculated that inter-cellular electric fields may induce MT orientation [47]. That such orientation is possible can be appreciated by noting that MTs are sensitive to fields $100mV/\mu m$, or $10^3 V/cm$, whereas fields measured across cellular membranes are $10^5 V/cm$ [47, 280].

Physical properties of MTs, e.g., ferroelectric properties, conditions for existence of solitary waves, coherent vibrational modes and electrical conductivity properties have been thoroughly investigated theoretically. A detailed model demonstrated how coherent lattice excitations propagate coupled electrostatic and elastic energy via solitary waves, which originate from the hydrolysis of a guanosine triphosphate (GTP) molecule bound to the β-tubulin monomer [318]. The model Hamiltonian contained effective dipole-dipole terms, the kinetic energy of the displaced dimers, the elastic restoring energy and the potential energy due to the environment and the effective field of the system. External electric fields of the order of magnitude consistent with the axonal action potential were shown capable of reordering these local dipoles. Energy loss-free transport along MTs has been shown possible [280,318,142] as a re-

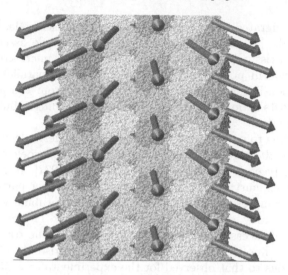

FIGURE 3.69: A microtubule structure showing the constituent dipoles. The arrows indicate the orientation of the permanent dipole moments of individual tubulin dimers with respect to the surface of a microtubule. Figure prepared using MolMol [215]. (After Tuszyński et al. [326].)

sult of "kink like excitations" (Fig. 3.70), or solitons, as an energy-transfer mechanism in MTs. In the model, this feature arises from non-linear coupling between the dielectric and elastic degrees of freedom of the tubulin dimers and from the inclusion of a viscous force representing the damping effect of the water. Several models exist regarding the nature of these excitation waves, but they all depend on the dipole moment of tubulin and its ability to flip, hence they are called "flip waves". They can be coupled to conformational states of protein (tubulin). Depending on the model and the parameters assumed, the speed of such waves has been estimated to be $10^{2\pm1} m/s$ [318].

Finally, water of hydration which decorates the surfaces of proteins and comprises almost 50% of the intracellular volume exhibits hexatic ordering of its dipole moments (Fig. 3.71) [203]. This indicates that ordered water can possess interesting polarization properties adding to an overall ferroelectric behavior of protein filaments in solution.

3.14.1.2 Voltage-dependent ion channels

Voltage-dependent ion channels (VDIC's) are commonly found in cell membranes and are involved with impulse conduction in nerve and muscle cells. These ion channels are glycoproteins embedded in the phospholipids bilayer of the plasma membrane that sharply alter their permeabilities to specific ions in response to changes in transmembrane voltage [188]. Understanding the

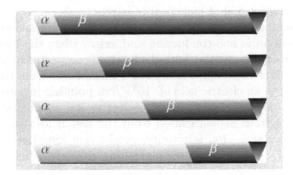

FIGURE 3.70: Conformational "kink" traveling along a MT. α represents a domain where the mobile electron is in the α sub-unit of the tubulin proteins, orienting the dipole in the direction of travel. (Refer to Fig. 3.67.) β represents a domain where the mobile electron is in the β sub-unit of the tubulin proteins, orienting the dipole roughly against the direction of travel. (After Trpisova et al. Ref. [318].)

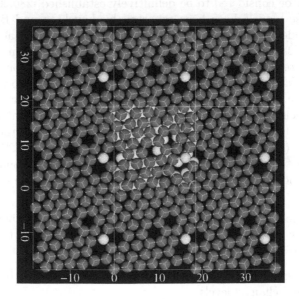

FIGURE 3.71: A schematic representation of water molecules bound to the surface of a protein. Note the hexagonal orientation. (After Luchko, [203].)

mechanisms that underlie the excitability of nerve and muscle membranes has been a long-standing goal of biological research.

These ion channels exist in two main conformations, closed and open. The closed conformation is non-conducting and exists when the membrane is at, near, or below its rest potential, with the internal compartment about $-70mV$ relative to the external compartment. For a membrane 5 nm thick, this voltage corresponds to an electric field of $10^5V/cm$ pointing inward. In the open position, the channel becomes a highly selective ion conductor, the opening process resulting from depolarization from the rest field to a value above a certain threshold [188]. Ion selectivity, conduction kinetics and reaction to pharmacological substances are distinguishing characteristics of these channels.

The study of ion channels has a long history. The early pioneering work of Hodgkin and Huxley [142] provided much of the insight, which is largely still applicable. They showed that (1) nerve impulse propagation could be understood if one assumed that the axon membrane contained separate ion channels for the conduction of Na^+ and K^+ ions, and (2) the channels have gates which respond to the depolarization of the membrane potential, opening on depoling. The movement of charged ions through the channel should create a short transient current, and such gating currents have been observed [201].

Various physical models for ion channel voltage gating have been proposed, but none can be considered to be definitively established [280,201]. Brief reviews of these models have been reported by Leuchtag and Bystrov [280]. One such model is a ferroelectric model for membrane excitation proposed by Leuchtag [280, 189] and specifically applied to the Na^+ ion channel. The model stipulates that the channel undergoes a phase transition from a ferroelectric state (closed) to a paraelectric (open) state with the open state acting as a superionic conductor. Leuchtag and Bystrov [280] further proposed that the ion channels are liquid crystalline components of the excitable membrane, and the phase transition which opens the channel is from a ferroelectric tilted chiral smectic (SmC^*) phase to a paraelectric non-tilted (SmA^*) phase.

To support their model, Leuchtag and Bystrov [280] have summarized experimental evidence of ferroelectricity in ion channels. The evidence is as follows:

- Excitable membranes have a heat-block temperature above which excitability fails. This feature is taken to be analogous to the Curie temperature in ferroelectrics.

- Hysteresis in the current-voltage relation is observed at both the membrane and channel levels.

- Thermal hysteresis in the opening/closing of the channel is taken as evidence for a first-order phase transition.

- Temperature-dependent currents suggest the existence of pyroelectricity.

- Na^+ channels have a surface charge density of $2.2\mu C/cm^2$, which is the range of the spontaneous polarization of known ferroelectrics.

- Excitable membranes swell in response to a voltage change, suggestive of piezoelectricity.

- Ion channels are known to be sensitive to pressure and membrane stretch consistent with the known stress dependence of the Curie temperature of ferroelectrics.

- Voltage dependent birefringence, a well-known effect in ferroelectrics, has been seen in excitable membranes.

Tokimoto et al. [313] presented a phenomological treatment of Leuchtag's ferroelectric hypothesis for the channel gating mechanism. The Gibbs free energy of the system, G, is written as a function of temperature, T, the electrical charge, q, and the electric displacement, D, which causes the rearrangement of charged molecules. It is found that the ground state of the membrane is a metastable (or rest) state analogous to a ferroelectric. An electrical or thermal stimulus drives the membrane to an excited paraelectric-like state in which matter and energy are exchanged through the membrane. This state is, however, unstable and the membrane returns to the rest state presumably by both Na^+ ion inactivation and the opening of the K^+ channels. Tokimoto et al. [313] provided a schematic representation of the structure and operation of the Na^+ and K^+ ion gates.

Although interesting and qualitatively capable of explaining many observed properties of voltage-dependent ion channels, Leuchtag's ferroelectric model is only one of a number of competing models, none of which has become generally accepted. One criticism of this model is that it does not take into account the possible involvement of membrane water in the gating process. Lu et al. [201] have developed a model in which the "gate" water is held in the channel by hydrogen bonding and by the fields induced by charges on the amino acids. In another development, Gordon et al. [124] proposed a modification of Leuchtag's hypothesis. The propagation of the action potential represents the polarization kink-soliton describing the motion of the interphase boundary between closed and open gate states associated with different values of the polarization. This is illustrated schematically in Fig. 3.72.

Undoubtedly the ferroelectric model will undergo considerable further evolution in the quest for a fuller understanding of voltage-dependent ion gates. The future of this work should be exciting.

Although the prevalence of piezoelectricity and pyroelectricity in biological materials naturally leads to the belief that ferroelectric mechanisms may be involved in a variety of biological functions, the occurrence of ferroelectricity in these materials is still a tenuous issue - one that is exciting, challenging and has the potential for strong impact in biology. At this time, theory appears to be well ahead of experiments in the quest for a definitive proof of bioferroelectricity. The current state of experimental work reflects the complexity

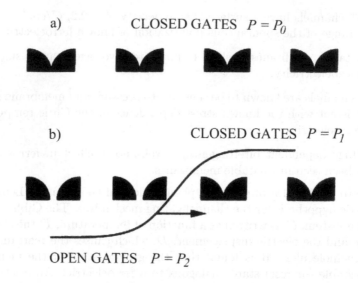

a)   CLOSED GATES $P = P_0$

b)   CLOSED GATES $P = P_1$

OPEN GATES $P = P_2$

FIGURE 3.72:   Schematic representation of the gating mechanism associated with the kink-soliton propagation of the polarization wave P. (a) closed gates; (b) the kink-soliton as a moving interphase boundary separating the closed gate and open gate states. (After Gordon et al. [124].)

of the inter- and intra-cellular materials and environments involved, the small length scale and the limitations of available tools. New advances in imaging, non-linear spectroscopies and other experimental diagnostics should improve the picture considerably.

Our emphasis in this section has been on (i) microtubules (MT's) and (ii) voltage-dependent ion channels (VDIC's). In both cases the most significant challenge is to definitively prove the role of ferroelectricity in their functions. Other challenges/opportunities that will undoubtedly contribute to resolving the main challenge include:

- Quantitative mapping of the electric field around MT's.

- Electric fields across cellular membranes are as large as 10^5 V/cm. How do these fields affect the orientation of MT's? Can these fields induce orientational order and array formation?

- Can the α - β building block (dimer) be utilized to demonstrate an FE tubulin-based system?

- Theory suggests that energy transfer along MT's occurs via "kink-like excitations," or solitons, and similarly solitons can open and close VDIC's. Such loss-free transport is exciting and has been hinted at in very recent experiments on the use of MT's to transport cargo in nano-scale laboratory experiments [64]. Definitive identification of the mechanisms involved would be a very significant advance.

- The role of water (hydration) in both MT and VDIC functions is critical and not yet completely understood. This role may be key to bioferro-electricity.

- A model [189] for VDIC's speculates that the inter-cellular Na^+ ion channel undergoes a phase transition from a ferroelectric (closed) to a paraelectric (open) state during its function, and, additionally, the "open" state is a superionic conductor. Experimental verification of these two mechanisms would be crucial accomplishments.

- A clear demonstration of FE (switchable) domains in small-scale biological materials would be a breakthrough.

Chapter 3 Questions and Problems

QUESTION 3.1 Can you describe a minimalistic membrane model? What are the main limitations of your model?

QUESTION 3.2 Membranes are a key system, but are nearly invisible - how could one experimentally investigate membranes?

PROBLEM 3.3 *Listeria monocytogenes* is a rod bacterium that has been implicated in several food poisoning epidemics. It moves by polymerizing actin behind itself, and its movement is controlled by the energy difference (the difference in critical concentration) between the two ends of the filaments. Suppose that listeria converts the change in Gibbs free energy of polymerization of actin into mechanical work with 100% efficiency. If the k_1 for the addition of one actin monomer is 12 /mM ·s and k_{-1} is 100 /s, how much work can be done by the polymerization of one actin molecule?

PROBLEM 3.4 Version 1: A water permeable membrane connects the bottoms of two beakers 5cm in radius. If beaker 1 contains 2L of water with 50mM glucose and beaker 2 contains 1L of 25mM KCl what is the equilibrium volumes of each beaker? Version 2: A water permeable membrane connects the bottoms of two beakers 5cm in radius. If beaker 1 contains 2L of water with 50mM glucose and beaker 2 contains 2.5×10^{-2}mol of KCl how much water must be in beaker 2 to maintain equilibrium?

PROBLEM 3.5 Calculate the net osmotic pressure for the Giant Sequoia and Red Mangrove.

TABLE 3.14: Concentration Data

Ion	Concentration (mmol/L)			
	Giant Sequoia		Red Mangrove	
	soil	sap	soil	sap
K^+	1	20	10	20
Na^+	2	0	450	0
Cl^-	30	2	530	2
NO_3^-	3	20	0	20
Ca^{2+}	7	1	10	1

(Based off *http://umech.mit.edu/6.021J/hw/hw3.pdf.*)

PROBLEM 3.6 Derive equation 3.4. Use the ideal gas approximation (e.g., $\mu_1 = RT\ln(c_i)$).

PROBLEM 3.7 The mechanical properties of microtubules play a significant role in processes such as cell division, cellular transport and cellular motility (in the case of cilia and flagella). One particular mechanical property that is studied is the elasticity of microtubules. Elastic properties have been studied by observing the bending of microtubules under force from thermal vibrations, optical tweezers, and hard surfaces. At the cellular level the bending of microtubules is well described by the continuum mechanics of elastic rods.

(a) Consider a microtubule as a solid rod anchored at one end with the other end free to move. If the energy to bend the rod is given by,

$$E = \frac{\xi_p L}{2\beta R_c^2}$$

where ξ_p is the persistence length, L is the length of the rod and R_c is the radius of curvature of the rod, find the energy required to move the free microtubule end by a small distance x perpendicular to the microtubule.

(b) Consider the above microtubule to be in thermal equilibrium. Find the bending amplitude from thermal vibrations.

(c) Equation 3.5 gives the persistence length of a thin solid cylinder. Microtubules however have a hollow core. The persistence length of a thin hollow cylinder is given by,

$$\xi_p = \beta Y \frac{\pi(R_o^4 - R_i^4)}{4}$$

where R_o is the outer radius and R_i the inner radius. Compare the energies required to bend a thin solid and a thin hollow cylinder of equal length the same distance. What can you conclude?

PROBLEM 3.8 *The size of cells and their components.* [Lehninger]

(a) If you were to magnify a cell 10000 fold (the typical magnification achieved using a microscope) how big would it appear? (assume you are viewing a typical eukaryotic cell with a diameter of $50\mu m$).

(b) If this cell were a muscle cell, how many molecules of actin could it hold assuming no other cellular components were present? (Actin molecules are spherical with a diameter of $3.6nm$; assume that the muscle cell is spherical.)

(c) If this were a liver cell of the same dimensions, how many mitochondria could it hold assuming there were no other cellular components present? (Assume mitochondria are spherical with a diameter of $1.5\mu m$.)

(d) Assuming glucose is present at a concentration of $1mM$, calculate how many molecules of glucose would be present?

PROBLEM 3.9 *Efficiency of membrane bound transporters*. It has been shown experimentally that E. Coli is a near perfect absorber of some kinds of molecules, to which the bacterial cell wall is impermeable. In this question we examine how this is achieved, without the entire membrane being devoted to the transport of one type of molecule.

Using Fick's laws of diffusion, it can be shown that the absorption current of a perfectly absorbing sphere is

$$I = 4\pi DsC_0$$

where D is the diffusion coefficient, s is the sphere radius and C_0 is the concentration difference. The absorption current of a perfectly absorbing disk, meanwhile, is given by

$$I = 4DsC_0$$

where s now refers to the radius of the disk. We wish to consider the absorption current to N disk like absorbers on the surface of a sphere. This problem is formally equivalent to the electrical problem of current flow through a medium of finite resisivity to N conducting patches on an insulating sphere, given the analogues to Ohm's law $C \sim V$, $I \sim I$.

(a) From the equations above, find expressions for the diffusion 'resistance' in each case, by posing analogous relations to Ohm's law, $I = V/R$.

(b) Find an expression for the total resistance posed by N disk like absorbers. (Hint: consider the analogous electrical problem).

(c) Find an expression for I_{sphere}/I_{Ndisks}.

(d) Considering a spherical cell with a radius of 2μ, disk absorbers with a radius of 10 Angstroms and the N needed for $I_{sphere}/I_{Ndisks} = 2$, find the average distance between absorbing units.

This question should convince you that many hundreds of different transporters or receptor proteins can be accommodated on the cell surface, each absorbing particles of a specific kind with an efficiency approaching that of a cell whose entire surface is devoted to such a task.

PROBLEM 3.10 Assume you know only that the effective spring constant k of DNA depends on the thermal energy $k_B T = 4 \times 10^{-21} J$, the strand length L_o and the persistence length ξ_P. Use dimensional analysis to derive

$$k = \frac{3k_B T}{4\xi_P L_o}$$

You may assume that k is inversely proportional to L_o where ξ_P is the persistence length, approximately equal to the size of the coil and L_o is the strand length.

PROBLEM 3.11 Animal cells have a typical size of about 10 microns. These cell surfaces cannot withstand osmotic pressure differences exceeding about 10^{-4} Pa. What do you conclude from this about the maximum elastic tension of animal cell walls?

PROBLEM 3.12 *ATP Synthase* The enzyme ATP synthase is located in the walls of mitochondria (small energy producing units inside cells). This enzyme transforms a low-energy chemical compound (ADP) into a high-energy compound (ATP) which requires about $7.3 kcal/mol$ of energy per ADP to ATP transformation. There is an H^+ concentration difference across the mitochondrial wall. Assume that the concentration ratio between the two sides of the wall is 1000. During ATP synthase activity, protons move through the enzyme from the high concentration side to the low concentration side. Compute the maximum osmotic work done by the solution during the movement of one proton. What is the minimum number of protons required per ADP-to-ATP transformation?

PROBLEM 3.13 Suppose that the electric potential outside a living cell is higher than that inside the cell by $0.070V$. How much work is done by the electric force when a sodium ion (charge $= +e$) moves from the outside to the inside?

PROBLEM 3.14 The inner and outer surfaces of a cell membrane carry a negative and positive charge, respectively. Because of these charges, a potential difference of about $0.070V$ exists across the membrane. The thickness of the membrane is $8.0 \times 10^{-9}m$. What is the magnitude of the electric field in the membrane?

PROBLEM 3.15 An axon is the relatively long tail-like part of a neuron, a nerve cell. The outer surface of the axon membrane (dielectric constant $= 5$, thickness $= 1 \times 10^{-8}m$) is charged positively, and the inner portion is charged negatively. Thus, the membrane is a kind of capacitor. Assuming that an axon can be treated like a parallel plate capacitor with a plate area of $5 \times 10^{-6}m^2$, what is its capacitance?

PROBLEM 3.16 The membrane that surrounds a certain type of living cell has a surface area of $5.0 \times 10^{-9}m^2$ and a thickness of $1.0 \times 10^{-8}m$.

Assume that the membrane behaves like a parallel plate capacitor and has a dielectric constant of 5.0.

(a) If the potential on the outer surface of the membrane is $+60.0mV$ greater than that on the inside surface, how much charge resides on the outer surface?

(b) If the charge in part (a) is due to K^+ ions (charge $+e$), how many such ions are present on the outer surface?

PROBLEM 3.17 DNA is a double stranded molecule wound in a helix. The two strands are held together by hydrogen bonds. The two strands have the same charge per unit length. In ion free water, DNA has a tendency to separate into two strands. This is not the case for DNA in an electrolyte under physiological conditions. Why?

PROBLEM 3.18 Here is a table of the ion concentrations of the most common ions in the cell interior and in the blood. Compute the conductivity of the cell interior and that of blood. Compare with sea water.

TABLE 3.15: Data for Problem 3.18

Type of Ion	Cell Interior (in milliMolar)	Blood (in milliMolar)
K^+	139	4
Na^+	12	145
Cl^-	4	116

Chapter 3 References

1. Abad, E. and Milke, A. *Ann. Phys.* **7**, 9, 1998.

2. Ajdari, A. *J. Phys. I (Paris)* **4**, 1577, 1994.

3. Altman, D., Sweeney, H. L. and Spudich, J. A. *Cell* **116**, 737-749, 2004.

4. Alberts, B., Bray, D., Lewis, J., Raff, M., Roberts, K., and Watson, J.D. *Molecular Biology of the Cell.* Garland Publishing, London, 1994.

5. Albrecht-Buehler, G.http://www.basic.nwu.edu/g-buehler/cellint.htm.

6. Albrecht-Buehler, G., *Cell Motility and the Cytoskeleton* **27**, 262, (1994).

7. Albrecht-Buehler, G., *Cell Motility and the Cytoskeleton* **32**, 299, (1995).

8. Albrecht-Buehler, G., *Experimental Cell Research* **236**, 43, (1997).

9. Alhenstaedt, H. *Naturwissenschaften* **48**, 465-472, 1961.

10. Ali, M.Y., Uemura, S., Adachi, K., Itoh, H., Kinosita, K. Jr. and Ishiwata, S. *Nat. Struct. Biol.* **9**, 464-467, 2002.

11. Ali, M.Y., Homma, K., Iwane, A.H., Adachi, K., Itoh, H., Kinosita, K., Yanagida, T., and Ikebez, M. *Biophys. J.* **86**, 3804-3810, 2004.

12. Alt, W. and Dembo, M. *Mathematical Biosciences* **156**, 207-228, 1999.

13. Amos, L.A. *Trends Cell Biol.* **5**, 48-51, 1995.

14. Amos, L.A. and Amos, W.B. *Molecules of the Cytoskeleton.* Macmillan Press, London, 1991.

15. Amos, L.A. and Cross, R.A. *Curr. Opin. in Struct. Biol.* **7**(2), 1997.

16. Amos, L.A. and Hirose, K. *Curr. Opin. in Cell Biol.* **9**, 4-11, 1997.

17. Applequist, D.E., Depuy, C.H., and Rinehart, K.L. *Introduction to Organic Chemistry*, John Wiley and Sons, New York, 1982.

18. Apweiler et al., *Nucleic Acids Res* **32** Database issue, D115-9, 2004.

19. Arizmendi, C. M. and Family, F. *Physica A: Statistical Mechanics and Its Applications* **269**, 285, 1999.

20. Arnal, I. and Wade, R. H. *Structure* **6**, 33-38, 1998.

21. Astumian, R.D., Bier, M. *Physical Review Letters* **72**, 1766, 1994.

22. Astumian, R. D. *Science* **276**, 917, 1997.

23. Astumian, R. D. and Derenyi, I. *Biophys. J.* **77**, 993, 1999.

24. Athenstead, H. *Ann NY Acad Sci* **238**, 68-94, 1974.

25. Bairoch, A. and Apweiler, R. *Nucleic Acids Res.* **26**, 38-42, 1998.

26. Baker, N.A. et al. *Proc. Natl. Acad. Sci.* **98** 10037-10041, 2001.

27. Baker, D. and Sali, A. *Science* **294(5540)**, 93-96, 2001.

28. Bakhshi, A. K. *Prog. Biophys. Molec Biol.* **61**, 187, 1994.

29. Balci, H., Ha, T., Sweeney, H.L. and Selvin, P.R. *Biophys. J.* **89**, 413-417, 2005.

30. Banerjee, A. and Luduena, R.F. *J. Biol. Chem.* **267**, 13335-9, 1992.

31. Bayley, P.M., Schilstra, M.J., and Martin, S.R. *J. Cell Sci.* **95**, 33, 1990.

32. Benedek, G. B., and Villars, F. M. H. *Physics With Illustrative Examples from Medicine and Biology*, Springer, Berlin, 2000.

33. Bergethon, P.R. and Simons, E.R. *Biophysical Chemistry. Molecules to Membranes*, Springer-Verlag, Berlin, 1990.

34. Bier, M. and Astumian, R. D. *Bioelectrochem. Bioenerg.* **39**, 67, 1996.

35. Blinc, R. and Zeks, B. *Soft Modes in Ferroelectics and Antiferroelectrics*, North Holland-American Elsevier, New York, 1974.

36. Block, S. M. *Trends Cell Biol.* **5**, 170, 1995.

37. Block, S. M., Goldstein, L. S., and Schnapp, B. J. *Nature* **348**, 348, 1990.

38. Bloom, G. S., Wagner, M. C., Pfister, K. K., and Brady, S. T. *Biochem* **27**, 3409, 1988.

39. Boal, D. *Mechanics of the Cell*, Cambridge University Press, 2002.

40. Boehm, K. J., Steinmetzer, P., Daniel, A., Vater, W., Baum, M., and Unger, E. *Cell Motil Cytoskeleton* **37**, 226, 1997.

41. Bolognani, L., Causa, F., Costato, M. and Milani, M. *Il Nuovo Cimento* **17**, 235, 1995.

42. Bolterauer, H. and Tuszyński, J. A. *International Journal of Nonlinear Sciences and Numerical Simulation (IJNSNS)* **3**, 185-190, 2002.

43. Bolterauer, H., Tuszyński, J. A., Unger, E. *Cell Biochemistry and Biophysics (CBB)* **42**, 95-119, 2005.

44. Bras, W. *PhD Thesis*, Liverpool John Moores University, October 1995.

45. Bras, W. *PhD Thesis*, University of Amsterdam.

46. Brown, J.A. and Tuszyński, J.A. *Physical Review E* **56**, 5834-5840, 1997.

47. Brown, J.A. and Tuszyński, J.A. *Ferroelectrics* **220**, 141, 1999.

48. Bruinsma, R., *Physics, 6A and 6B*, International Thomson Publishing, 1998.

49. Brunori, M. and Wilson, M. T. *Biochimie* **77**, 668, 1995.

50. Burgess, S., Walker, M., Wang, F., Sellers, J.R., White, H.D., Knight, P.J. and Trinick, J. *J. Cell Biol.* **159**, 983-991, 2002.

51. Burr, H.S. and Northrop, F.S.C. *Proc. Nat. Acad. Sci. U.S.A.* **25**, 284, 1939.

52. Buss, F., Spudich, G. and Kendrich-Jones, J. *Ann. Rev. Cell Dev. Biol.* **20**, 649-676, 2004.

53. Buttiker, M. *Z. Phys. B* **68**, 161, 1987.

54. Buzan, J. M. and Frieden, C. *Proc. Natl. Acad. Sci. U.S.A.* **93**, 91-95, 1996.

55. Canters, G. W. and Dennison, C. *Biochimie* **77**, 506, 1995.

56. Carlier, M. F., Pantaloni, D., and Korn, E. D. *J. Biol. Chem.* **259**, 9987-9991, 1984.

57. Carlier, M.F., Melki, R., Pantaloni, D., Hill, T.L., and Chen, Y. *Proc. Natl. Acad. Sci. U.S.A.* **84**, 5257-5261, 1987.

58. Carpenter, E. J., Huzil, J. T., Lirdueña, R. F., and Tuszyński, J. A. *European Journal of Biophysics* vol. **36(1)**, 2006.

59. Case, R.B., Pierce, D.W., Hom-Booher, N., Hart, C.L., and Vale, R.D. *Cell* **90**, 959-966, 1997.

60. Casella, J. F., Barron-Casella, E. A., and Torres, M.A. *Cell Motil. Cytoskel.* **30**, 164-170, 1995.

61. Cassimeris, L. *Cell. Motil. Cyto.* **26**, 275-281, 1993.

62. Cassimeris, L., Gard, D., Tran, P. T., and Erickson, H. P. *J. Cell Sci.* **114**, 3025, 2001.

63. Chen, C.S., Mrksich, M., Huang, S., Whitesides, G.M., and Ingber, D.E. *Science* **276**, 1425-1428, 1997.

64. Chou, K.-C., Zhang, C.-T., and Maggiore, G.M. *Biopolymers* **34**, 143, 1994.

65. Chrétien, D., Flyvbjerg, H., and Fuller, S. D. *Eur. Biophys. J.* **27**, 490, 1998.

66. Chrétien, D. and Wade, R.H. *Bio. Cell* **71**, 161-174, 1991.

67. Chrétien, D., Metoz, F., Verde, F., Karsenti, E., and Wade, R.H. *J. Cell Biol.* **117**, 1031-1040, 1992.

68. Civelecoglu, G. and Edelstein-Keshet, L. *Bull, Math. Biol.* **56**, 587 - 616, 1998.

69. Cole, D. G. and Scholey, J. M. *Trends Cell Biol.* **5**, 259, 1995.

70. Cole, K.S. *Trans. Faraday Soc.* **33**, 966, 1937.

71. Colli, L., Facchini, U., Guidotti, G., Lonati, R.D., Orsenigo, M., and Sommariva, O. *Experientia* **11**, 479-481, 1955.

72. Connolly, M.L. *Science* **221**, 709-13, 1983.

73. Consta, S. and Kapral, R. *J. Chem. Phys.* **101**, 10908, 1994.

74. Cooper, J. A., Buhle, E. L. Jr., Walker, S. B., Tsong, T. Y. and Pollard, T.D. *Biochemistry* **22**, 2193-2202, 1983.

75. Coppin, C. M., Pierce, D. W., Hsu, L., and Vale, R. D. *Proc. Natl Acad. Sci U.S.A.* **94**, 8539, 1997.

76. Coy, D. L., Wagenbach, M., and Howard, J. *J. Biol. Chem.* **274**, 3667, 1999.

77. Cramer, L.P. *Biochem. Soc. Symp.* **65**, 173-205, 1999.

78. Crevel, I. M., Lockhart, A., and Cross, R. A. *J. Mol. Biol.* **273**, 160, 1997.

79. de Gennes, P.-G. *Introduction to Polymer Dynamics.* Cambridge University Press, Cambridge, 1990.

80. De La Cruz, E. M., Ostap, E. M. and Sweeney, H. L. *J. Biol. Chem.* **276**, 32373-32381, 2001.

81. de Pablo, P. J., Schaap, I. A. T., MacKintosh, F. C., and Schmidt, C. F. *Physical Review Letters* **91**, 098101, p4, 2003.

82. Derenyi, I. and Vicsek, T. *Proc. Natl. Acad. Sci. U.S.A.* **93**, 6775, 1996.

83. De Robertis, E.D.P. and De Robertis, E.M.F. *Cell and Molecular Biology*, Saunders College, Philadelphia, 1980.

84. Doering, C.R., Horsthemke, W., and Riordan, J. *Physical Review Letters* **72**, 2984, 1994.

85. Diaz, J.F., Pantos, E., Bordas, J. and Andreu, M.J. *J. Mol. Biol.* **238**, 214-225, 1994.

86. Doi, M. *J. Physiol. (Paris)* **36**, 607-617, 1975.

87. Dombeck, D.A., Kasischke, K.A., Vishwasrao, H.D., Ingelsson, M., Hyman, B.T., and Webb, W.W. *Proc. Natl. Acad. Sci. U.S.A.* **100**(12), 7081-7086, 2003.

88. Dreyer, J. L. *Experientia* **40**, 653, 1984.

89. Ducharme, S., Palto, S.P. and Fridkin, V.M. in *Ferroelectric and Dielectric Thin Films*, edited by H.S. Nalwa, Academic Press, San Diego, 2002.

90. Duke, T. and Leibler, S. *Biophys. J.* **71**, 1235, 1996.

91. Dustin, P. *Microtubules.* Springer-Verlag, Berlin, 1984.

92. Edelstein-Keshet, L. *Eur. Biophys. J.* **27**, 521-531, 1998.

93. Einarsdóttir, O., Georgiadis, K. E., and Sucheta, A. *Biochemistry* **34**, 496, 1995.

94. Elbaum, M., Fygenson, D., and Libchaber, A. *Phys. Rev. Lett.* **76**, 4078, 1996.

95. Eley, D. D. *Mol. Cryst. Liq. Cryst.* **171**, 1, 1989.

96. Endow, S. A. *Guidebook to the Cytoskeletal and Motor Proteins*, chapter C-terminal Motor Kinesin Proteins. Oxford University Press, Oxford, 1999.

97. Endow, S. A. *Molecular Motors*, chapter Molecular Motor Directionality. Wiley-VCI, Weinheim, 2003.

98. Erickson, H. P. *J. Molec. Biol.* **206**, 465-474, 1989.

99. Felgner, H., Frank, R., and Schliwa, M. *J. Cell Sci.* **109**, 509, 1996.

100. Feynman, R.P., Leighton, R.B., and Sands, M. *The Feynman Lectures on Physics*, Addison-Wesley, Reading MA, 1969.

101. Fisher, M. E. and Kolomeisky, A. B. *Proc. Natl. Acad. Sci. U.S.A.* **96**, 6597-6602, 1999.

102. Fletterick, R.J. *Nature* **395**, 813-816, 1998.

103. Flory, P.J. *Statistical Mechanics of Chain Molecules*. Wiley, New York, 1969.

104. Flyvbjerg, H., Holy, T.E., and Leibler, S. *Phys. Rev. Lett.* **73**, 2372, 1994.

105. Flyvbjerg, H., Holy, T.E., and Leibler, S. *Phys. Rev. E* **54**, 5538-5560, 1996.

106. Fong, P. *Bull. Americ. Phys. Soc.* **13**, 613, 1968.

107. Forkey, J.N., Quinlan, M.E., Shaw, M.A., Corrie, J.E.T. and Goldman, Y.E. *Nature* **422**, 309-404, 2003.

108. Frey, E., Kroy, K., and Wilhelm, J. *Adv. Struct. Biol.* **5**, 135-168, 1998.

109. Frieden, C. *Proc. Natl. Acad. Sci. U.S.A.* **80**, 6513-6517, 1983.

110. Frieden, C., and Goddette, D. *Biochemistry* **22**, 5836-5843, 1983.

111. Friesner, R. A. Current Biology Ltd, Minireview, *Structure* **2**, 339, 1994.

112. Froehlich, H. *Adv. in Electronics and Electron Physics* **53**, 85-152, 1980.

113. Fukuda, E., Ueda, H., and Rinaldi, R. *Biophys. J.* **16**, 911-918, 1976.

114. Furukawa, T. *Phase Transitions* **18**, 143 1989.

115. Fygenson, D. K., Elbaum, M., Shraiman, B., and Libchaber, A. *Phys. Rev. E* **55**, 850, 1997.

116. Fygenson, D.K., Braun, E., and Libchaber, A. *Phys. Rev. D* **50**, 1579-1588, 1994.

117. Gao, Q., Sheinbeim, J.I., and Newman, B.A. *J. Polym. Sci. B: Polym. Phys.* **37**, 3217, 1999.

118. Geeves, M. A. and Holmes, K. C. *Annu. Rev. Biochem.* **68**, 687-728, 1999.

119. Gennis, R. B., *Biomembranes: Molecular Structure and Function*, Springer, 1989.

120. Gibbons, F., Chauwin, J. F., Desposito, M., and Jose, J. V. *Biophys. J.* **80**, 2515-2526, 2001.

121. Gittes, F., Mickey, B., Nettleton, J., and Howard, J. *J. Cell Biol.* **120**, 923, 1993.

122. Gilmanshin, R. I. and Lazarev, P. I. *Material Science* **13**, 71, 1987.

123. Goodby, J.W. et al., editors. *Ferroelectric Liquid Crystals*, Gordon and Breach Science Publishers, Philadelphia, PA, 1991.

124. Gordon, A., Vugmeister, B.E. , Rabits, H. , Dorfman, S., Felsteiner, J., and Wyder, P. *Ferroelectrics* **220**, 291, 1999.

125. Gradsteyn, I. S., and Ryzhik, I. M. *Tables of Integrals, Series and Products*, Academic Press, New York, 1965.

126. Gu, Q. and Popp, F.-A. *Experientia* **48**, 1069, 1992.

127. Gurwitsch, A.A. *Experientia* **44**, 545, 1988.

128. Hackney, D. D. *Nature* **376**, 215, 1995.

129. Hackney, D. D. *Annu. Rev. Physiol.* **58**, 731, 1996.

130. Han, Y., Sablin, E. P., Nogales, E., and Downing, R.J.F.K.H. *J. Struct. Biol.* **128**, 26-33, 1999.

131. Hancock, W. O. and Howard, J. *J. Cell Biol.* **140**, 1395, 1998.

132. Hancock, W. O. and Howard, J. *Molecular Motors*, chapter Kinesin: Processivity and Chemomechanical Coupling., 243-269. Wiley-VCI, Weinheim, 2003.

133. Hardman and Limbird. *Goodman and Gilman, The Pharmacological Basis of Therapeutics*, 9th ed., 1228, 1257-1261, 1603, 1996.

134. Hari et al. *Cell Motil Cytoskeleton* **56**, 45-56, 2003.

135. Hays, T.S. and Salmon, E.D. *J. Cell Biol.* **110**, 391-404, 1990.

136. Hess, B. and Mikhailov, A. *Science* **264**, 223, 1994.

137. Higgins and Sharp, *Gene* **73**, 237-44, 1988.

138. Hinner, B., Tempel, M., Sackmann, E., Kroy, K. and Frey, E. *Phys. Rev. Lett.* **81**, 2614–2618, 1998.

139. Hirokawa, N., Shiomura, Y., and Okabe, S. *The Journal of Cell Biology* **107**, 1449-1459, 1998.

140. Hirokawa, N., Pfister, K. K., Yorifuji, H., Wagner, M. C., Brady, S. T., and Bloom, G. S. *Cell* **56**, 867, 1989.

141. Hirose, K., Henningsen, U., Schliwa, M., Toyoshima, C., Shimizu, T., Alonso, M., Cross, R. A., and Amos, L. A. *EMBO* **19**, 5308-14, 2000.

142. Hodgkin, A.L. and Huxley, A.F. *J. Physical* (London 116, 449, 473 and 497) (1952; 117, 500), 1952.

143. Hol, W.G. *Prog. Biophys. Mol. Biol.* **45**, 149 , 1985.

144. Horio, T. and Hotani, H. *Nature London* **321**, 605-607, 1986.

145. Howard, J. *Mechanics of Motor Proteins and the Cycloskeleton.* Sinauer Associated Inc., 2001.

146. Howard, J., Hudspeth, A. J., and Vale, R. D. *Nature* **342**, 154, 1989.

147. Houchmandzadeh, B. and Vallade, M. *Phys. Rev. E* **6320**, 53, 1996.

148. Hua, W., Chung, J., and Gelles, J. *Science* **295**, 844, 2002.

149. Humphrey, W., Dalke, A., and Schulten, K. *J. Molecular Graphics* **14**, 33 , 1996.

150. Huxley, H. E. *Science* **164**, 1356-1366, 1969.

151. Huxley, A. F. *Biophys. Chem.,* **7**, 257, 1957.

152. Hyman, A.A., Salser, S., Dreschel, D.N., Unwin, N., and Mitchison, T.J. *Molec. Biol. Cell* **3**, 1155-1167, 1992.

153. Hyman, A.A., Chrétien, D., Arnal, I., and Wade, R.H. *J. Cell. Biol.* **128**, 117-125, 1995.

154. Ichinose, S.-I. and Minato, T. *Phys. Cond. Matt.* **5**, 9145, 1993.

155. Ingber, D.E. *J. Cell Sci.* **104**, 613, 1993.

156. Ingber, D.E. *Ann. Rev. Physiology* **59**, 575-599, 1997.

157. Ionov, S.P. and Ionova, G.V. *Dok 1. Biophys.* **202**, 22, 1972.

158. Inclán and Nogales, *J. Cell Sci.* **114**, 413-22, 2001.

159. Inoue, Y., Iwane, A. H., Miyai, T., Muto, E., and Yanagida, T. *Biophys. J.* **81**, 2838, 2001.

160. Iwata, S., Lee, J.W., Okada, K., Lee, J.K., Iwata, M., Rasmussen, B., Link, T.A., Ramaswamy, S. and Jap, B.K. *Science* **281**, 64-71, 1998.

161. Janmey, P. A., Euteneuer, U., Traub, P., and Schliwa, M. *J. Cell Biol.* **113**, 155 , 1991.

162. Janmey, P. A., Hvidt, S., Käs, J., Lerche, D., Maggs, A., Sackmann, E., Schliwa, M. , and Stossel, T. P. *J. Biol. Chem.* **269**, 32503-32513, 1994.

163. Janosi, I. M., Chretien, D., and Flyvbjerg, H. *Eur. Biophys. J.* **27**, 501, 1998.

164. Jelinek, F., Pokorny, J., Saroch, J., Trkal, V., Hasek, J., and Palan, B. *Bioelectrochemistry and Bioenergetics* **48**, 261-266, 1999.

165. Jülicher, F., Adjari, A., and Prost, J. *Rev. Mod. Phys.* **69**, 1269-1281, 1997.

166. Käs, J., Strey, H., Tang, J. X., Finger, D., Ezzell, R., Sackmann, E. and Janmey, P. A. *Biophys. J.* **70**, 609-625, 1996.

167. Keller, D. and Bustamante, C. *Biophys.J.,* **78**, 541, 2000.

168. Kepler, R.G. and Anderson, R.A. *Adv. in Phys.* **41**, 1992.

169. Khan and Luduena. *Invest New Drugs* **21**, 3-13, 2003.

170. Kikkawa, M., Sablin, E. P., Okada, Y., Yajoma, H., Fletterick, R. J., and Hirokawa, N. *Nature* **411**, 439-445, 2001.

171. Kimura, K. and Inokuchi, H., *J.Phys.Soc. Japan*, 2218, 1982.

172. King, R.W.P. and Wu, T.T. *Phys. Rev. E* **58**, 2363-2369, 1998.

173. Kinosian, H. J., Selden, L.A., Estes, J. E. and Gershman, L. C. *Biochem.* **32**, 12353-12357, 1993.

174. Kis, A., Kasas, S., Babic, B., Kulik, A. J., Benoit, W., Briggs, G. A. D., Schonenberger, C., Catsicas, S., and Forro, L. *Phys. Rev. Lett.* **89**, 248101, 2002.

175. Klotz, I.M. *Protein Sci.* **2**, 1992-1999, 1993.

176. Kobayasi, S., Asai, H., and Oosawa, F. *Biochim. Biophys. Acta* **88**, 528-540, 1964.

177. Koradi, R., Billeter, M., and Wuthrich, K. *J. Mol. Graphics* **14**(1), 51-55 and 29-32, 1996.

178. Kozielski, F., Sack, S., Marx, A., Thormahlen, M., Schonbrunn, E., Biou, V., Thompson, A., Mandelkow, E.-M., and Mandelkow, E. *Cell* **91**, 985-994, 1997.

179. Kraulis, J. per., *Journal of Applied Crystallography* **24**, 946-950, 1991.

180. Kubisz, L., Jozwiak, G., Jaroszyk, F., Tuliszka, M. and Kudynski, R. *Acta Physiol. Pol.* **35**(5-6), 571-6, 1984.

181. Kull, F.J., Sablin, E.P., Lau, R., Fletterick, R.J. and Vale, R.D. *Nature* **380**, 550-555, 1996.

182. Kurashi, M., Hoshi, M., and Tashiro, H. *Cell Motil. Cytoskeleton* **30**, 221, 1995.

183. Kuznetsov, S. A. and Gelfand, V. I. *Proc. Natl. Acad. Sci. U.S.A.* **83**, 8530, 1986.

184. Landauer, R. *J. Stat. Phys.* **53**, 233, 1988.

185. Lang, S.B. *Modern Bioelectricity*, A.A. Marino (ed.), Marcel Dekker, Inc., New York, 243, 1998.

186. Ledbetter, M.C. and Porter, K.R. *J. Cell Biol.* **19**, 239-250, 1963.

187. Leibler, S. and Huse, D.A. *J. Cell Biol.* **121**, 1357-1368, 1993.

188. Leuchtag, H.R. and Bystrov, V.S. *Ferroelectrics* **220**, 157, 1999.

189. Leuchtag, R. *Ferroelectrics* **236**, 23, 2000.

190. Li, H., DeRosier, D.J., Nicholson, W.V., Nogales, E., and Downing, K.H. *Structure* **10**, 1317-1328, 2002.

191. Lindahl, E., Hess, B., and van der Spoel, D. *J. Mol. Model.* **7**, 306-317, 2001.

192. Lipowsky, R. and Harms, T. *Eur. Biophys.J.* **29**, 542, 2000.

193. Lister, I., Schmitz, S., Walker, M., Trinick, J., Buss, F., Veigel, C., and Kendrick-Jones, J. *EMBO J.* **23**, 1729-1738, 2004.

194. Liu et al. *J. Surg. Res.* **99**, 179-86, 2001.

195. Lodish, H., Berk, A., Matsudaira, P., Kaiser, C.A., Krieger, M., Scott, M.P., Zipursky, L., and Darnell, J. *Molecular Cell Biology*, 5th ed., W. H. Freeman and Company, 2004.

196. Lovinger, A.J. *Science* **220**, 1115, 1983.

197. Lowe, J. and Amos, L. A. *Nature* **391**, 203-206, 1998.

198. Löwe, J., Li, H., Downing, K.H., and Nogales, E. *J. Mol. Biol.* **313(5)**, 1045-1057, 2001.

199. Lu, Q., and Luduena, R.F. *J. Biol. Chem.* **269**, 2041-7, 1994.

200. Lu, Q., Moore, G.D., Walss, C., and Luduena, R.F. *Adv. Struct. Biol.* **5**, 203-227, 1998.

201. Lu, J., Yin, J., and Green, M.E. *Ferroelectrics* **220**, 249, 1999.

202. Luby-Phelps, K. *Curr. Opin. Cell Biol.* **6**, 3-9, 1994.

203. Luchko, T., personal communication, 2004.

204. Luduena, R.F. *Int. Rev. Cytol.* **178**, 207-75, 1998.

205. Ma, L., Xu, J., Coulombe, P. A., and Wirtz, D. *J. Biol. Chem.* **274**, 19145, 1999.

206. Mader, Sylvia. *Biology*, 6th edition. William C. Brown, Dubuque, IA, 1996.

207. Magnasco, M. O. *Phys. Rev. Lett.* **71**, 1477, 1993.

208. Magnasco, M. O. *Phys. Rev. Lett.* **72**, 2656, 1994.

209. Mandelkow, E.M., Mandelkow, E., and Milligan, R. *J. Cell Biol.* **114**, 977-991, 1991.

210. Mandelkow, E.-M. and Mandelkow, E. *Cell Motil. and Cytoskel.* **22**, 235-244, 1992.

211. Mandelkow, E. and Johnson, K. A. *Trends Biochem. Sci.* **23**, 429, 1998.

212. Mandelkow, E. and Hoenger, A. *Curr. Opinion in Cell Biol.* **11**, 34, 1999.

213. Maniotis, A., and Ingber, D.E. *Science*, 1997.

214. Maniotis, A.J., Chen, C.S. and Ingber, D.E. *Proc. Natl. Acad. Sci. U.S.A.*, **94**, 849-854, 1997.

215. Martí-Renom, M.A., Stuart, A.C., Fiser, A., Sánchez, R., Melo, F., and Šali, A. *Annu. Rev. Biophys. Biomol. Struct.* **29**, 291-325, 2000.

216. Marx, A., and Mandelkow, E. *Eur. Biophys. J.* **22**, 405, 1994.

217. Mathur, S.C., Scheinbeim, J.I. and Newman, B.A. *J. Appl. Phys.* **56**, 2419 , 1984.

218. Mehta, A.D., Rock, R.S., Rief, M., Spudich, J.A., Mooseker, M.S. and Cheney, R.E. *Nature* **400**, 590-593, 1999.

219. Meréchal, Y. *Proton Transfer in Hydrogen-Bonded Systems*, Edited by T. Bountis, Plenum Press, New York, 1992.

220. Mershin, A., Kolomenskii, A. A., Nanopoulos, D.V. and Schuessler, H. A. *Biosystems* **77**(1-3), 73-85, 2004.

221. Meurer-Grob, P., Kasparian, J. and Wade, R. H. *Biochemistry* **40**, 8000, 2001.

222. Mickey, B. and Howard, J. *J. Cell Biol.* **130**, 909-917, 1995.

223. Mirny, L.A., and Shakhnovich, E.I. *J. Mol. Biol.* **264**, 1164-79, 1996.

224. Mitchell, P. *Chemiosmatic Coupling in Oxidative and Photosynthetic Phosphorylation*, Glynn Research UK 1961.

225. Mitchison, J.M. *Biology of the Cell Cycle*. Cambridge University Press, Cambridge, 1973.

226. Mitchison, T. and Kirschner, M. *Nature London* **312**, 237-242, 1984.

227. Mizushima-Sugano, J., Maeda, T. and Miki-Noumura, T. *Biochimica et Biophysica Acta* **755**, 257, 1983.

228. Mogilner, A., Fisher, A. J., and Baskin, R. J. *J. Theor. Biol.* **211**, 143, 2001.

229. Moore, J.R., Krementsova, E.B., Trybus, K.M. and Warshaw, D.M. *J. Muscle Res. Cell Motil.* **25**, 29-35, 2004.

230. Morowitz, H. J. *Am. J. Physiol.* **235**, R99, 1978.

231. Morris, C.A., Wells, A.L., Yang, Z., Chen, L.-Q., Baldacchino, C.V. and Sweeney, H.L. *J. Biol. Chem.* **278(26)**, 23324-23330, 2003.

232. Moser, C. C., Page, C. C., Farid, R. and Dutton, P. L. *J. Bioenergetics and Biomembranes* **27**, 263, 1995.

233. Mozzetti et al. *Clin. Cancer Res.* **11**, 298-305, 2005.

234. Murphy, D. B., Gray, R. O., Grasser, W. A. and Pollard, T. D. *J. Cell Biol.* **106**, 1947-1954, 1988.

235. Nagy, I. Z. *Experimental Gerontology* **30**, 327, 1995.

236. Nalwa, H.S., editor. *Ferroelectric Polymers*, Marcel Dekker, New York, 1995.

237. Nanopoulos, D.V., Mavromatos, N.E. and Zioutas, K. *Advances in Structural Biology* **5**, 127-137, 1998.

238. Nedelec, F. *J. Cell Biol.* **158**, 1005, 2002.

239. Némethy, G., Scheraga, H.A. 1962. *J. Phys. Chem.* **66**, 1773- 1789.

240. Nguyen, H. and Higuchi, H. *Nature Structural and Molecular Biology* **12**, 127-132, 2005.

241. Nicholls, A., Sharp, K., and Honig, B. *Structure, Function and Genetics* **11**, 281, 1991.

242. Nicklas, R.B. and Ward, S.C. *J. Cell Biol.* **126**, 1241, 1994.

243. Nicklas, R.B., Ward, S.C., and Gorbsky, G.J. *J. Cell Biol.* **130**, 929, 1995.

244. Nishikawa, S. (and 11 others) *Biochem. Biophys. Res. Commun.* **290**, 311-317, 2002.

245. Nogales, E., Wolf, S.G., and Downing, K.H. *Nature London.* **391**, 199-203, 1998.

246. Nogales, E. *Cell. Mol. Life Sci.* **56**, 133, 1999.

247. Oakley et al., *Cell* **61**, 1289-301, 1990.

248. Okada, Y. and Hirokawa, N. *Proc. Natl. Acad. Sci. U.S.A.* **97**, 640-5, 2000.

249. Oosawa, F., and Asakura, S. *Thermodynamics of the Polymerization of Protein.* Academic Press, London; New York, 1975.

250. Owicki, J.C., Springgate, M.W., and McConnell, H.M. *Proc. Nat. Acad. Sci U.S.A.* **75**, 1616, 1978.

251. Panda et al. *Proc. Natl. Acad. Sci. U.S.A.* **91**, 11358-62, 1994.

252. Peskin, C. S. and Oster, G. *Biophys. J.* **68**, 202, 1995.

253. Peyrard, M. ed. *Nonlinear Excitations in Biomolecules.* Springer-Verlag, Berlin, 1995.

254. Pink, D. A. *Theoretical Models of Monolayers, Bilayers and Biological Membranes in Biomembrane Structure and Function* Ed. D. Chapman, McMillan Press, London, 319-354, 1984.

255. Pokorny, J., Jelinek, F., and Trkal, V. *Bioelectrochemistry and Bioenergetics* **45**, 239-245, 1998.

256. Pollard, T.D. *J. Cell Biol.* **103**, 2747-2754, 1986.

257. Press, W., Teukolsky, S., Vetterling, W., Flannery, B. *Numerical Recipes in C: The Art of Scientific Computing.* Cambridge University Press, Cambridge, 1992.

258. Proc. 3rd and 4th International Conferences on Ferroelectric Liquid Crystals, *Ferroelectrics* **121**, (1991), 147 1993.

259. Procopio, J. et al., *Phys. Rev. E* **55**, 6285, 1997.

260. Prost, J., Chauwin, J. F., Peliti, L., and Ajdari, A. *Phys. Rev. Lett.* **72**, 2652-2655, 1994.

261. Purcell, T.J., Morris, C., Spudich, J.A. and Sweeney, H.L. *Proc. Natl. Acad. Sci. U.S.A.* **99**, 14 159-14 164, 2002.

262. Ranganathan et al. *Cancer Res.* **56**, 2584-9, 1996.

263. Ray, S., Meyhoefer, E., Milligan, R. A., and Howard, J. *J. Cell Biol.* **121**, 1083, 1993.

264. Richards et al. *Mol. Biol. Cell.* **11**, 1887-903, 2000.

265. Rickard, J. E. and Sheterline, P. *J. Mol. Biol.* **201**, 675-681, 1988.

266. Rieder, C. L., and Salmon, E. D. *J. Cell Biol.* **24**, 223-233, 1994.

267. Risken, H. *The Fokker-Planck Equation.* Springer-Verlag, Berlin, 1989.

268. Roach et al. *Cell Motil. Cytoskeleton* **39**, 273-85, 1998.

269. Roberts, R., Lister, I., Schmitz, S., Walker, M., Veigel, C., Trinick, J., Buss, F. and Kendrick-Jones, J. *Phil. Trans. R. Soc. B* **359**, 1931-1944, 2004.

270. Rock, R. S., Rice, S. E., Wells, A. L., Purcell, T. J., Spudich, J.A. and Sweeney, H. L. *Proc. Natl Acad. Sci. U.S.A.* **98**, 13655-13659, 2001.

271. Roger. *PhD thesis*, 1996.

272. Rolfe, D.F.S. and Brown, G.C. *Physiol. Rev.* **77**, 731-758, 1997.

273. Rosenfeld, S.S. and Sweeney, H.L. *J. Biol. Chem.* **279**, 40100-40111, 2004.

274. Rosenfeld, S.S., Houdusse, A. and Sweeney, H.L. *J. Biol. Chem.*, **280**(7), 6072-6079, 2005.

275. Ruppel, K.M., Lorenz, M., and Spudich, J.A. *Curr. Opin. Struct. Bio.* **5**, 181-186, 1995.

276. Ruth, B., Popp, F.A. *Zeitschrift fur Naturforschung* **31c**, 741-745, 1976.

277. Saenger, W. *Principles of Nucleic Acid Structure.* Springer-Verlag, New York, 1984.

278. Sakamoto, T., Wang, F., Schmitz, S., Xu, Y., Xu, Q., Molloy, J.E., Veigel, C., and Sellers, J.R. *J. Biol. Chem.* **278**, 29 201-29 207, 2003.

279. Samara, G.A. in *Solid State Physics*, edited by H. Ehrenreich and F. Spaepen, Academic Press, New York, Vol. 56, p. 240, 2001.

280. Sataric, M.V., Tuszyński, J.A. and Zakula, R.B. *Phys. Rev. E* **48**, 589-597, 1993.

281. Sato, M., Schwartz, W. H., Selden, S. C., and Pollard, T. D. *J. Cell Biol.* **106**, 1205, 1988.

282. Schafer, D., Jennings, P., Cooper, J. *J. Cell Biol.* **135**, 169-179, 1996.

283. Schuessler, H.A., Mershin, A., Kolomenskii, A.A., and Nanopoulos, D.V. *J. Modern Optics* **50**, 2381-2391, 2003.

284. Schulz, G.E. and Schirmer, R.H. *Principles of Protein Structure.* Springer-Verlag, Berlin, 1979.

285. Scott et al. *Arch. Otolaryngol. Head Neck Surg.* **116**, 583-9, 1990.

286. Semënov, M.V. *J. theor. Biol.* **179**, 91-117, 1996.

287. Sept, D. Models of Assembly and Disassembly of Individual Microtubules and their Ensembles, *PhD thesis*, University of Alberta, 1997.

288. Sept, D., Limbach, H.-J., Bolterauer, H., and Tuszyński, J.A. *J. theor. Biol.* **197**, 77-88, 1999.

289. Sept, D., Baker, N. A., and McCammon, J. A. in *46th annual meeting of the Biophysical Society*, San Francisco, CA, 2002.

290. Sept, D., Baker, N.A. and McCammon, J.A. *Protein Science* **12**, 2257-2261, 2003.

291. Schliwa, M., editor *Molecular Motors.* Wiley-VCH, Weinheim, 2003.

292. Schnitzer, M. J., Visscher, K., and Block, S. *Nature Cell Biol.* **2**, 718, 2000.

293. Scholey, J. M., Heusner, J., Yang, J. T., and Goldstein, L. S. *Nature* **338**, 355, 1989.

294. Schoutens, J.E. *J. Biol. Phys.* **31**, 35, 2005.

295. Sinden, R. *DNA Structure and Function*, Academic Press, San Diego, 1990.

296. Singer, S.J., and Nicolson, G.L. *Science* **175**, 720, 1972.

297. Sirenko, Y., Stroscio, M., and Kim, K. *Phys. Rev. E* **54**, 1816, 1996.

298. Small, J. V, Stradal, T., Vignal, E., and Rottner, K. *Trends in Cell Biol.* **12**, 112-120, 2002.

299. Snyder, G.E., Sakamoto, T., Hammer, J.A. (III), Sellers, J.R., and Selvin, P.R. *Biophysical Journal* **87**, 1776-1783, 2004.

300. Song, Y. H., Marx, A., Muller, J., Woelke, G., Schliewa, M., Krebs, A., Hoenger, A., and Mandelkow, E. *EMBO J.* **15**, 6213-25, 2001.

301. Spurck, T.P. and Pickett, H.J. *J. Cell Biol.* **105**, 1691-1705, 1987.

302. Stamenovic, D., Mijailovich, S. M., Tolic-Norrelykke, I. M., Chen, J., and Wang, N. *Am. J. Physiol. Cell Physiol.* **282**, 617, 2002.

303. Stracke, R., Boehm, K.J., Wollweber, L., Unger, E. and Tuszyński, J.A. *Biochemistry and Biophysics Research Communications* **293**, 602-609, 2002.

304. Stryer, L. *Biochemistry.* W.H. Freeman and Co. San Francisco, 1981.

305. Sucheta, A., Ackrell, B. A. C., Cochran, B. and Armstrong, F. A. *Nature* **356**, 361, 1992.

306. Svoboda, K., and Block, S.M. *Cell* **77**, 773, 1994.

307. Sweeney, H., and Houdusse, A. *Phil. Trans. R. Soc. B* **359**, 1829-1841, 2004.

308. Tabony, J. and Job, D. *Nature London.* **346**, 448-451, 1990.

309. Tanaka, H., Homma, K., Iwane, A.H., Katayama, E., Ikebe, R., Saito, J., Yanagida, T. and Ikebe, M. *Nature* **415**, 192-195, 2002.

310. Tanford, C., *The Hydrophobic Effect: Formation of Micelles and Biological Membranes*, 2nd ed., Wiley, 1980.

311. Tarusov, B.N., Polivoda, A.I., and Zhuravlev, A.I. *Biophysics* **6**, 83-85, 1961.

312. Tobacman, L. S., and Korn, E. D. *J. Biol. Chem.* **258**, 3207-3214, 1983.

313. Tokimoto, T., Shirane, K., and Kushibe, H. *Ferroelectrics* **220**, 273, 1999.

314. Tolomeo, J. A., and Holley, M. C. *Biophys. J.* **73**, 2241, 1997.

315. Tomishige, M., Klopfenstein, D. R., and Vale, R. *Science* **297**, 2263-7, 2002.

316. Tran, P.T., Walker, R.A., and Salmon, E.D. *J. Cell Biol.* **138**, 105-117, 1997.

317. Tributsch, H. and Pohlmann, L. *J.Theor. Biol.* **165**, 225, 1993.

318. Trpisova, B. and Tuszyński, J.A. *Phys. Rev. E* **55**, 3288-3302, 1997.

319. Tsiavaliaris, G., Fujita-Becker, S. and Manstein, D. J. *Nature* **427**, 558-561, 2004.

320. Tsong, T.Y., Liu, D.-S., Chauvin, F., Gaigalas, A. and Astumian, R.D. *Bioelectrochemistry and Bioenergetics* **21**, 319-331, 1989.

321. Tuszyński, J.A. and Dixon, J.M. *Physical Review E* **64**, 51915-1-51915-7, 2001.

322. Tuszyński, J.A., Brown, J.A., and Sept, D. *Journal of Biological Physics* **29**, 401-428, 2003.

323. Tuszyński, J.A., Carpenter, E.J., Crawford, E., Brown, J.A., Malinski, W., and Dixon, J.M. *Proceedings of ICMENS , International Conference on MEMS, NANO and Smart Systems*, Banff, Ed. W. Badawy and W. Moussa, IEEE Computer Society, Los Alamitos, California, 55-61, 2003.

324. Tuszyński, J.A., Luchko, T., Carpenter, E., and Crawford, E. *Theoretical and Computational Nanoscience* **1**, 392-397, 2004.

325. Tuszyński, J.A., Luchko, T., Portet, S., and Dixon, J.M. *European Journal of Physics E: Soft Condensed Matter* **17** (No.1), 29-35, 2005.

326. Tuszyński, J.A., Brown, J.A., Crawford, E., Carpenter, E.J., Nip, M.L.A., Dixon, J.M., and Sataric, M.V. *Mathematical and Computer Modelling* **41**, 1055-1070, 2005.

327. Tuszyński, J.A., Carpenter, E.J., Huzil, J.T., Malinski, W., Luchko, T. and Ludueńna, R.F. *IJDB* (special issue) **50**, 341-358, 2006.

328. Vale, R. D., Reese, T. S., and Sheetz, M. P. *Cell* **42**, 39, 1985.

329. Vale, R.D., Coppin, C.M., Malik, F., Kull, F.J. and Milligan, R.A. *J. Biol. Chem.* **269**, 23769-23775, 1994.

330. VanBuren, V., Odde, D. J., and Cassimeris, L.*Proc. Natl. Acad. Sci. U.S.A.* **99**, 6035, 2002.

331. van Gunsteren, W. F., Billeter, S. R., Eising, A. A., Hünenberger, P. H., Kruüger, P., Mark, A. E., Scott, W. R. P., and Tironi, I. G. *Biomolecular Simulation: The GROMOS96 Manual and User Guide*, vdf Hochschulverlag, Zürich, Switzerland. 1996.

332. Vassilev, P.M., Dronzine, R.T., Vassileva, M.P., and Georgiev, G.A. *Biosci. Rep.* **2**, 1025-1029, 1982.

333. Vater, W., Stracke, R. Böhm, K.J., Speicher, C., Weber, P., and Unger, E. *The Sixth Foresight Conference on Molecular Nanotechnology.* Santa Clara, CA (U.S.A.) 1998.

334. Veigel, C., Wang, F., Bartoo, M.L., Sellers, J.R., and Molloy, J.E. *Nature Cell Biol.* **4**, 59-65, 2002.

335. Venier, P., Maggs, A.C., Carlier, M.-F., and Pantaloni, D. *J. Biol. Chem.* **269**, 13353-13360, 1994.

336. Verhey, K. J., Lizotte, D. L., Abramson, T., Barenboim, L., Schnapp, B. J., and Rapoport, T. *J. Cell Biol.* **143**, 1053, 1998.

337. Veselovskii, V.A., Sekamova, Y.N., and Tarusov, V.N. *Biophysics* **8**, 147-150, 1963.

338. Visscher, K., Schnitzer, M. J., and Block, S. M. *Nature* **400**, 184, 1999.

339. Vladimirov, Y.A., and L'vova, O.F. *Biophysics* **9**, 548-550, 1964.

340. Vogel, Gretchen. *Science* **279**, 1633-1634, 1998.

341. Volkenstein, M.V. *General Biophysics.* Academic Press, San Diego

342. von Massow, A., Mandelkow, E. M., and Mandelkow, E. *Cell Motil. Cytoskeleton* **14**, 562, 1989.

343. Wagner, O., Zinke, J., Dancker, P., Grill, W., and Bereiter-Hahn, J. *Biophys. J.* **76**, 2784, 1999.

344. Walker, M.L., Burgess, S.A., Sellers, J.R., Wang, F., Hammer 3rd, J.A., Trinick, J., and Knight, P.J. *Nature* **405**, 804-807, 2000.

345. Warshaw, D.M., Kennedy, G.G., Work, S.S., Krementsova, E.B., Beck, S. and Trybus, K.M. *Biophys. J.:Biophys. Lett.* L30-L32, 2005.

346. Wegner, A. and Savko, P. *Biochemistry* **21**, 1909-1913, 1982.

347. Weitzel, E.K., Tasker, R., and Brownell, W.E. *J. Acoust. Soc. Am.* **114**, 1462-1466, 2003.

348. Wells, A. L., Lin, A. W., Chen, L. Q., Safer, D., Cain, S. M., Hasson, T., Carragher, B.O., Milligan, R. A., and Sweeney, H. L. *Nature* **401**, 505-508, 1999.

349. Westerhoff, H. V., Tsong, T. Y., Chock, P. B., Chen, Y.-D., and Astumian, R. D. *Proc. Nat. Acad, Sci. U.S.A.* **83**, 4734, 1986.

350. Woehlke, G., Ruby, A.K., Hart, C.L., Ly, B., Hom-Booher, N., and Vale, R.D. *Cell* **90**, 207-216 , 1997.

351. Yildiz, A., Forkey, J.N., McKinney, S.A., Ha, T., Goldman, Y.E., and Selvin, P.R. *Science* **300**, 2061-2065, 2003.

352. Yildiz, A., Park, H., Safer, D., Yang, Z., Chen, L.Q., and Selvin, P.R. *J. Biol. Chem.* **279**, 37 223-37 226, 2004.

353. Yoshimura, K., Batiza, A. and Kung, C. *Biophys. J.* **80**, 2198-2206, 2001.

354. Young, E. C., Mahtani, H. K., and Gelles, J. *Biochemistry* **37**, 3467, 1998.

355. Zhuravlev, A.I., Tsvylev, O.P., and Zubtova, S.M. *Biophysics* **18**, 1101-1105, 1973.

356. Zuckermann, M. J., Georgallas, A., and Pink, D. A. *Can. J. Phys.* **63**, 1228-1234, 1985.

Chapter 4

What Are Life Processes?

4.1 Oxidative Phosphorylation

Chemical reactions within the body are responsible for storing, releasing, absorbing and transferring the energy humans need to move, breathe and pump blood etc. Reactions that require energy are called endothermic reactions while those which release energy are termed exothermic reactions. The energy required for muscle contraction is brought about through a network of chemical reactions within which there are two ingredient types (input and output ingredients). Input ingredients come from the air we breathe and the food we eat.

The two outputs include the carbon dioxide we exhale and the water we produce as a by-product. To be more specific in the lungs, the body removes oxygen from inhaled air and transports it to muscle cells via hemoglobin into the bloodstream. The body digests food in the mouth, stomach and intestines and processes some of it into glucose ($C_6H_{12}O_6$), part of which is transported to muscle cells. Here oxygen and glucose combine to form water, carbon dioxide and an exothermic reaction to also produce energy, E_{out}. Thus, in terms of a chemical reaction

$$C_6H_{12}O_6 + 6O_2 \rightarrow 6CO_2 + 6H_2O + E_{out} \qquad (4.1)$$

One can measure the net oxygen inhaled and/or the net carbon dioxide exhaled and the energy released can then be deduced from the glucose oxidation reaction.

Necessarily, time is required for the transport of oxygen and glucose to muscle cells, but the muscle fibers need not wait until glucose oxidation provides the energy required. This is because the body has the ability to store energy. We have already mentioned the role of ATP, ADP and AMP in this process and as long as ATP is available immediate muscle contraction can take place. When the power demand of muscles increases abruptly the energy required cannot be supplied by oxidation since time is required to move to a new equilibrium state. This energy gap is provided by energy stored in the muscle cells in the form of existing ATP in a form not requiring oxidation. After the transition time, respiration will gradually increase ATP until a new steady state is attained. During exercise and depending on the form of physical activity, for approximately the first two minutes energy is supplied anaerobically (without oxygen) and aerobically thereafter via respiration using readily available (not stored) ATP. When the body uses up energy faster than respiration can support ATP production available energy declines - a condition called oxygen debt. Thus, at the end of vigorous exercise, when the demand for energy is returning to normal the "debt" is repaid when, for example, the sprinter gasps for air.

Almost all metabolic energy of animals and humans comes from the conversion of adenosine triphosphate (ATP) into ADP (adenosine diphosphate) and AMP (adenosine monophosphate). The energy to convert ADP and AMP back to ATP is supplied by the oxidation of carbohydrates, fats and proteins. Metabolism requires a constant supply of oxygen which predominantly determines the metabolic rate. Oxygen is supplied by blood pumped by the heart at a variable rate depending on the type of activity pursued. For example, a person completely at rest consumes 15 ℓ of oxygen per hour. The energy production is linked to the oxygen supply via the relationship

$$E = 2 \times 10^4 J/(\ell \text{ of oxygen supplied}) \qquad (4.2)$$

Thus, at rest the metabolic power generated is approximately

$$P = \frac{15\ell}{h} \cdot \frac{2 \times 10^4 J}{\ell} \cdot \frac{1h}{3600s} = 83W \qquad (4.3)$$

Food energy is the energy stored by chemical bonds and has traditionally been measured in food calories but in science joules are used as a measure of energy. One food calorie is equivalent to 1000 physics calories or 4186 joules. To measure the energy content of food, the food is burnt in oxygen and the measured heat that is produced is the energy content. The human body combines food with oxygen, i.e., it burns food, to produce CO_2.

As shown above, the basic metabolic rate of an average person, when resting, is about 80 to 100 watts (W). Walking at 2.5 km per hour (h) takes another 100 J/s and walking faster at 5 km/h takes another 100 J/s. About $\frac{3}{4}$ of all the energy generated is converted to heat and the remainder can be used for activities like walking and running. Taking 100 W as an average metabolic rate for an adult translates into a dietary requirement of 2600 food cal a day. During vigorous physical exercise such as cycling the metabolic rate may increase five-fold to over 400 W or even ten-fold under extreme conditions.

4.1.1   The Biochemical Energy Currency- The ATP Molecule

All life processes require an energy supply, which originally comes from the electromagnetic energy of visible light, emitted by the sun and absorbed by pigments of photosynthetic units. The absorbed light energy is transformed, accumulated and transmitted to other parts of the cell, to other parts of the plant, and finally to other organisms which are not able to obtain energy by photosynthesis directly. For this purpose, the energy of molecular excitation is transformed to the chemical energy of high-energy compounds. The most common storage molecule of chemical energy in the cell is adenosine triphosphate (ATP), formed in the process of photosynthesis and used in nearly all processes of energy conversion in other cells. ATP is a complex nano-machine that is the most widely distributed high-energy compound within the human body. All living organisms produce ATP, which in turn powers virtually every activity of the cell and organism.

The hydrolysis of ATP, which produces adenosine diphosphate (ADP) and is catalyzed by special enzymes called ATPases, allows the use of this stored energy for ionic pumps, molecular synthesis, the production of mechanical energy and many other applications. The amount of energy, stored by the $ADP \rightarrow ATP$ reaction in the cell, however, is limited due to osmotic stability. Therefore, other molecules, such as sugars and fats, are used for long-term energy storage. The free energy of ATP, however, is used to synthesize these molecules. Subsequently, in the respiratory chain, these larger storage molecules are decomposed, whereas ATP is continually recycled.

Approximately 7.3 kcal/mole of energy is released in the ATP hydrolysis reaction whose chemical reaction is given by the equation

$$ATP^{4-} + H_2O \rightarrow ADP^{3-} + HPO_4^{2-} + H^+ \qquad (4.4)$$

There are four negatively charged oxygen atoms in close proximity on the ATP molecules and as these repel one another the molecule is under considerable strain. The ADP molecule contains only three negatively charged oxygen ions and thus in the conversion, the overall electrostatic energy is reduced. Under physiological conditions the ATP molecule is metastable. When an ATP molecule binds to certain enzymes, however, the activation barrier is lowered and the molecule moves down the potential energy curve releasing a phosphate group and turning into ADP. When the ATP converts to ADP, the ATP is said to be "spent" having done some form of work in the process. Then the ADP is usually immediately recycled in the mitochondria where it is recharged and comes out again as ATP. At any instant each cell contains about one billion ATP molecules satisfying the cell's needs for only a few minutes and hence the supply must be rapidly recycled. The total human body content of ATP is only about 50 g. However, the average daily intake of 2,500 food calories translates into a turnover of 180 kg of ATP.

4.1.1.1 The Structure of ATP

ATP contains the purine base adenine and the sugar ribose which together form the nucleoside adenosine. The basic building blocks in the formation of ATP are carbon, hydrogen, nitrogen, oxygen and phosphorus which are assembled in a complex whose mass is 500 Daltons. One phosphate ester bond and two phosphate anhydride bonds hold the three phosphates (PO_4) and the ribose together (see Fig. 4.1)

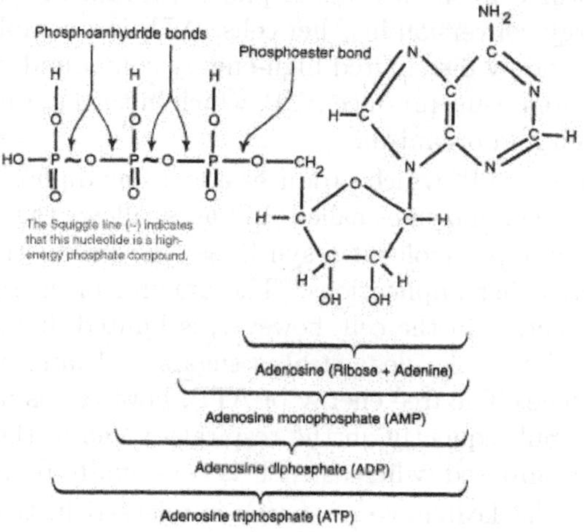

FIGURE 4.1: The chemical structure of ATP.

The high-energy bonds of ATP are rather unstable hence the energy of ATP is readily released when ATP is hydrolyzed in cellular reactions. ATP is not excessively unstable, but it is useful because its hydrolysis is slow in the absence of a catalyst. This insures that its stored energy is released rapidly only in the presence of the appropriate enzyme. Note that ATP is an energy-coupling agent and not a fuel. When ATP is produced by one set of reactions, it is almost immediately consumed by another.

4.1.1.2   The Function of ATP

The ATP is used for many cell functions including transport of substances across cell membranes, mechanical work, supplying the energy needed for muscle contraction, etc. It supplies energy to the heart and skeletal muscles as well as to membrane potentials and flagella. A major role of ATP is in chemical work, supplying the needed energy to synthesize the various macromolecules. ATP is also used as an on-off switch both to control chemical reactions and to send messages. It can bond to one part of a protein molecule, causing another part of the same molecule to change its conformation, thereby inactivating it. Subsequent removal of ATP causes the protein to return to its original shape making it again functional. Both phosphorylation and dephosphorylation can be used for this purpose, i.e., a phosphorylated protein can be either the biologically active or inactive form.

ATP is manufactured as a result of several cell processes including fermentation, respiration and photosynthesis. Most commonly the cells use ADP as a precursor molecule and then add a phosphate group to it. In eukaryotes this can occur either in the cytosol or in mitochondria. Charging ADP to form ATP in the mitochondria is called chemi-osmotic phosphorylation. This process occurs in specially constructed chambers located in the mitochondrion's inner membranes. The mitochondrion itself functions to produce an electrical chemical gradient by accumulating hydrogen ions in the space between the inner and outer membrane. This energy comes from the estimated 10,000 enzyme chains in the membranous sacks on the mitochondrial walls.

Most of the energy in cells is converted from food energy released by the electron transport chain. Cellular oxidation in the Krebs cycle causes an electron build-up that is used to push H^+ ions outward across the inner mitochondrial membrane. As the charge builds up, it provides an electrical potential that causes a flow of hydrogen ions across the inner membrane into the inner chamber. This process is coupled to an enzyme attached to ADP which catalyzes the addition of a third phosphorus to form ATP. In the case of eukaryotic cells the energy comes from food which is converted to pyruvate and then to acetyl coenzyme A (acetyl CoA). Acetyl CoA then enters the Krebs cycle which releases energy that results in the conversion of ADP back into ATP. When the repulsion due to a high proton concentration reaches a certain threshold, the hydrogen ions are forced out of a revolving-door-like structure mounted on the inner mitochondria membrane called ATP synthase complexes. This enzyme functions to reattach the phosphate groups to the ADP molecules. Each revolution of ATP synthase requires the energy of about nine hydrogen ions returning into the mitochondrial inner chamber. Located on the ATP synthase are three active sites, each of which converts ADP to ATP with every turn of the wheel. Under maximum conditions, the ATP synthase wheel turns at a rate of up to 200 revolutions per second, producing 600 ATP molecules in the process.

4.1.1.3 A Double Energy Packet

Although ATP contains the amount of energy necessary for most metabolic reactions, at times more energy is required. The solution is for ATP to release two phosphates instead of one, producing an adenosine monophosphate (AMP) plus a chain of two phosphates called a pyrophosphate (PPi). An intricate enzyme called adenylate kinase is able to transfer a single phosphate from an ATP to the AMP, producing two ADP molecules. The two ADP molecules can then enter the normal Krebs cycle designed to convert ADP into ATP. The main energy carrier the body uses is ATP, but other energized nucleotides are also utilized such as thymine, guanine, uracil and cytosine for making RNA and DNA, and for use in many important cell signaling pathways. The Krebs cycle only recharges ADP, but the energy contained in ATP

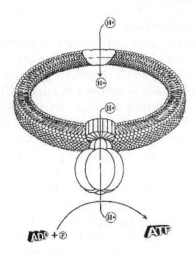

FIGURE 4.2:   A schematic diagram of the functioning of the ATPase.

can be transferred to one of the other nucleosides by means of an enzyme
called nucleoside diphosphate kinase. This enzyme transfers the phosphate
from a nucleoside triphosphate (ATP) to a nucleoside diphosphate such as
guanosine diphosphate (GDP) to form guanosine triphosphate (GTP). The
nucleoside diphosphate kinase works by one of its six active sites binding
nucleoside triphosphate and releasing the phosphate which is bonded to a
histidine residue on the kinase. Then the nucleoside triphosphate, which is
now a diphosphate, is released, and a different nucleoside diphosphate binds
to the same site and as a result the phosphate that is bonded to the enzyme
is transferred, forming a new triphosphate. Scores of other enzymes exist in
order for ATP to transfer its energy to the various places where it is needed.
Each enzyme must be specifically designed to carry out its unique function.

4.1.1.4   The Methods of Producing ATP

A crucial difference between prokaryotes and eukaryotes is the means they
use to produce ATP. All life forms produce ATP by three basic chemical
methods: oxidative phosphorylation, photo-phosphorylation and substrate-
level phosphorylation. In prokaryotes ATP is produced both in the cell wall
and in the cytosol by glycolysis. In eukaryotes most ATP is produced in
chloroplasts (for plants), or in mitochondria (for both plants and animals).

Mitochondria produce ATP in their internal membrane system called the
cristae. Since bacteria lack internal membrane systems, they must produce
ATP in their cell membrane which they do by two basic steps. The bacterial
cell membrane contains a unique structure designed to produce ATP and no
comparable structure has been found in any eukaryotic cell.

Note that ATPase conventionally refers to $ATP \rightarrow ADP$ activity, reverse ATPase to $ADP \rightarrow ATP$ activity. In bacteria, the reverse ATPase and the electron transport chain are located inside the cytoplasmic membrane between the hydrophobic tails of the phospholipid's membrane inner and outer walls. The breakdown of sugar and other food causes the positively charged protons on the outside of the membrane to accumulate to a higher concentration than they are on the membrane inside. This creates an excess positive charge on the outside of the membrane and a relatively negative charge on the inside. The result of this charge difference is a dissociation of H_2O molecules into H^+ and OH^- ions. The H^+ ions that are produced are then transported outside of the cell and the OH^- ions remain on the inside. This results in a potential energy gradient whose force, called a proton motive force, can accomplish a variety of cell tasks including converting ADP into ATP. In some bacteria such as Halobacterium this system is modified by use of transfer of protons across the plasma membrane to the periplasm. The energy temporarily changes rhodopsin from a trans to a cis form. The trans to cis conversion causes deprotonation and the transfer of protons across the plasma membrane to the periplasm. The resulting proton gradient is used to drive ATP synthesis through the use of the reverse ATPase complex.

4.1.1.5 Chloroplasts

Chloroplasts are double membraned ATP-producing organelles found only in plants. Inside their outer membrane is a set of thin membranes organized into flattened sacs stacked up like coins called thylakoids. The disks contain chlorophyll pigments that absorb solar energy. The chloroplasts first convert the solar energy into ATP, which is then used to manufacture storage carbohydrates that can be converted back into ATP when energy is needed. The chloroplasts also possess an electron transport system for producing ATP. The electrons that enter the system are taken from water. During photosynthesis, carbon dioxide is reduced to a carbohydrate by energy obtained from ATP. Photosynthesizing bacteria (cyanobacteria) use yet another system where they use chlorophyll bound to cytoplasmic thylakoids.

4.2 Diffusion Processes

One of the most important transport processes for ions and biomolecules in cells is their diffusion. It allows, due to entropic processes, for an equitable distribution of solutes within accessible space.

4.2.1   Translational Diffusion

Diffusion is usually associated with a gradient of concentration as is shown in Fig. 4.3.

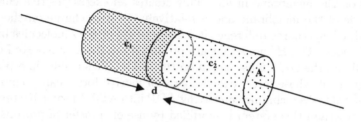

FIGURE 4.3:   An illustration of the diffusion process due to a concentration gradient in a cylindrical container.

In contrast to the mass flow of liquids, diffusion involves the random spontaneous movement of individual molecules. This process can be quantified by a constant known as the diffusion coefficient, D, of the material, given in general by the Stokes-Einstein equation

$$D = \frac{k_B T}{f} \tag{4.5}$$

where k_B is the Boltzmann constant, T is the absolute temperature in K and f is a frictional coefficient. If the diffusing molecule is spherical, is in low concentration, is larger than the solvent molecules and does not attract a layer of solvent molecules to itself then the frictional coefficient is

$$f = 6\pi\eta r \tag{4.6}$$

where r is the Stokes radius of the diffusing molecule in m, and η is the coefficient of viscosity of the solvent expressed in Ns/m^2. Since $D = k_B T/f$ and $f = 6\pi\eta r$, then

$$D = \frac{k_B T}{6\pi\eta r} \quad \text{or} \quad r = \frac{k_B T}{6\pi\eta D} \tag{4.7}$$

The diffusion constant of a particular molecular species thus depends on the nature of the molecule and the solvent. Large molecules have smaller diffusion constants. The diffusion constant D depends also on temperature. In Table 4.1, diffusion constants for several molecules at room temperature in water are listed.

TABLE 4.1: Diffusion Constants at Room Temperature in Water

Molecule	Diffusion Constant D (in m^2/s)
water	2×10^{-9}
oxygen	8×10^{-10}
glucose	6×10^{-10}
tobacco mosaic virus	3×10^{-12}
DNA (molar mass: 5 million g)	1×10^{-12}
protein (order of magnitude)	1×10^{-10}
hemoglobin	6.9×10^{-12}

Note that the volume of a sphere is $V = \frac{4}{3}\pi r^3$, the mass of a molecule is $m = \rho V$ and the molar mass M is $M = mN_A$, where N_A is Avogadro's number $(6.02 \times 10^{23}$ molecules/mol). Combining the above equations, we obtain

$$M = \rho \frac{4}{3}\pi \left(\frac{k_B T}{6\pi \eta D} \right)^3 N_A \tag{4.8}$$

With all the quantities in this equation in SI units, the molar mass will be in kg/mol.

As can be seen from Table 4.1 for biologically important molecules diffusing through water at room temperatures values of the diffusion constant, D, range from 1 to 100 $\times 10^{-11} m^2/s$, the corresponding range of molecular weights being about 10^4. The diffusion constant may be shown to be related to the temperatures and viscosity of the liquid by (Eq. 4.7). In (Eq. 4.8) the radius of a sphere is proportional to the cube root of its mass and therefore we conclude that D is inversely proportional to the cube root of the mass; see (Eq. 4.7). This explains why, for a wide range of molecular weights, the values of D are in a comparatively small range. In gases this result does not hold and D becomes inversely proportional to the square root of the mass of solute particles.

An important characteristic of diffusion processes is the proportionality of the average value of the displacement, R, squared of a diffusing particle to the time, t, elapsed of the diffusion process.

$$\bar{R}^2 = 6Dt \tag{4.9}$$

This equation helps us to determine whether some cellular processes such as transcription and translation are physically feasible. For example, one might ask whether or not the rates of diffusion of $D = 2 \times 10^{-9} m^2/s$ are sufficient to allow 50 amino acids per second to be made into protein at the ribosome. Taking the distance to be the length of a bacterial cell (i.e., $3\mu m$) gives

$$t = \frac{\bar{R}^2}{6D} = \frac{(3 \times 10^{-6})}{6 \times 2 \times 10^{-9}} = 7.5 \times 10^{-4}s \tag{4.10}$$

The process in a real bacterial cell would not be quite this rapid since the cytoplasm will be about five times as viscous as water. This will decrease the diffusion constant to one-fifth of the value used, with the result that the time will be increased by a factor of five to 3.8×10^{-3} s which is still very fast. Thus diffusion, while a slow process in a bulk liquid, is a very fast process of a cell.

Fick's Law states that the rate of diffusion per unit area in a direction perpendicular to the area is proportional to the gradient of concentration of solute in that direction. The concentration is the mass of solute per unit volume, and the gradient of concentration is the change in concentration per unit distance. If the concentration changes from c_1 to a lower value c_2 over a short length d of the pipe, then the mass m of the solute diffusing down the pipe in time t is given by

$$\frac{m}{t} = DA\frac{c_1 - c_2}{d}$$
(4.11)

This is a simplified version of Fick's Law whose differential form is given by

$$J = -D\frac{dc}{dx}$$
(4.12)

This equation states that J, the flux of particles (number of particles passing through an imaginary normal surface of unit area per unit time), is related to the force which is pushing them $\left(-\frac{dc}{dx}\right)$.

Osmosis is usually defined as the transport of molecules in a fluid through a semipermeable membrane due to an imbalance in its concentration on either side of the membrane. Osmosis may be by diffusion, but it may also be a bulk flow through pores in the membrane. In either case, water moves from a region of high concentration to a region of low concentration.

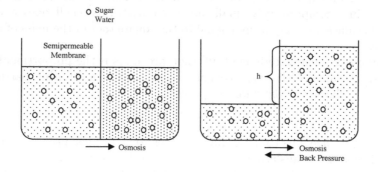

FIGURE 4.4:   An illustration of the process of osmosis.

In Fig. 4.4, the pressure on the right is then greater than the pressure on the left by an amount $h\rho g$, where ρ is the density of the liquid on the right, and is called the relative osmotic pressure.

The general formula for the osmotic pressure of a solution containing m moles of solute per volume V of solvent is

$$\Pi = \frac{m}{V}RT = cRT \qquad (4.13)$$

The net osmotic pressure exerted on a semipermeable membrane separating the two compartments is thus the difference between the osmotic pressure of each compartment. This equation is known as van't Hoff's Law and it looks exactly like the ideal gas law ($pV = nRT$) but osmotic pressure refers to the pressure exerted on a semi-permeable membrane by an aqueous solution, while the ideal gas law refers to the pressure exerted on the wall of a container by an enclosed gas. A detailed study of osmosis began in the middle of the 19th century with experiments on plant cells. If a plant cell is put into a concentrated solution of sugar, for example, the living portion of the cell (the protoplasts) withdraws from the walls. If the cells thus treated are removed and placed in pure water the protoplasts expand again. This phenomenon is easily observed under a microscope and is known as plasmolysis.

The osmotic pressure, Π, can be found from experiments on weak solutions to be proportional to the concentration of solute, i.e., inversely proportional to the volume of the solution, and also proportional to the absolute temperature. Hence the law of osmosis may be written

$$\Pi V = R'T \qquad (4.14)$$

where R' is a constant depending only on the mass of the solute present. It transpires that

$$R' = nR \qquad (4.15)$$

where R is the universal gas constant and n is the number of moles of solute present. The solute behaves as if it were a perfect gas so osmotic pressure arises from the bombardment of the walls of the container by the molecules of sugar in the solution. At higher concentrations (Eq. 4.14) is not valid for reasons analogous to those that cause the breakdown of simple gas laws (which were subsequently reformulated using the Van der Waals equation).

Reverse osmosis may take place in situations where an external pressure is applied to one compartment such that it exceeds the osmotic pressure (see Fig. 4.5).

Osmosis between roots and groundwater is thought to be responsible for the transfer of water into many plants. Groundwater has a higher water concentration than sap, so osmosis moves water into the roots. Water in sap is then transferred by osmosis into cells, causing them to swell with increased pressure. This pressure is called turgor pressure and is partially responsible for the ability of many plants to stand up. Relative osmotic pressure is not large enough to cause sap to rise to the top of a tall tree, however. To do this, the sap would have to contain more dissolved materials than is found to be the case.

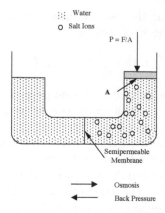

FIGURE 4.5: Reverse osmosis occurs when a pressure greater than the osmotic pressure is applied.

4.2.2 Diffusional Flow Across Membranes

Biological membranes are highly complex, dynamic structures that regulate the flow of compounds into the cell. Structurally, the conventional model (the so-called Singer-Nicolson model) describes the membrane as a bipolar, lipid layer, interspersed with regions of ion pumps, globular proteins and pores. A complete analysis of particle flow across the membrane would have to include the influence of pressure gradients, Coulomb attraction and of concentration differences. However, the degree of influence of each driving force varies. For example, the flow of water across the membrane, called osmosis, is chiefly explained as the result of a pressure gradient (see above). This flow is influenced by the other two mechanisms, but they are not as dominant. Meanwhile, the phenomenon of active transport is often described as resulting from Coulomb attraction between the solute and the membrane. This phenomenon occurs frequently; however, the primary mechanism for transport of solute is a difference of concentration across the membrane. The flow resulting from this difference, diffusion, dominates cellular processes such as metabolism (energy-yielding processes essential for life) and respiration (the flow of oxygen into the cell and release of CO_2). Because of its predominance in solute flow, an analysis of the process of diffusion across the membrane is worthwhile.

The discussion here is divided into two subsections: cells without a source of the diffusing particles, and cells with a source. In the first part, there is an emphasis on the essential features of transport across the membrane. As the second part, the emphasis is on limiting effects of the cell, due to diffusion, cell size and concentration. Specifically, from rather a crude approximation, a region, about the center of the cell, in which there is no consumption of oxygen is mathematically predicted, and experimentally verified.

The fundamental equations describing the process of diffusion are the continuity equation, Fick's Law, and the diffusion equation. In cells without a source, these are written as:

$$\frac{\partial C}{\partial t} = -\nabla j \qquad j = -D\nabla C \qquad \frac{\partial C}{\partial t} = D\nabla^2 C \qquad (4.16)$$

In cells with a source, the diffusion equation is altered to accommodate a (presumed) constant source Q:

$$\frac{\partial C}{\partial t} = D\nabla^2 C + Q \qquad (4.17)$$

4.2.2.1 Cells Without a Source

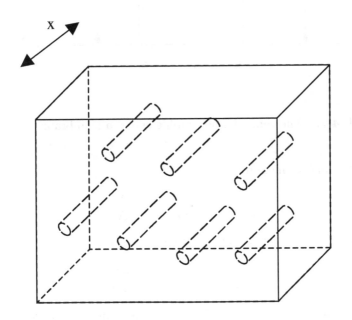

FIGURE 4.6: A schematic illustration of a membrane with pores.

The porous membrane illustrated in Fig. 4.6 is assumed to be of the simplified shape of a thin rectangular slab, interspersed with cylindrical pores. In order to calculate the flow rate through this membrane, two further assumptions are required. First, that the concentrations on either side of the membrane are uniform. This is accurate if the fluids are either stirred or circulated, as is the case for blood plasma in a capillary. The second assumption is that the profile of the concentration in the pore is linear. This is the quasi-stationary situation, for if $\frac{\partial C}{\partial t} \simeq 0$, then $\frac{\partial^2 C}{\partial x^2} \simeq 0$ (from the diffusion equation), and so, $\frac{\partial C}{\partial x} \simeq \text{const}$. The concentration profile, with these assumptions, is

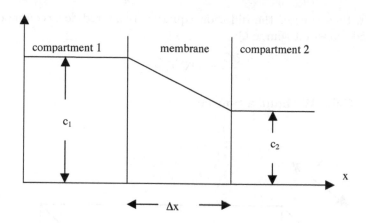

FIGURE 4.7: The concentration profile across a membrane.

With this, inside the pore,

$$\frac{\partial C}{\partial x} = -\frac{C_1 - C_2}{\Delta x} = -\frac{\Delta C}{\Delta x} \tag{4.18}$$

Using Fick's law, the flow of solute through a single pore is

$$\frac{j(x,t)}{\text{pore}} = -D\frac{\partial C}{\partial x} = D\frac{\Delta C}{\Delta x} \tag{4.19}$$

Commonly, the flow rate is described by J, the number of particles flowing per second. The relation between the current density $j(x,t)$ and J is simply

$$J = j(x,t) \cdot (\text{cross-sectional area}) = j(x,t) \cdot \pi a^2 \tag{4.20}$$

(for cylindrical pores). Therefore, for n pores,

$$J = n\pi a^2 \frac{D}{\Delta x}\Delta C \quad \text{or} \quad J = \wp \Delta C \quad \text{where} \quad \wp = n\pi a^2 \frac{D}{\Delta x} \tag{4.21}$$

Thus, the flow rate is proportional to the difference of concentration, where the constant of proportionality, $\wp$, is called the permeability. Essentially, it is this constant which dictates whether a solute will pass through the membrane or not. As may be expected, the permeability is dependent upon the size of the pores, the thickness of the membrane and the diffusion constant.

Now that we have a rough idea of what the flow rate is, we can determine the time required for the concentrations to equilibrate. Consider the flow across a membrane separating two compartments of varied volumes and concentrations (as diagrammed in Fig. 4.8).

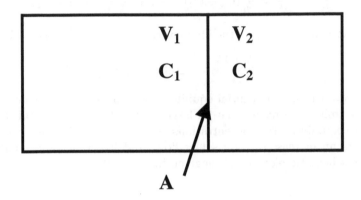

FIGURE 4.8:   Sketch of two compartments separated by a membrane.

Let us assume that $C_1 > C_2$, and let N_1 be the number of particles in (1) and N_2 be the number of particles in (2). The total solute flow per second for the entire membrane is (J A), where A), where J is defined above, and A is the membrane area. From the conservation of particles, the particle decrease corresponds to a particle increase in (2). That is,

$$\frac{dN_1}{dt} = -\frac{dN_2}{dt} = -J \cdot A = -A\wp(C_1 - C_2) \tag{4.22}$$

But,

$$N_1(t) = C_1(t)V_1, \qquad N_2(t) = C_2(t)V_2 \tag{4.23}$$

so

$$\frac{dN_1}{dt} = V_1\frac{dC_1}{dt} = -A\wp(C_1 - C_2), \qquad \frac{dN_2}{dt} = V_2\frac{dC_2}{dt} = +A\wp(C_1 - C_2) \tag{4.24}$$

With particle conservation, $N_1 + N_2 = $ constant, and

$$\frac{dN_1}{dt} + \frac{dN_2}{dt} = 0 \tag{4.25}$$

Substituting (Eq. 4.24) into this,

$$V_1\frac{dC_1}{dt} + V_2\frac{dC_2}{dt} = \frac{d}{dt}(V_1C_1 + V_2C_2) = 0 \tag{4.26}$$

That is, $(V_1C_1 + V_2C_2)$ is a constant in time. Let the initial concentrations be $C_1(0)$ and $C_2(0)$, and the final, common concentration be C_∞. So,

$$C_\infty = \frac{V_1}{V_1 + V_2}C_1(0) + \frac{V_2}{V_1 + V_2}C_2(0) \tag{4.27}$$

Using (Eq. 4.24),

$$\frac{dC_1}{dt} - \frac{dC_2}{dt} = \frac{d}{dt}(C_1 - C_2) = -A\wp\frac{V_1 + V_2}{V_1 V_2}(C_1 - C_2) \tag{4.28}$$

Defining the constant, τ_0, as the characteristic time for concentration equilibrium

$$\tau_0^{-1} = A\wp\frac{V_1 + V_2}{V_1 V_2}, \qquad \Delta C(t) = \Delta C(0)e^{(-\frac{t}{\tau_0})} \tag{4.29}$$

we have obtained an exponential equilibration process for the concentration function of time. This means that for large values of τ_0, a great deal of time is required to equilibrate the concentrations. As a check, as the area is increased one expects an increase of total solute flow, and hence, a lower equilibrium time. The schematic plot of $C(t)$ is given in Fig. 4.9.

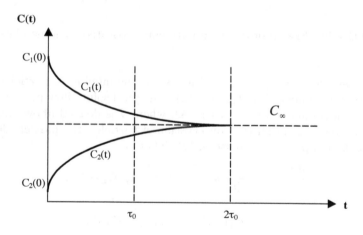

FIGURE 4.9: Plot of C(t).

Solving for $C_1(t)$ and $C_2(t)$, we finally obtain

$$C_1(t) = C_\infty + [C_1(0) - C_\infty]e^{(-\frac{t}{\tau_0})} \qquad C_2(t) = C_\infty + [C_2(0) - C_\infty]e^{(-\frac{t}{\tau_0})} \tag{4.30}$$

Thus, the two concentrations exponentially approach a common value, depending on the magnitude of τ_0.

4.2.2.2 Cells With a Source

The cell is now assumed to have a constant source and is described by the 'modified diffusion equation':

$$\frac{\partial C}{\partial t} = \nabla^2 C + Q \tag{4.31}$$

where, if $Q > 0$, the solute is produced and if $Q < 0$, the solute is consumed. In order to simplify the analysis, the cell will be assumed to be spherical. This is admittedly a crude approximation but one which gives us a good insight into the nature of the diffusion processes taking place in the cell. In addition, a quasi-stationary situation will, again, be assumed, so that $\frac{\partial C}{\partial t} \simeq 0$.

Let the concentration and the diffusion constant inside the cell be C_i and D_i, respectively, and those outside be C_e and D_e, respectively (as diagrammed in Fig. 4.10). In addition, let the solute concentration a great distance from the cell be C_o. By this, we assume the cell is placed in a medium of concentration C_o, so that near the membrane, the concentration varies, but nowhere else. It may be any non-negative value, including zero.

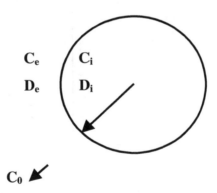

FIGURE 4.10: Diagram showing the concentration distribution in space at the beginning of the diffusion process.

Consequently, the diffusion equations are:

$$D_i \nabla^2 C_i + Q = 0, \qquad D_e \nabla^2 C_e = 0 \tag{4.32}$$

With the condition that whatever enters (leaves) the cell, leaves (enters) the external environment, at $r = r_0$ (where r_0 is the radius of a spherical cell),

$$D_i \frac{dC_i}{dr} = D_e \frac{dC_e}{dr} \quad \text{and} \quad -D_i \frac{dC_i}{dr} = k\Delta C = k(C_i - C_e) \tag{4.33}$$

where $k = \wp A$. By making the substitution,

$$C_i^* = C_i r \quad \text{and} \quad C_e^* = C_e r \tag{4.34}$$

in the diffusion equations (for spherical cells) we obtain

$$\frac{d^2 C_i^*}{dr^2} = -\frac{Q}{D_i} r, \quad \frac{d^2 C_e^*}{dr^2} = 0 \tag{4.35}$$

Integrating the above equations, and solving for C_i and C_e subject to the boundary conditions we find:

$$C_i = C_0 + \frac{Q r_0}{3k} + \frac{Q}{6 D_i} \left(r_0^2 - r^2 \right) + \frac{Q r_0^2}{3 D_e} \quad C_e = C_0 + \frac{Q r_0^3}{3 D_e} \frac{1}{r} \tag{4.36}$$

The variations of the concentrations C_i and C_e both for $Q > 0$, and for $Q < 0$ are represented graphically in Fig. 4.11.

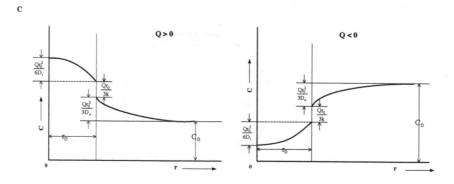

FIGURE 4.11:   Concentration profiles as a function of the spherical radius for both $Q > 0$ (left panel) and $Q < 0$ (right panel).

Notice that if $Q < 0$, and $r = 0$, then $C_i = 0$, when

$$C_0 = C^* = |Q| \left(\frac{r_0}{3k} + \frac{r_0^2}{6 D_i} + \frac{r_0^2}{3 D_e} \right) \tag{4.37}$$

However, if $C_0 < C^*$, then $C_i < 0$ and $r = 0$ which is physically impossible! So, to alleviate this problem, assume that there is a region, at a radius r_1, in which no consumption occurs. That is, for $r = r_1$

$$\frac{dC_i}{dr} = 0, \quad C_i = 0 \tag{4.38}$$

For $r > r_1$, the same conditions apply as before. Re-solving the diffusion equations with these new conditions, we now find

$$C_i = C_0 + \frac{Qr_0}{3k} + \frac{Q}{6D_i}\left(r_0^2 - r^2\right) + \frac{Qr_0^2}{3D_e} - \left(\frac{1}{D_e r_0} - \frac{1}{D_i r_0} + \frac{1}{kr_0^2}\right)Qr_1^3 - \frac{Qr_1^3}{3D_i}\frac{1}{r}$$

$$C_e = C_0 + \frac{Q}{3D_e}\left(r_0^3 - r^3\right)\frac{1}{r} \tag{4.39}$$

Using the boundary conditions at $r = r_1$, a cubic equation in r_1 is determined.

$$C_0 - C^* - \frac{Qr_1^2}{2D_i} - \frac{1}{3}\left(\frac{1}{D_e r_0} - \frac{1}{D_i r_0} + \frac{1}{kr_0^2}\right)Qr_1^3 = 0 \tag{4.40}$$

Solving this equation for r_1, and introducing it into (Eq. 4.39), the diffusion equation will, thus, be solved. This leads to a lengthy formula of no real practical use. Instead, a comparison with experimental data is made.

For $C_0 < C^*$, the total rate of consumption is given by

$$Q_{tot} = \frac{4}{3}\pi\left(r_0^3 - r_1^3\right)Q \tag{4.41}$$

since all of the consumption is assumed to take place within a shell of volume

$$\frac{4}{3}\pi\left(r_0^3 - r_1^3\right) \tag{4.42}$$

For $C_0 > C^*$, Q_{tot} is equal to

$$Q_{tot} = \frac{4}{3}\pi r_0^3 Q \tag{4.43}$$

With C_0 decreasing from C^* to zero, Q_{tot} also decreases to zero. A theoretical comparison of solute consumption to C_0 is possible with the following definition:

$$\gamma = \frac{r_i}{r_0} = \left[1 - \frac{Q_{tot}}{Q_0}\right]^{1/3} \tag{4.44}$$

For now, the cubic equation in r_1 becomes:

$$C_0 = C^* + \frac{Qr_0^2}{2D_i}\gamma^2 - \left(C^* + \frac{Qr_0^2}{2D_i}\right)\gamma^3 \tag{4.45}$$

The following in Fig. 4.12 is a graphical comparison of data, obtained in the study of oxygen consumption in *Arbacia* eggs and bacteria, to this theoretical prediction. There is remarkable accuracy, considering the assumptions made.

The case of cell respiration is quite detailed, but shows similar results as the diagrams in Fig. 4.13 indicate. As a particular example consider oxygen consumption by aerobic bacteria. Oxygen is supplied by diffusion for the metabolism of aerobic bacteria. Consider a single bacterium in an aqueous

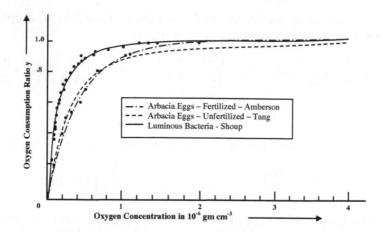

FIGURE 4.12:   A graphical comparison of data, obtained in the study of oxygen consumption in Arbacia eggs and bacteria, to the theoretical prediction in Eq. (4.45).

solution which contains a certain amount of dissolved oxygen. Let us suppose that the concentration of oxygen far from the bacterium has its equilibrium value, C_0; for oxygen dissolved in water it is about $0.2 mol/m^3$. Suppose also in analogy to our earlier discussion that the bacterium is spherical, has a radius R, is constantly consuming oxygen, and that at the surface of the bacterium oxygen molecules are adsorbed very efficiently. As a consequence, the oxygen concentration just outside the bacterium must be close to zero. The concentration difference between the surface of the bacterium and infinity will set up a diffusion current I of oxygen from infinity to the bacterium. The oxygen concentration $C(r)$, where $r = 0$ is the origin of the coordinate system, must depend only on the distance r from the origin. At the surface of the bacterium, where $r = R$, the concentration must vanish so $C(R) = 0$. The oxygen molecule diffusion current density $J(r)$ also can only depend on r. To find out what the functions $C(r)$ and $J(r)$ are, we note that the number of oxygen molecules is constant provided the product of $J(r)$ and the area, A, of a spherical surface at r is constant. This is, in fact, just the diffusion current I since $J = I/A$. Therefore

$$J(r) = -\frac{I}{4\pi r^2} \tag{4.46}$$

the minus sign indicating that current is flowing inwards from large to small r. The concentration, $C(r)$, may be also written as

$$J(r) = -D\frac{dC}{dt} \tag{4.47}$$

Equating (Eq. 4.46) and (Eq. 4.47) we find that

$$\frac{dC(r)}{dr} = \frac{I/D}{4\pi r^2} \tag{4.48}$$

(Eq. 4.48) may be solved by separation of variables to give the concentration $C(r)$ as

$$C(r) = -\frac{I/D}{4\pi r} + \text{constant} \tag{4.49}$$

We find the constant in (Eq. 4.49) by recalling that far from the bacterium i.e., $C(r) = C_0$. Hence

$$C(r) = -\frac{I/D}{4\pi r} + C_0 \tag{4.50}$$

Recall that the oxygen concentration at the surface of the bacterium, i.e., at $r = R$, must be zero. Thus in (Eq. 4.50), putting $C(R) = 0$ gives

$$I = 4\pi RDC_0 \tag{4.51}$$

(Eq. 4.51) determines the maximum number of oxygen molecules which can be taken in by the bacterium per second. The number of oxygen molecules required per unit volume by a bacterium is called the metabolic rate M and for bacteria is about $20 mol/m^3 s$. The incoming diffusion current, I, must equal or exceed M times the volume, $V = 4/3\pi R^3$, of the bacterium for it to function normally. Hence

$$I > \frac{4}{3}\pi R^3 M \tag{4.52}$$

(Eq. 4.51) combined with (Eq. 4.52) gives an inequality for the size of R. That is

$$R < \sqrt{\frac{3DC_0}{M}} \tag{4.53}$$

which provides an upper bound for the size of the bacterium. Inserting values for the parameters one finds that $R < 10$ m. Bacteria large compared to this size will simply not get enough oxygen by pure diffusion. This limit is, in fact, about the size of a typical spherical bacterium. Larger cells can be produced by making them very long keeping the diameter small, i.e., cylindrical. This greatly increases the diffusion current, and diffusion imposes no size limitation on the length of the cylindrical bacteria. Large bacteria are indeed cylindrical rather than spherical.

In conclusion, several rather interesting results are obtained, by limiting analysis to only the concentration difference mechanisms. For cells without a source, the permeability characteristic time governs the diffusion process while the pore size, the membrane thickness, the diffusion constant and the membranous area were shown to be factors in these descriptions. For cells with a source, the existence of a region, in the cell, of no oxygen consumption was theoretically predicted, and experimentally verified. Not all cells have

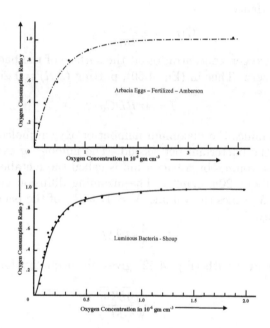

FIGURE 4.13: Diagrams of oxygen consumption versus oxygen concentration for the cells of Arbacia Eggs (left panel) and luminous bacteria (right panel).

this property, but for those that do, a study of the diffusion process reveals a possible explanation.

A more in-depth analysis of the transport of particles across a membrane should include the other two mechanisms: the pressure gradient, and the Coulomb attraction. In addition, the non-stationary case should be investigated with and without the other two mechanisms. Also, since chaos theory is becoming more prominent in biological applications, a closer study of the interplay of these mechanisms would most likely show signs of chaotic dynamics. Nonetheless, limiting the analysis to just diffusion resulted in rather interesting conclusions.

4.3 Proton Transport and Bioenergetics

Protonic conduction differs from electronic conduction in several respects. First, it is a positive charge which is three orders of magnitude more massive than an electron. Secondly, it is abundant in water complexes that surround all bio-polymers.

4.3.1 Proton Transport

Protons constitute mobile units in the commonly encountered hydrogen-bonded structures in peptides, proteins and DNA. Finally, protons are freed in the hydrolysis reaction of the energy-giving molecules of ATP and GTP. We expect, therefore, to find protonic conduction in biochemical processes at a sub-cellular level. However, as far as we know it, protonic conduction usually manifests itself through mechanisms of fault and defect migration in linear and closed (ring-like) organic polymers (see Fig. 4.14).

Two types of faults in Hydrogen bond systems are found. D faults possess an extra proton and L faults have a deficiency of a proton. This is shown in Fig. 4.16

In Table 4.2 we have summarized the potential applications of bio-molecular complexes as future devices in the fast developing area of bio-electronics.

4.3.2 Bioenergetics: The Davydov Model

In general, the following mechanisms of intermolecular energy transfer must be considered in the context of cellular activities:

- energy transfer by radiation (which will be discussed later)

- energy transfer by charged waves (such as calcium waves or action potential propagation)

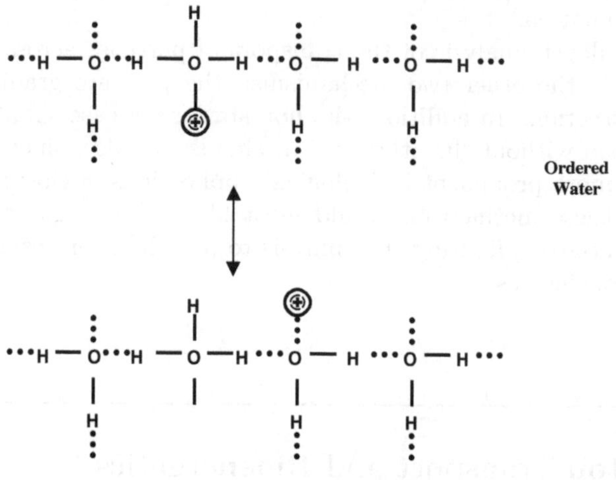

FIGURE 4.14:   Schematic representation of protonic conduction in chain-like and ring-like organic polymers.

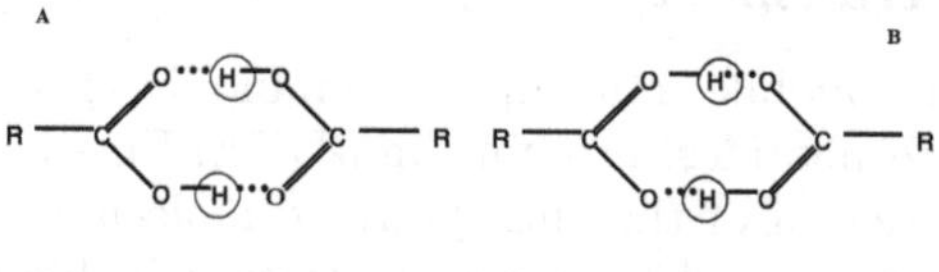

FIGURE 4.15:   Dimer rings.

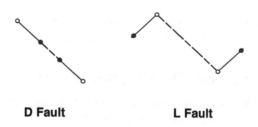

D Fault   **L Fault**

FIGURE 4.16:   Faults in the hydrogen bond system.

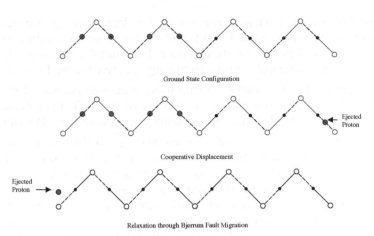

Ground State Configuration

Cooperative Displacement

Relaxation through Bjerrum Fault Migration

FIGURE 4.17: Proton migration in hydrogen bonded chains.

TABLE 4.2: Potential Functions of Molecular Complexes as Bioelectronic Materials

Function	Molecular Complexes
Wiring	Polyene anitibiotics, conductive polymers: actin, MT's.
Storage	Bacteriorhodopsin, reaction centres (PSII) cytochromes, blueproteins, ferritin, collagen, DNA.
Gates and Switches	Bacteriorhodopsin, photosynthetic systems, cell receptors, ATPase.
Input/Output Devices	Photosensitive proteins, enzymes, receptors, metal-protein complexes.

- energy transfer by hopping of individual charges (which we will discuss later)

- energy transfer by elementary excitations such as phonons or excitons that will be briefly described in this section.

It is generally accepted that energy transfer processes in living matter involve proteins. While the structure of the proteins involved in these processes has been quite well understood, the same cannot be said about their functions, in particular energy transduction. It is known, however, that the currency of biochemical energy is the molecule ATP or its analogs such as GTP. ATP binds to a specific site on a protein, reacts with a water molecule and releases the energy of 0.48 eV in a process called hydrolysis. While single energetic events pose no serious challenge to the theoretician, energy transport on length scale of protein filaments or DNA strands is still very much an open problem. A.S.

Davydov [60] developed a non-linear theory of biological energy transfer which focuses on the so-called amide I bond of $C = O$ in a peptide chain containing the $H - N - C = O$ group and its vibrational and dipolar coupling. A review of potential applications of solitons to biology was written by Lomdahl [165].

Now suppose that the amide - I vibration is excited on one of the three spines of an helix (for example by the energy of the ATP hydrolysis). The oscillating $C = O$ dipole with dipole moment interacts with the dipoles of the neighboring peptide groups of the spine and $\frac{2|\vec{d}|^2}{R^3}$ is the interaction energy between two parallel dipoles positioned at a distance R from each other. Thus the energy of the initial excitation does not remain localized; it propagates through the system. However, the excitation of the amide - I vibration results in the deformation of the $H - O$ bond since it is much weaker than the covalent bonds of the α helix. The deformation of the hydrogen bond can be described as the deformation of a spring of length K. The initial deformation (the excitation of the amide - I vibration) of a spring will cause the deformation of neighboring springs of the same strength and the disturbance will propagate along the chain of coupled springs (H - O binds) as a longitudinal sound wave. According to Davydov the longitudinal sound wave couples non-linearly to the amide - I vibration and thus acts as a potential well trapping the energy of the amide - I vibration. The energy remains localized and travels along the α helix chain as a soliton. The non-linear relation between the amide - I vibration and the sound wave can be expressed by the parameter

$$\chi = \frac{dE}{dR} \qquad (4.54)$$

where E is the amide - I excitation energy.

Davydov [60] used a semiclassical approach to model solitons traveling along the single chain of hydrogen-bonded peptide groups. The amide - I vibrations were treated quantum mechanically while the longitudinal sound wave was treated classically. In the continuum approximation this leads to the non-linear Schrödinger equation

$$i\hbar\frac{\partial a}{\partial t} + J\frac{\partial^2 a}{\partial^2 x} - E_0 a + \kappa|a|^2 a = 0 \qquad (4.55)$$

where $\kappa = \frac{4\chi^2}{K(1-s^2)}$, $s = \frac{v}{v_s}$, $v_s = R\sqrt{\frac{K}{M}}$ is the velocity of sound, $|a(x,t)|^2$ is the probability amplitude that the excitation will occur at a position x in the chain of hydrogen bonded peptide groups at time t, $J = \frac{2|\vec{d}|^2}{R^3}$, $E_0 = E - 2J + \frac{1}{2}\int_{-\infty}^{+\infty}\left[M\left(\frac{\partial u}{\partial t}\right)^2 + \left(K\frac{\partial u}{\partial x}\right)^2\right]dx$ is the excitation energy, M is the mass of the peptide group, E the amide - I excitation energy, u the displacement of the peptide group at position x.

For a stationary (i.e., non-moving) soliton, the solution of (Eq. 4.55) is represented by

$$a(x,t) = \frac{\chi}{\sqrt{(2KJ)}} \sec h \frac{\chi^2}{KJ}(x - x_0)exp\left[-\frac{i}{\hbar}\left(E_0 - \frac{\chi^2}{K^2J}\right)t\right] \quad (4.56)$$

where x_0 is the position of maximum probability of amide - I excitation along the chain. The energy of the stationary soliton is

$$E_{sol} = E - 2J - \frac{\chi^4}{3K^2J} \quad (4.57)$$

In (Eq. 4.56) the amplitude a(x,t) represents a "bump-like" soliton. In case of an absolutely rigid chain, i.e., for $K \to \infty$, the non-linear Schrödinger equation reduces to the linear Schrödinger equation whose solutions have the form of plane waves [245]. That means that the energy of the amide - I vibration is not localized, but is uniformly distributed along the whole chain. According to (Eq. 4.57), the energy in the system is then larger than the soliton energy. Since the system tends to occupy the state with a lower energy, the situation with the energy localized in the soliton is more probable.

In reality the three spines of an α helix interact with each other through transverse dipole-dipole interactions. This problem of α-helix interaction can be solved by means of three coupled non-linear Schrödinger equations.

On the basis of Davydov's ideas it has been suggested that the energy transport in proteins is carried out by localized pulse-like excitations. Some investigations done on proteins support these considerations. Numerical computations of Davydov solitons on α helices performed by J.M. Hyman, D.W. McLaughlin and A.C. Scott [120] revealed that solitons can arise only above some threshold value of the χ parameter. In other words the non-linear interaction between the dipole vibration of a peptide group of a spine of an α helix and its deformation caused by the vibration must be sufficiently strong. The results of the calculations are shown in Fig. 4.18. It can be clearly seen how a "soliton-like" object arises at the value of χ of about $0.45 \times 10^{-10}N$.

Another fact which supports the Davydov soliton theory has been provided by studies done on the crystalline polymer ACN (acetanilide). Interest in this chemical compound has arisen because of the great similarity of its structure to the structure of the α helix. ACN consists of hydrogen bonded peptide chains which are held together by Van der Waals forces. In 1984 Careri et al. found in the infrared absorption spectrum of ACN an anomalous line at $1650cm^{-1}$ shifted only $15cm^{-1}$ from the amide - I line at $1665cm^{-1}$. Attempts to explain this phenomenon in a conventional way have all failed. However, Careri et al. [34] suggested that solitons could be responsible for this experimental observation. It was assumed that the line was caused by non-linear coupling between the amide - I vibration and an out-of-plane displacement of the hydrogen-bonded proton. Such coupling allows for a soliton excitation by direct electromagnetic radiation. The α helix soliton cannot be excited in this

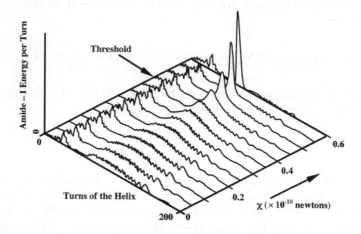

FIGURE 4.18: Total amide - I energy (summed over all three spines of the α helix) as a function of χ.

way since the displacement of the whole and thus the much heavier peptide group takes place there.

A.C. Scott developed his model [246] in analogy to that of Davydov for the α helix and obtained acceptable values of χ. However, numerical simulations of Lomdahl et al. [165] have shown that the value of the coupling constant may be too small to support a soliton. Some other arguments have been found which support the explanation of the anomalous line in the ACN absorption spectrum proposed by Careri and Scott. Experiments with polarized light appear to provide evidence for the existence of solitons in the ACN crystal. If the line at $1650 cm^{-1}$ is due to a soliton, it should behave in the same manner as the amide - I $1665 cm^{-1}$ line, e.g., it should have the same polarization. Observations have confirmed these expectations.

Attempts have also been made to apply the soliton model of biochemical energy transfer to globular proteins. While globular proteins lack translational symmetry, α-helix protein chains may still allow solitons to travel. It has been argued that the Davydov model can be generalized to this case, taking into account the full geometry of the protein. Promising results of model calculations performed for globular proteins have been obtained. However, so far experimental evidence still remains elusive.

4.4 Electronic and Ionic Conductivities of Microtubules and Actin Filaments and Their Consequences for Cell Signaling - from (275)

The prevailing view of neural information processing is based on passive properties of the membrane derived from the application of linear cable theory to dendrites, e.g., [223]. Recent studies, however, have suggested nonlinear models that accommodate new experimental evidence that appears to be inconsistent with this classical view. In particular, conventional theories do not satisfactorily explain: a) significant fluctuations of synaptic efficacy over short periods of time in response to a recent burst of activation, e.g., [107], b) a variety of active active ion-channels capable of affecting the local membrane properties, and c) non-linear responses localized to specific dendritic sites, pointing to highly specialized mechanisms correlated with specific inputs. On the other hand, non-linearities inherent in the new models, e.g., [248; 148], give rise to a wider repertoire of (computational) capabilities such as multiplication, fast correlation, etc. Moreover, experimental and theoretical support for conductive properties of protein filaments and the ionic clouds around them is becoming available which indicates a far more complex picture of signaling an information processing phenomena in neurons and other cells.

In this section we suggest that functional electrodynamic interactions between cytoskeletal structures and ion-channels are central to the neural information processing mechanism. These interactions are supported by long-range ionic wave propagation along microtubule networks and actin filaments and exhibit subcellular control of ionic channel activity, hence impact the computational capabilities of the whole neural function. Cytoskeletal biopolymers, including actin filaments (AFs) and microtubules (MTs), constitute the backbone for wave propagation, and in turn interact with membrane components to modulate synaptic connections and membrane channels. Indeed, only recently clear functional interactions between these cytoskeletal structures have become apparent. Association of MTs with AFs in neuronal filopodia appears to guide microtubule growth, and plays a key role in neurite initiation [63]. This is further evidenced by the presence in neurons of proteins that interact with both MTs and F-actin, and proteins that can mediate signaling between both types of filaments. This is likely used to control microtubular invasion. The microtubule associated proteins MAP1B and MAP2, for example, are known to interact with actin in vitro, e.g., cross-linking, MAP2 and/or MAP1B is highly probable to associate with both types of filaments contributing to the guidance of MTs along AF bundles. Extensive evidence exists confirming a direct interaction between AFs and ion channels and a regulatory functional role associated with actin. Thus it is clear that the cytoskeleton has a direct connection to membrane components, in particular channels and synapses.

Below, we demonstrate how each individual cytoskeletal component is capable of supporting both ionic and electronic wave propagation. In particular, we present a molecular dynamics model of the dendritic microtubule network (MTN) where arrays of MTs are interconnected by MAP2s. In this model, ionic waves propagate along the MTN and interact with C-termini of MTs to generate collective modes of behavior. Also in this section, a biophysical model of non-linear ionic wave propagation along AFs is presented, supported by experimental evidence. Finally, we describe a model for a direct regulation of ion-channels and synaptic strength by AFs and networks of MTs that control the electrical response of the dendrite in particular, and the neuron in general. We begin by providing a general overview of the structure and function of a nerve cell, emphasizing the role of cytoskeletal filaments, ion channels and their interactions.

4.4.1   The Neuron

The neuron is the quintessential communicating cell. A great variety of neural subtypes can be found interconnected to one another in complex electrochemical circuits via multiple synaptic connections. Thus, the amount of information processing by a neural circuit depends on the number of such connections or synapses. Neurons are highly polarized cells, whose level of morphological complexity increases during maturation into distinct subcellular domains. A typical neuron possesses three functional domains. These are the cell body or *soma* containing the nucleus and all major cytoplasmic organelles, a single axon, which extends away from the soma and takes cable like properties, a variable number of dendrites, different in shape and complexity, which emanate from the soma. The axonal terminal region, where the synapse contacts it with other cells, displays a wide range of morphological specialization (see Fig. 4.19).

Two major types of synapses can be found: (a) asymmetric, responsible for transmission of excitatory impulses, and (b) symmetrical, present in inhibitory synapses. Turning morphology into a functional feature, the soma and dendritic tree are the major domains of receptive inputs. Thus both the dendritic arborization and the axonal ramifications confer a high level of subcellular specificity in the localization of particular synaptic contacts on a given neuron. The extent to which a neuron may be interconnected largely depends on the three-dimensional spreading of the dendritic tree. Dendrites contain a large number of dendritic spines, which are small sac-like organelles, projecting from the dendritic trunk. Dendritic spines are more abundant in highly arborized cells such as pyramidal neurons and scarcer in lowly interconnected, smaller-sized dendritic tree inter-neurons.

Spines are the dendritic regions where most excitatory inputs are found, such that each spine contains one asymmetric synapse. The number of excitatory inputs can be correlated with the number of spines present in the dendrite. Dendritic spines are complex saccular enlargements, containing ri-

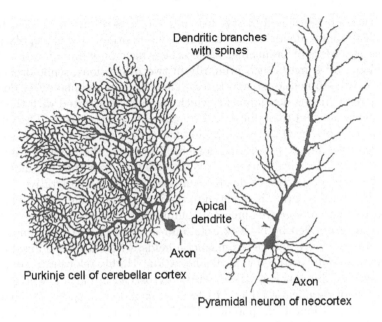

FIGURE 4.19:   Two prominent examples of nerve cells.

bosomes, and cytoskeletal structures, including actin, and α- and β-tubulin. Electrically speaking, a single neuronal circuit is closed such that different inputs reaching the soma and the dendritic tree are transferred to the axon at the other pole of the cell which, in turn, is responsible for transmitting efferent neural information. Both the morphology of the axon and its course through the nervous system are correlated with the type of information processed by a particular neuron, and by its interconnectivity with other neurons. While the axon leaves the soma with a small swelling known as the axon hillock, some neurons protrude the axonal projection from the main dendrite instead. Cytoskeletal components are highly relevant in the structure/function of the neuron. The axonal hillock contains large "parallel" bundles of MTs. Axons are also located with other cytoskeletal structures, such as neurofilaments, which are more abundant in axons than in dendrites. The dynamics in cytoskeletal structures is central to our contention that information can be processed and "delivered" to the synaptic function by changes in the cytoskeletal structures.

Dendrites are the principal element responsible for synaptic integration and for the changes in synaptic strengths that take place as a function of neuronal activity. While dendrites are the principal sites for excitatory synaptic input, little is still known about their function. The activity patterns inherent to the dendritic tree-like structures, such as in integration of synaptic inputs, are

likely based in the wide diversity of shapes and sizes of the dendritic arboriza-
tions [224]. Some dendrites are unipolar, while others are multipolar; some
have many orders of branching, while others have only one or two orders of
branching; and some branch primarily in two dimensions, while others have
complex three-dimensional structures. The size and complexity of dendritic
trees increase during development, which has been associated with the ability
of the neural system to organize and process information [100]. The majority
of synapses, both excitatory and inhibitory, terminate on dendrites, so it has
long been assumed that dendrites somehow integrate the numerous inputs to
produce single electrical outputs. It is increasingly clear that the morphological functional properties of dendrites are central to its integrative function.
Branching in this context is essential to the increasingly complex ability of the
neuron to respond to developmental complexity. Nerve branching and complexity are controlled by both external and internal cues. Of current interest
to this are substrate and electromagnetic fields, while also recognized are the
chemical gradients involved in local adaptation. Highly relevant to adaptation
is cytoskeletal dynamics, as the dendritic cytoskeleton plays a central role in
the process of ramification, filopodia formation and more specialized neuronal
activity such as long-term potential (LTP) and long-term depression (LTD)
circuit dynamics, including memory formation.

The electrotonic or passive properties of dendrites became of great importance to our understanding of dendritic integration of synaptic signals. Early
hypotheses were supported by work on motor neurons [70] and sensory receptors [75, 138] - in that action potentials were initiated in the axon hillock
region as a result of the graded or algebraic summation of EPSPs and IP-
SPs occurring in various parts of the neuron. Because of the relatively short
space constant calculated for motor neurons [50], it was predicted that excitatory post-synaptic potentials from synapses on the distal portions of dendrites
would be ineffective in firing a neuron. Rall [222] revolutionized the field of
dendrite physiology by using cable theory for the study of neurons and their
dendrites. Action potentials initiated in dendrites by synaptic stimulation
were later observed propagating in both directions at velocities of around 0.3
m/s [13, 55, 87]. Direct intracellular recordings from dendrites of cerebel-
lar Purkinje cells [164], cortical neurons [115] and hippocampal neurons [284]
confirmed that action potentials could be initiated and propagated in dendrites. Very little is known, however, about the active, or voltage-dependent,
properties of dendrites and many questions remain unanswered such as the
types of ion channels present in dendrites, location, and how active properties
alter synaptic integration.

Firing of action potentials elicits characteristic M-shaped Ca^{2+} profiles
across the neuron. The Ca^{2+} rise is small in the soma, highest in the proximal
apical and basal dendrites, and small again in the distal apical dendrites [128,
196, 226, 227, 228]. The spatial distribution of Ca^{2+} channels explains the
distinct Ca^{2+} rise in various cell locations but rather reflects the propagation
into the dendrites of trains of action potentials. Imaging with a Na^+-sensitive

dye has yielded similar patterns, suggesting that Ca^{2+} influx into the dendrites is driven by Na^+-dependent action potentials [128,]. Ca^{2+}-dependent action potentials have been demonstrated in hippocampal neurons and dendrites [244]. Most of the action potential induced Ca^{2+} entry is mediated by voltage-gated Ca^{2+} channels [196]. Multiple types of voltage-gated Ca^{2+} channels have been demonstrated in hippocampal neurons [72, 79, 139 , 198]. Based on the addition of specific Ca^{2+} channel blockers, the presence of L-, N-, P-, Q-, R- and T-type Ca^{2+} channels has been reported [48]. This channel distribution may contribute to spike-triggered Ca^{2+} entry into the soma and apical dendrites. Both low-voltage activated (LVA) and at least one high-voltage activated (HVA) channels seem to be most predominant in regions of the dendrite at least a 100 mm from the soma. Experimental results strongly suggest a heterogeneous distribution of different types of voltage-gated Ca^{2+} channels within the soma and dendrites of hippocampal neurons [283]. The presence of Na^+ channels in distal dendrites helps explain the back-propagation of action potentials well into the dendritic tree [171, 172, 258, 259, 263]. Action potential propagation in dendrites is rather low, and susceptible to the effects of cable filtering.

A most prominent feature of dendritic ion channel regulation is the potential boosting and enhancement of distal synaptic events. Voltage-gated Na^+ and Ca^{2+} channels may play an important role in amplifying the magnitude of EPSPs towards the soma. However, channel regulation may also be important for dendritic interactions in the immediate vicinity of the synaptic input. Voltage-gated channels may alter the local input resistance and time constant, which in turn would influence both spatial and temporal summation of EPSPs and IPSPs, thus conveying highly non-linear interactions; signal amplification [131,256]. Thus, the non-linear properties of dendrites may serve as electronic devices. In addition to providing an output, the action potentials elicited by synaptic input, whether initiated at the local synaptic sites or at the axon hillock and back-propagated into the dendrites, would essentially reset the dendritic membrane potential. Action potentials might also provide a feedback or an associative signal to other synapses active just before or just after the action potential.

Back-propagating action potentials may also provide a feedback to dendritic synapses after a neuronal output has occurred. Ca^{2+} signals are implicated in post-synaptically-induced synaptic plasticity such as short-term potentiation (STP) [176], LTP and LTD [23, 47]. Actin has long been postulated to play important roles in neuronal morphogenesis. High concentrations of actin have been found in the leading edge of the growth cones of developing neurons and in synapses of mature neurons. Actin filament dynamics also plays a role in a number of related phenomena, including axon initiation, growth, guidance and branching, in the morphogenesis and stability of dendrites and dendritic spines, in synapse formation and stability. Neurons conduct two types of cytoplasmic growth processes, in axons and dendrites, both of which are functionally and morphologically distinct, but are developmentally led from the growth cone. Growth cones are composed of finger-like filopodia and veil-like lamellipodia that help the growing processes explore their environment. Axons travel long distances, upon which reaching their targets, produce terminal branches where the growth cones are converted into pre-synaptic terminals. Dendrites usually do not extend such long distances, but often branch extensively, giving rise to dendritic trees. Upon proper contact with axons, post-synaptic dendritic specializations create functional synapses by opposition with the pre-synaptic terminals of axons. These are localized to small protrusions on their dendritic trees called dendritic spines, which are the site of most excitatory synapses. In the mammalian brain these multiple synaptic connections are believed to play important roles in learning and memory. Thus, neural morphogenesis requires changes in both the underlying cytoskeleton and specific interactions with the plasma membrane. To generate a membrane protrusive structure such as an axon, coordinated changes in actin and microtubular organization are required. Growth cones of developing neurons, for example, rapidly extend and retract filopodia by concentred changing in cytoskeletal dynamics, namely bundled AFs in filopodia and composed of a cross-linked actin network in lamellipodia [e.g., 286, 159, 160]. In this section we have outlined the main features of signaling taking place within a neuron. Major emphasis was placed on the role played by the cytoskeleton in the neuron's structure, growth and internal signaling. The next section discusses the main biophysical properties of the cytoskeleton.

4.4.2   The Cytoskeleton

The cytoskeleton is a major component of all living cells. It is made up of three different types of filamental structures, including actin-based microfilaments (MFs), intermediate filaments (IFs) (e.g., neurofilaments, keratin) and tubulin-based microtubules (MTs). All of them are organized into networks, which are interconnected through numerous particular proteins, and which have specific roles to play in the functioning of the cell. The cytoskeletal networks are mainly involved in the organization of different directed movements in cell migration, cell division or in the internal transport of materials. Polymerization of MFs is responsible for cell migration and for the remodeling of the leading edge of cells. Molecular motors are protein complexes which are associated with the cytoskeleton and drive organelles along MTs and MFs in a "vectorial" transport.

MTs are some of the most basic and most important cytoskeletal elements in the morphological scheme of neurons. As for the case of cytochalasins in actin structures, addition of microtubular modifying agents was first used to assess a role of MTs in axon elongation. Microtubular disrupting agents affect axons first and have no early effects on growth cones and filopodia, suggesting that MTs provide support for the axon. Conversely, microtubular invasion may be a critical element in the formation of neurite projections. Initially expanded lamellipodia first undergo segmentation at certain spots where MTs accumulate in an ordered array [266, 287, 62] and gradually "migrate away" from the cell body. These lamellipodia transform into nascent growth cones. Neurite elongation takes place as the MTs in the shaft become compressed into a narrower bundle. In cultured hippocampal and cortical neurons this process most likely implies the coordinated control of the major cytoskeletal components, including actin, MTs and associated proteins.

Thus, MTs are essential for neurite development. One particular aspect of microtubular involvement in neurite outgrowth is vesicular transport. Most membrane insertion occurs at the growth cone of axons [56, 58]. MAP-stabilized MTs form parallel or antiparallel arrays, which might act as compression resistant struts inside neurites [124]. Because MTs constantly invade the actin cytoskeleton in lamellipodia of epithelial [238] and neuroblastoma cells [62], such a mechanical role for MTs could also be important during de novo neurite initiation. Neurite-like protrusions, for example, can be induced in non-neuronal cells by exogenous expression of stabilizing proteins such as MAP2 and tau [158, 146, 288]. Spontaneous neurite initiation is dependent on both the presence and the dynamic properties of MTs.

A number of proteins associate with MTs, including MT motors, kinases, so-called "structural" MAPs, which alter microtubule structure, and specialized MAPs, which bind to microtubule plus or minus ends. Members of the MAP2/tau family have long been proposed to be important regulators of neurite behavior [282, 30, 97]. MAP2 and tau stabilize MTs by reducing catastrophe and promoting rescues, leading to prolonged growth periods and thus enhanced net microtubule accumulation [68, 91, 149]. MAP2 and tau stabilize MTs by binding along the outer ridges of protofilaments [4].

F-actin and MTs cytoskeletal networks are typically thought to fulfill separate, independent cellular roles. Highly dynamic actin networks are known for their role in cell spreading and contraction. The more stable MT cytoskeleton is known for its importance in cell division and organelle trafficking. However, recent studies provide important roles for the actin cytoskeleton in cell division and trafficking and for MTs in the generation and plasticity of cellular morphology. A direct physical association between both cytoskeletons has been suggested, because MTs often preferentially grow along actin bundles and transiently target actin-rich adhesion complexes.

Association of MTs with actin cables in filopodia appears to guide microtubule growth along the most efficient path anti-parallel to retrograde flow [243]. Thus, specific coupling between MTs and actin bundles presumably promotes MT advance. Proteins that are able to interact with both MTs and F-actin, or proteins that can mediate signaling between both cytoskeletons likely control the regulation of MT invasion. This is further strengthened by specific microtubule association of the microtubule associated proteins MAP1B and MAP2, both known to interact with actin in vitro [214, 270, 242, 249, 53]. Furthermore, MAP2 binds to F-actin and efficiently induces actin bundle formation [242, 101]. It is likely that by crosslinking, MAP2 and/or MAP1B associated with both cytoskeletons could be involved in guidance of MTs along actin filament bundles. Alternatively, MAPs could shuttle from MTs to actin and could alter F-actin behavior by actively crosslinking AFs.

The effect of changes in the cytoskeleton on ion channel activity has been the focus of recent attention. Most studies have employed drugs that selectively stabilize or destabilize either AFs or MTs, with resulting effects on specific ion channel activity, either as changes in whole cell conductance or single channel activity by the patch clamping techniques. Although direct binding of specific cytoskeletal proteins to individual channel proteins remains to be demonstrated, recent work in which exogenous purified actin has been added to the cytoplasmic side of channels excised from the cell membrane has in numerous cases directly confirmed an effect of cytoskeletal filaments. A body of earlier evidence suggested that various cytoskeletal components, including actin and actin-associated proteins, anchor, co-localize and regulate both the spatial stability as well as the function of ion transport proteins. It is possible that actin either binds directly to the channels, or that actin may first interact with actin-binding proteins, which in turn regulate by binding or other indirect interaction- ion channel function. Independent studies demonstrated that membrane-resident CFTR is functionally regulated by AF organization, which may also involve a direct interaction with actin. Thus, ion channel may be controlled by direct and indirect cytoskeletal interactions.

Before we present an integrated view of the conductive properties of the active and passive structural blocks of the cytoskeleton, we should review the current state of knowledge regarding their conductive properties.

4.4.3   Overview of Biological Conductivity

Cope[52] has presented a detailed review of the applications of solid state physics concepts to biological systems. He divided the discussion in seven groups. The first involves semiconduction of electrons across enzyme particles as a rate-limiting process in cytochrome oxidase evidenced by kinetic patterns of enzymes and microwave Hall measurements. Secondly, pn-junction conduction electrons were suggested by kinetics of photobiological free radicals in the eye and photosynthesis. Their I-V characteristics conform to the diode equation:

$$I = I_0 exp(V/kT) \qquad\qquad (4.58)$$

Thirdly, there may be a connection between superconduction and growth in nerves. Fourthly, phonons and polarons are involved in mitochondrial phosphorylation. Next, piezoelectricity and pyroelectricity may be involved in growth of nerves and bone structures. Then, infrared electromagnetic waves may transmit energy in lipid bilayers of nerves and mitochondria. Finally, complexed sodium and potassium ions in structured cell water may be analogous to valence band electrons in semiconductors, free cations being analogous to conduction electrons. Ionic process in cell water resemble electronic conduction in seminconductors. For dried proteins or DNA, Szent-Gyorgi [264] noticed that their conductivities obeyed the relationship:

$$\sigma = \sigma_0 exp(-E_a/kT) \tag{4.59}$$

with activation energies E_a on the order of 1 eV. The presence of water may cause considerable changes in conductivities; for example in cytochrome oxidase this may be reduced to 0.3 eV.

Adessi et al. [1] have provided a theoretical treatment of transmission through a poly(g)DNA molecule. They found that there was a modification of the rise of a B-DNA form which can induce a shift of the conduction channel towards a valence one. They found that deformation of the backbone of the molecule has a significance on the hole transport. Furthermore, the presence of ionic species (e.g., Na) in the surrounding of the molecule can create new conduction channels. In photo-excitation experiments hole transport occurs through coherent transport at short distance and long-range incoherent or hopping transport at large distances. Fink et al. [78] studied the metallic behavior in λ DNA using an electron projection technique. Superconducting behavior has been observed by Kasumov et al. [137]. Porath et al. [217] observed semiconductor behavior with a poly(g)-poly(c)DNA molecule. A scanning force microscope has been used by de Pablo et al. [61] to observe insulating behavior. This section models a single poly(g) DNA lying between electrodes and transmission was computed with a Landauer-Buettiker formalism based on Green's functions. The electronic properties of O_2-doped DNA have been studied by Mehrez et al. [185] who suggested that DNA conduction is affected by O_2-doping, in particular hole doping. However, ab initio atomic structure calculations of O_2-doped DNA indicate negligible charge transfer. The absence of dc conductivity in λ DNA has been investigated by de Pablo et al. First principles electronic structure calculations indicate a minimum DNA resistance of $10^{16}\Omega$ per molecule and a minimum resistivity of $10^4\Omega m$. However, lower energy electron bombardment may dramatically increase conductivity, and with electron energies between 50 and 200 eV may lead to metallic conduction with a resistance of $2 \times 10^8\Omega$. Fink et al. [78] discussing DNA and conducting electrons stated that I-V characteristics of DNA indicate ohmic-like behavior with a resistance of 2 $M\Omega$ for a 600 nm-long DNA rope corresponding to a $\rho = m\Omega$ cm. These authors speculate that DNA may be semiconducting, insulating or metallic simultaneously depending on the arrangement of molecules in the structure. Endres et al. [73] discussed the quest for high-conductance DNA saying that DNA conduction is an unsolved problem first enunciated by Eley and Spivey [71]. Experimental outcomes range from insulating to semiconducting to even superconducting and the uncertainties depend on: (a) contacts between the electrodes and the DNA and (b) differences in the molecules (sequence, length, arrangement, preparation) within the DNA. It appears that drying DNA can lead to conformations with localized electronic high states although hole doping of the backbone by counter-ions might also be possible. Wet DNA may support electrical current due to solvent impurity states in the large high-II* energy gap. Davies and Inglesfield [59] discussed the embedding method for conductance of DNA calculations. It is believed that the pathway for transport in DNA runs through the bases in the center of the double helix molecule. The dominant transport mechanism for DNA appears to be coherent hopping. Calculations show extreme sensitivity to the choice of contact orbitals.

Marechal [181] discussed the transfer of protons as a third fundamental property of hydrogen bonds. The importance of the cooperative resonance types in cyclic structures is stressed. Proton transfers along H-bonds have been discussed as a mechanism of vital importance for chains, helices and in cyclic structures such as rings. Consta and Kapral [49] discussed proton transfer in mesoscopic molecular transfers. Proton transfer rates and mechanics are studied theoretically with solvent molecules in solution. The proton transfer occurs as a result of orientational fluctuations on the cluster's surface. The environment influences solvent effects and reaction rates. The rate constant was established to be 1/59 ps at 260 K and for the bond length fixed at 2 Å. Morowitz [199]discussed proton semiconductors and energy transduction in biological systems. He presented the possibility of proto-chemistry paralleling electro-chemistry. ATP synthesis was discussed as coupled to proton transport assuming a gated protein semiconductor across a membrane. Protein conduction in biological structures involves the water of hydration and protein molecules.

Ripoll et al. [231] investigated ionic condensation and signal transduction. Ion condensation on biopolymers at a critical value of the charge density may lead to intra-cellular signaling due to the diffusion of condensed counter-ions in the near region along cytoskeletal filaments. A feedback mechanism was proposed between condensation/decondensation of calcium and the activation of calcium-dependent enzymes. This is shown to create coherent patterns of protein phosphorylation. It was postulated that ion condensation operates in signal transduction. In the case of polymer ionics [225], the conductivity depends on temperature via the formula:

$$\sigma T = A exp[-B/(T - T_o)] \qquad (4.60)$$

indicating activated hopping. The concept of the Fermi level in solution has been extended by Bockris and Khan [24] through the introduction of a reversible potential for a redox couple.

Unfortunately, measurements of the conductivity of biopolymers (including MTs) are fraught with difficulties due to their structural heterogeneity, strong dependence on the environmental conditions (pH, ionic concentration, temperature, etc.) and the inherent liquid state of the sample. Nonetheless, attempts have been made recently to measure resistance values of MTs, both intrinsically and as ionic conduction cables. Fritzsche et al. [85] made electrical contacts to single MTs following dry-etching of a substrate containing gold microelectrodes. Their results indicate intrinsic resistance of a 12 micron-long

MT to be in the range of 500 megaohms giving a value of resistivity of approximately 40 mega-ohms/micron in their dry state. The same group performed measurements on gold-coated MTs [86] which were covered with a 30-nm layer of gold following sputtering. The resistance of the thus metallized MT's was estimated to be below 50 ohms, i.e., originated entirely due to the metallic coating. Minoura and Muto [192] measured the ionic conductivity of MT's using an electro-orientation method under an alternating electric field. The ionic MT conductivity was calculated as 1.5 +/- 0.5 mS/m which is approximately 15 times greater than that of the buffer solution. This was attributed to the counter-ion polarization and is consistent with another recent study of ionic conductivity along MT's. Furthermore, Goddard and Whittier [96] reported their measurements of RF reflectance spectroscopy of samples containing the buffer solution, free tubulin in buffer, MTs in buffer, and finally, MTs with MAP's in buffer. The sample size in each case was 5 ml. The concentration of tubulin was 5 mg/ml and in the last case the concentration of MAP2 and tau protein was 0.3 mg/ml. The average DC resistance reported by these authors was: (a) 0.999 $k\Omega$ (buffer), (b) 0.424 $k\Omega$ (tubulin), (c) 0.883 $k\Omega$ (MTs) and (d) 0.836 $k\Omega$ (MTs+ MAPs). It's not straightforward to translate these results into resistivity of MTs without making assumptions about their geometrical arrangement and connectivity. However, assuming all tublin being polymerized in case (c) and a uniform distribution of MT's forming a combination of parallel and series networks, one can find the resistance of a 10 mm-long MT, forming a basic electrical element in such a circuit, to have approximately a 8 $M\Omega$ resistance. This compares very favorably to an early theoretical estimate of MT conductivity based on the Hubbard model with electron hopping between tubulin monomers. This model predicted the resistance of a 1-mm MT to be in the range of 200 $k\Omega$, hence a 10-mm MT would be expected to have an intrinsic resistance of 2 $M\Omega$, within the same order of magnitude as the result reported by Goddard and Whittier [96]. While direct experimental analysis of intrinsic conductivity of the various cytoskeletal components is still nebulous, theoretical modeling of these effects is very well developed. The next section presents our current understanding of the intrinsic conductivity of MT's.

4.4.4 Intrinsic Electronic Conductivity of Microtubules

Microtubules (MT's) are ubiquitous in eukaryotic cell biology. Their primary role is to serve as the cellular scaffold and to serve as a corridor for the transport of vesicles through the cell by motor proteins. The interest in MT's stems from their involvement in a number of crucial cellular processes including mitosis and, more recently, the discovery of their role in the communication between the exterior of the cell and the nucleus [94, 180]. The special arrangement of MT's just prior to chromosome separation makes them an ideal candidate for mediating a signal of some sort which would cause the co-ordinated breakage of sister chromatids. In the axons of nerve cells, the

MT's are arranged in bundles, parallel to the axon [112]. These groups of MT's span the entire length of the axon and might serve to carry signals from the cell body to the nerve terminus or vice versa. No mechanism for this sort of direct feedback is known. We demonstrate here that the MT may be able to conduct a current along its length through a process of electron hopping between binding sites on each dimer.

The MT is made up of a protein called tubulin. Each sub-unit contains α-tubulin and β-tubulin monomers joined together to form a dimer. The dimers are connected end to end to form a protofilament. These protofilaments, usually 13 in number, are wrapped up to form a hollow tube, the MT. The dimers are themselves 5 nm in diameter and have a length of about 8 nm. A MT has an outer diameter of 25 nm and an inner diameter of 15 nm.

The tubulin dimer of which the MT is composed has at least two different conformational states [118]. That is, while the amino-acid chain of the protein remains fixed, its three-dimensional shape may change. It is believed that these conformational changes are the result of hydrolysis and an associated charge movement from one binding site within the molecule to another. It is precisely this charge movement which may lead to the conduction through the polymer.

One may take a semi-classical view of the movement of electrons along the protofilament. We ignore the fact that the protofilament is nearly one-dimensional and consider its *bulk* properties. Assuming two conduction electrons for each tubulin dimer, the conduction electron density is:

$$n_0 = 2/(8nm \times 5nm \times 5nm) \tag{4.61}$$
$$= 1.0 \times 10^{19}/cm^3.$$

In comparison with other materials, this value is slightly higher than typical semi-conductor concentrations, usually $10^{13} - 10^{17}/cm^3$, but is much less than typical metals which have a conduction electron density of $10^{22} - 10^{23}/cm^3$. However, the electrons are expected to have relatively low mobilities unless they are excited from the ground state configuration. The excitation energy is a function of both the number of electrons within the protofilament and its size. Suppose that the mobility of excited electrons is at least similar to the mobility of electrons within semi-conductors such as Si where the electron mobility, μ_e, is about 1300 cm^2 V^{-1} s^{-1}. The electrical conductivity σ is given in a semi-classical model by the well-known formula

$$\sigma = ne\mu_e \tag{4.62}$$

where e is the electron charge and n is the conduction electron density given by

$$n = n_0 e^{-\Delta/\kappa T} \tag{4.63}$$

where n_0 represents the total number of electrons. If we assume that the gap between valence and conduction bands, Δ, is about 0.40 eV then at physiological temperature, some of the carriers will be excited and should be available to conduct. Applying the above formula (4.63), a conductivity of about 10^{-2} (in SI units) is expected for the protofilament.

$$\sigma \sim (10^{25} e^{-0.40/0/026})(1.6 \times 10^{-19})(0.13) \sim 0.04 \Omega^{-1} m^{-1} \qquad (4.64)$$

However, if the hybridization of electronic orbitals lowers the gap to a value closer to zero, the expected conductivity would be significantly higher:

$$\sigma \sim (10^{25})(1.6 \times 10^{-19})(0.13) \sim 2 \times 10^{5} \Omega^{-1} m^{-1} \qquad (4.65)$$

These values should be compared to values in excess of 10^{8} for copper and other good conductors, 10^{7} for lesser conductors and semi-metals. Typical semi-conductors such as Si and Ge have intrinsic conductivities of $10^{-3} - 10^{+1}$ at room temperature but when heavily doped, the conductivity may rise to between 10^{4} and 10^{5}.

In order to model the MT as a semiconductor, the individual tubulin dimer must first be physically analyzed. The $\alpha\beta$-tubulin dimer is comprised of about 890 amino acids which represents about 13000 nuclei. A gross simplification which we introduce is to treat each monomer as having an effective site energy for electronic binding. Fluctuations of the site energy on scales less than the monomer spacing are therefore washed out. It is convenient to model this system as two quantum wells separated by a potential barrier. The two wells represent two binding sites for an electron which may hop between the monomers resulting in the change of molecular conformation. We can estimate the relevant parameters of the well depth and the well width from kinetic studies and with a knowledge of the molecule's overall geometry. Once the bound states and their energies are known for this system, this knowledge may be applied to a second-quantization scheme. This procedure allows us to make progress without specifically requiring us to solve for the wavefunction of the system.

A 1D chain of quantum double wells is pictured in Fig. 4.20. For simplicity, we are initially making each binding site have the same energy. If this is related to GTP hydrolysis, they may differ by about 0.17 eV (4 kcal/mol) [121]. This is roughly the quantity of energy that is stored in the lattice when GTP is hydrolyzed to GDP [122, 33]. The barrier height between dimers has been estimated from kinetic data. The bond strength must be somewhat larger than the hydrolysis energy delivered to the lattice or else hydrolysis of the dimers would cause immediate disassembly. Hopping along the protofilament by electrons presumably has an energy comparable to this binding energy. We have estimated 0.4 eV for intra-dimer hopping and 1.0 eV for the inter-dimer hopping. What may be varied in the model is the central barrier which must be overcome for each tubulin dimer to change conformation. If it is nearly as large as the barrier to electron movement between dimers, then the protofilament will look like a chain of single quantum wells rather than double quantum wells because the difference between the α and β monomers will not be distinguished. However, if the central barrier is too small, the chain will again look like a series of single quantum wells since the effect of the barrier would be negligible. The depth of the deeper well, f_1, has been selected as 0.4 eV initially. Given the value, we assumed a 25% difference in well depths. If f_1 is much larger, the dimer nature of the protofilament is lost because the wells are effectively the same depth.

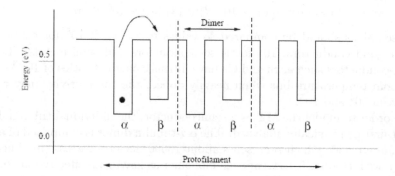

FIGURE 4.20:   A one-dimensional chain of quantum wells representing the energy profile along a MT protofilament where the α and β monomers have alternating site energies.

The results for the single electron eigenstates of a double quantum-well may be found using a transfer matrix method for the time-independent 1D Schrodinger equation in each interval in which the piece-wise constant potential is defined

$$\hat{H}\psi = \frac{\hat{p}^2}{2m}\psi + \hat{V}\psi = E\psi \qquad (4.66)$$

The oscillatory solution

$$\psi(x) = A\cos(kx) + B\sin(kx) \qquad (4.67)$$

applies to $E > V$ where

$$k = a\sqrt{2m(E-V)}/\hbar \qquad (4.68)$$

while for $E > V$, the solutions are exponentially damped

$$\psi(x) = Ae^{\kappa x} + Be^{-\kappa x} \qquad (4.69)$$

where

$$k = \sqrt{2m(V-E)}/\hbar \qquad (4.70)$$

Now, matching conditions for the wave function and its derivative are applied at each interface. This gives a 2×2 matrix relating the coefficients in the wave function expansion according to

$$\begin{pmatrix} A_2 \\ B_2 \end{pmatrix} = \begin{pmatrix} {}^{2,1}M_{11} & {}^{2,1}M_{12} \\ {}^{2,1}M_{21} & {}^{2,1}M_{22} \end{pmatrix} \begin{pmatrix} A_1 \\ B_1 \end{pmatrix} \qquad (4.71)$$

In the above equation, the superscript to the left of the M denotes which coefficients are being linked. The subscripts indicate the matrix element. In an iterated fashion, one eventually arrives at the following:

$$\begin{pmatrix} A_n \\ B_n \end{pmatrix} = \begin{pmatrix} {}^{n,1}M_{11} & {}^{n,1}M_{12} \\ {}^{n,1}M_{21} & {}^{n,1}M_{22} \end{pmatrix} \begin{pmatrix} A_1 \\ B_1 \end{pmatrix} \qquad (4.72)$$

Now, for bound states $A_1 = B_n = 0$ so that the wave-function decays outside the region of the quantum wells. Therefore,

$$B_n = 0 \quad \Rightarrow \quad {}^{n,1}M_{22} = 0 \qquad (4.73)$$

Given the choices for our quantum double well: width, $a = 2$ nm and height, $V = 1.0$ eV, one finds that there are two bound states for this dimer problem. As the number of dimers is increased, one resorts to numerical analysis. In Fig. 4.21, the energy of the two bound states is plotted as the energy of the central barrier is varied. The lowest energy state is the symmetric wavefunction and increases in energy sharply as the central barrier is increased. The second bound state is the anti-symmetric wavefunction and has a node within the barrier and hence its energy increases only marginally.

The energy calculation for a sequence of n such dimers leads to the development of bands as hybrid orbitals develop across the polymer. Roughly speaking, one band is of hybridized symmetric orbitals and the higher lying band of hybridized anti-symmetric orbitals. The separation of the two bands and the spread of each individual band determine the values of the parameters which shall be used in the tight binding Hubbard model used to model the polymer.

The second-quantized Hamiltonian for a many-electron multidimer system becomes

$$\hat{H} = \sum_{i,\sigma} \epsilon_i \hat{c}^\dagger_{i\sigma} \hat{c}_{i\sigma} + \sum_{i \neq j, \sigma} t_{ij} \hat{c}^\dagger_{i\sigma} \hat{c}_{j\sigma} \qquad (4.74)$$

where the operators $\hat{c}^\dagger_{i\sigma}$ and $\hat{c}_{i\sigma}$ are the usual fermion creation and annihilation operators, and where i runs from 1 to $2n$ representing the two sites on each of the n dimers. The ϵ_i represent the potential energy at each site and the kinetic terms t_{ij} account for hopping from one site to another. There are two different kinds of interactions depending on whether an electron hops within a dimer, or between dimers. Hence, this kinetic parameter has two different values

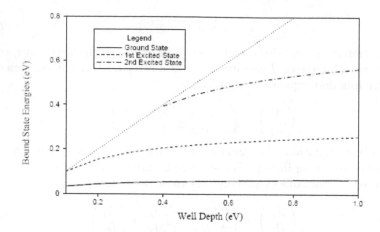

FIGURE 4.21: Variation of the bound state energies with the depth of the double quantum wells, $f_1 = f_2$. The straight line indicates where the bound state energy is equal to the well depth.

along the chain. Since there are no interactions between the states with $+\frac{1}{2}$ spin and those with $-\frac{1}{2}$ spin, the state space may be divided into two identical subspaces labelled by the electron spin. In each of these subspaces, the matrix corresponding to this chain of dimers is tri-diagonal and may be diagonalized to determine the itinerant eigenstates of the system. The coefficient t_d which represents the movement of an electron within a dimer causes the eigenstates to be split into two bands by an amount, $2t_d$. Within each of these bands, the coefficient t_l of the inter-dimer hopping causes these two single electron bands to broaden each to about $2t_l$. Now, although the above description is qualitative, clearly whenever $t_d \sim t_l$, these bands will overlap and transitions between the states will be important.

Considering the band formation in the quantum well formulation, we can assign values to t_d and t_l. The two bands are separated by about 0.40 eV, so $t_d \sim 0.20$ eV and using the bottom band to estimate t_l it widens until it is about 0.10 eV thick, so $t_l \sim 0.05$ eV.

The Hubbard model is the simplest second quantized model to capture all of the essential features of a real system of interacting particles [161]. In this model, electrons are the fundamental excitations. All states may be built up from the vacuum using the fermion creation operators. Once the number of sites is selected, n, and the number of electrons is determined, e, there are

$$\binom{2n}{e} = \frac{2n!}{e!(2n-e)!} \tag{4.75}$$

distinct electronic states. These states are built up by applying in normal order each of the combinations of e-electron states.

The second quantized Hamiltonian for the Hubbard model with electron-electron interactives is conventionally written as

$$\hat{H} = \sum_{i,\sigma} \epsilon_i \hat{c}_{i\sigma}^\dagger \hat{c}_{i\sigma} + \sum_{i \neq j,\sigma} t_{ij} \hat{c}_{i\sigma}^\dagger \hat{c}_{j\sigma} + \sum_i U_i \hat{c}_{i\uparrow}^\dagger \hat{c}_{i\uparrow} \hat{c}_{i\downarrow}^\dagger \hat{c}_{i\downarrow} \qquad (4.76)$$

The last term accounts for electron-electron repulsion now that we have gone to a many electron system. In our simple model, this is limited to an on-site interaction. To this point, we have made each of the site energies equivalent ($\epsilon_i = \epsilon, \forall i$). In what follows, $U = 2.0$ eV which is the energy associated with two bare electrons lying about 0.7 nm apart. This choice seems reasonable within a well of width 2.0 nm. We try to solve the problem by considering the electron-electron term as a perturbation to study how it affects the electronic states of the dimer. The Hamiltonian in the absence of the final U-term may be diagonalized by the following unitary transformation:

$$\alpha_{1/\sigma}^\dagger = \frac{1}{\sqrt{2}}(c_{1\sigma}^\dagger + c_{2\sigma}^\dagger) \qquad (4.77)$$

$$\alpha_{x/\sigma}^\dagger = \frac{1}{2\sqrt{2}}(c_{1\sigma}^\dagger - c_{2\sigma}^\dagger) \qquad (4.78)$$

The linear combination of operators is simple because we have made the site energies ϵ_1 and ϵ_2 equal. These new operators are easily shown to obey the usual anti-commutation relations for fermions and are said to be in the itinerant or distributed electron basis. The Hamiltonian may now be rewritten in terms of these new operators, as

$$\hat{H} = \sum_\sigma (\epsilon + t)\alpha_{1\sigma}^\dagger \alpha_{1\sigma} + \sum_\sigma (\epsilon - t)\alpha_{2\sigma}^\dagger \alpha_{2\sigma} + \frac{1}{2} \overset{\sum}{\underset{\sigma \neq \sigma\prime}{i,j,l,m}} V_{ijlm} \alpha_{i\sigma}^\dagger \alpha_{j\sigma\prime}^\dagger \alpha_{l\sigma\prime} \alpha_{m\sigma}$$

$$(4.79)$$

the vacuum state is represented by $|\Omega>$. Acting with our Hamiltonian on this basis gives us the matrix form of the Hamiltonian,

$$\hat{H} = \begin{pmatrix} 2(\epsilon+t) + \frac{(U_1+U_2)}{4} & 0 & \frac{(U_1-U_2)}{4} & \frac{(U_2-U_1)}{4} & 0 & \frac{(U_1+U_2)}{4} \\ 0 & 2\epsilon & 0 & 0 & 0 & 0 \\ \frac{(U_1-U_2)}{4} & 0 & 2\epsilon + \frac{(U_1+U_2)}{4} & -\frac{(U_1+U_2)}{4} & 0 & \frac{(U_1-U_2)}{4} \\ \frac{(U_2-U_1)}{4} & 0 & -\frac{(U_1+U_2)}{4} & 2\epsilon + \frac{(U_1+U_2)}{4} & 0 & \frac{(U_2-U_1)}{4} \\ 0 & 0 & 0 & 0 & 2\epsilon & 0 \\ \frac{(U_1+U_2)}{4} & 0 & \frac{(U_1-U_2)}{4} & \frac{(U_2-U_1)}{4} & 0 & 2(\epsilon-t) + \frac{(U_1+U_2)}{4} \end{pmatrix}$$

The Hamiltonian is then diagnolized in order to determine its eigenstates. In the absence of the Coulomb repulsion ($U = 0$), the electrons are localized on the $2'$ site with energy $2(\epsilon - t)$, the gap to the first excited state being $2t$. However, as U is increased, the effect of the Coulomb interaction raises the energy of the ground state relative to the excited singlet states which do not feel the Coulomb repulsion. Consequently, the gap between the ground state and the first excited state is actually reduced through the introduction as shown in Fig. 4.22. Thus, when conductivity is being considered, it is important to bear in mind that the gap is reduced by the presence of the electron-electron interactions.

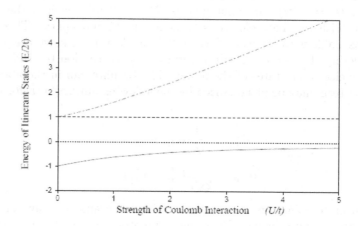

FIGURE 4.22:   Energy of the itinerant states as a function of U/t with the energy plotted in units of $2t$.

To this point, we have made each of the site energies equivalent, $(\epsilon_i = \epsilon)$, the electron repulsion the same at all sites $(U_i = U)$. The lone complication lies within the kinetic term, where the hopping parameter will depend on the two sites between which the electron hops. However, since we first consider the MT protofilament, it shall initially be a constant as well, $t = 0.4$ eV, where the choice comes from the separation of the energy bands in the 1D Schrödinger picture.

Since the Hamiltonian conserves both the particle number and the overall spin of each state, we are free to work in subspaces enumerated by the spin of the states it contains. This reduces the problems slightly since it is possible to consider individually the spin $+1$, spin 0 and spin -1 systems separately in a two-electron system. In addition, the symmetry implies that the spin $+1$ and spin -1 systems have the same conductive properties. We also find that the ground state is always a state consisting of the lowest possible total spin.

When the inter-dimer hopping term is small relative to the intra-dimer hopping term $t_{inter} < t_{intra}$, the form of the ground state is independent of the number of dimers. In each case, the ground state is a linear combination of singlet states on each dimer. The singlet states are those where two electrons are paired on an individual binding site, or where the state consists of the anti-symmetric combination of an up-electron and a down-electron on neighbouring binding sites of the dimer. When the inter-dimer hopping parameter is increased, the density of states becomes more uniform between the ground state and the highest energy eigenstate with the gap between the ground state and the first excited state of the system becoming diminished. Since the hopping term within a dimer is largest, the system prefers to have two electrons with the same spin within the same dimer in the ground state.

For electronic conductivity the band formation importantly depends upon the hopping parameter between dimers. When this parameter is small relative to the intra-dimer hopping, $t_{inter}/t_{intra} \sim 0.1$, there remains a rather large gap between the ground state and the first excited state of about $2t_{intra}$ which is reduced as the repulsion term increases. However, when the hopping between dimers is equal to the hopping within dimers, $t_{inter} \sim t_{intra}$, the gap is minimized. In the event that $t_{inter} > t_{intra}$, we would effectively have to relabel the dimers and we would then return to the picture where $t_{inter} < t_{intra}$.

Kubo was the first to derive formulae for the electrical conductivity in solids and his derivation is applied to our case in the present section [151]. In the Coulomb gauge we write the vector potential for a uniform electric field along the direction of the chain, $A(t)$, as

$$A(t) = \frac{cE(t)}{-iw}e^{iwt}. \tag{4.80}$$

Adding a field-coupling term to the Hubbard Hamiltonian and using a canonical transformation gives [174]

$$\hat{H}_0' = -t\sum_{i\sigma}[\hat{c}_{i\sigma}^\dagger \hat{c}_{i+1\sigma}e^{-ieA(t)/c} + \hat{c}_{i\sigma}^\dagger \hat{c}_{i-1\sigma}e^{ieA(t)/c}] \tag{4.81}$$

where a is the lattice spacing and c is the speed of light. Expanding to second order in the vector potential, we find

$$\hat{H}_A = \frac{A}{c}\hat{j} - e\frac{A^2}{c^2}a^2\hat{H}_0, \tag{4.82}$$

where

$$\hat{j} = iEat\sum_{i\sigma}(\hat{c}_{i\sigma}^\dagger \hat{c}_{i+1\sigma} - \hat{c}_{i\sigma}^\dagger \hat{c}_{i-1\sigma}). \tag{4.83}$$

So, the energy of the original system is corrected to this order and the interaction term is proportional to the current.

Suppose a time-dependent external electric field is applied to a solid,

$$E_\alpha^{\text{ext}}(r,t) = \Xi_\alpha^{\text{ext}} e^{iq\cdot r - i\omega t}, \tag{4.84}$$

where α represents the Cartesian coordinate directions. In a linear response theory, the induced current is proportional to the applied electric field:

$$J_\alpha(r,t) = \sum_\beta \sigma\prime_{\alpha\beta}(q,\omega) \Xi_\beta^{\text{ext}} e^{iq\cdot r - \omega t} \tag{4.85}$$

where $\sigma\prime_{\alpha\beta}$ is a parameter relating the observed current density to the applied field's direction, its periodicity and frequency. However, the symbol appearing in (4.85) is not the conductivity we seek. Rather, we want the conductivity which represents the response to the total electric field in the solid, a quantity that can be measured. This conductivity takes into account all of the currents, induced by the external fields, that create their own electric fields. Thus we seek $\sigma_{\alpha\beta}$ that relates the macroscopic electric field to the currents of the system.

$$J_\alpha(r,t) = \sum_\beta \sigma_{\alpha\beta}(q,\omega) E_\beta e^{iq\cdot r - i\omega t} \tag{4.86}$$

$$E_\alpha(r,t) = \Xi_\alpha e^{iq\cdot r - i\omega t} \tag{4.87}$$

$$\sigma_{\alpha\beta} = Re(\sigma_{\alpha\beta}) + iIm(\sigma_{\alpha\beta}) \tag{4.88}$$

We write the Hamiltonian of the system as $\hat{H} + \hat{H}\prime$ where the latter term contains the electric field which we shall introduce as a time-dependent perturbation. The evolution of the operators is given as follows [66]

$$\hat{j}(x,t) = e^{i\hat{H}t} \hat{j}(x) e^{-i\hat{H}t} \tag{4.89}$$

where $\hat{H}$ is the unperturbed Hamiltonian. Thus the interaction picture is adopted. The Kubo formula for electrical conductivity gives the result in terms of a current-current correlation function:

$$\sigma_{\alpha\beta}(q,w) = \frac{1}{\omega} \int_0^\infty dt e^{i\omega t} < \Omega | \hat{j}_\alpha^\dagger(q,t)\hat{j}_\beta(q,0) - \hat{j}_\beta(q,0)\hat{j}_\alpha^\dagger(q,t) | \Omega > + i\frac{n_0 e^2}{m\omega}\delta_{\alpha\beta} \tag{4.90}$$

where $|\Omega>$ is the ground state of the system which has not been perturbed by the electric field. The first term in (4.90) is known as the incoherent contribution to the conductivity. Now, we use (4.89) in the above equation and use the fact that the basis we are using is that of the eigenstates. Consequently, with proper normalization the new expression for the real part of the conductivity is [173]

$$\sigma(\omega) = \frac{\pi}{Z}\frac{(1 - e^{-\beta\omega})}{\omega} \sum_{n,m} e^{-\beta E_m} |<n|\hat{j}_\alpha|m>|^2 \delta(\omega - (E_n - E_m)) \tag{4.91}$$

The DC conductivity is calculated by a procedure [174, 173] which employs the sum rule

$$\int_0^\infty \sigma'(\omega)d\omega = -\frac{\pi e^2}{2}a^2 < K >$$ (4.92)

where

$$\sigma'(\omega) = D\delta(\omega) + \sigma_{kubo}(\omega)$$ (4.93)

and $< K >$ is the expectation value of the kinetic energy term of the Hamiltonian. The strategy of computing $\sigma(\omega)$ and subsequently the Drude contribution at either zero or finite temperature is reasonably straightforward. The first step is to construct the matrix corresponding to the Hamiltonian (eq. 4.76) in a subspace which is defined by the geometry of the lattice, the number of electrons and the total electron spin. This leads to serious computational difficulties since for a moderately sized lattice of say ten dimers and eight electrons of arbitrary spin, the corresponding space is about 77 million states. A matrix of dimension 77 million square has about 5.9×10^{15} elements. Obviously, working on a moderately sized lattice, we are restricted to a low concentration of electrons. Due to electron-hole symmetry, we are also able to study the nearly filled situation. It is useful to keep in mind the relationship

$$\lambda = \frac{1240nm \cdot eV}{\text{Energy}}$$ (4.94)

So in the plots following, the visible light range is between 1.75eV and 3.10 eV. Energies larger than this range correspond to ultraviolet and smaller to infrared wavelengths. Since cells respond to light in the near-infrared [8], and centrioles which are comprised of MTs are proposed to be the site of light detection [9], absorption in this energy range is of great interest to cell biology. These energies should also be compared to the energy available from ATP and GTP hydrolysis which are 0.49 eV and 0.22 eV, respectively. These values provide limits for the interactions of chemical energy in the conduction process.

At zero temperature, the linear polymer results are summarized concisely by saying that there is no DC conductivity which agrees with the previous modelling of Hubbard systems [89].

The following results for the 1D chain of tubulin dimers represent the AC conductivity. There are peaks in $\sigma(\omega)$, also known as the optical conductivity, which correspond to excitations between electronic states of the polymer. Electronic excitation results in electron redistribution and hence the development of a current. We have selected $\epsilon = 0.02$ eV to present our results which is roughly kT and is smaller than the three energy scales in the Hubbard model, t, U and $4t^2/U$ [200] so we expect to be able to pick out features corresponding to specific transitions in the predicted optical spectra.

We shall begin with two dimers and $t = 0.4$ eV, $U = 2.0$ eV and equal site energy on the α and β monomer sites. This system can support up to eight electrons. The results for these systems are presented in Fig. 4.23. The predicted spectra for the two and three electron cases are quite similar. The differences include a new but small peak in the three electron spectrum at an energy just below the major peak and variations in the detail of the absorption band centered about 2.6 eV. However, the qualitative similarity of the two spectra is remarkable in terms of band locations. Contrast these two cases with the four-electron system which corresponds to half-filling. The difference is that the peak in the optical conductivity spectrum increases dramatically from about 0.48 eV to 1.64 eV. This change indicates how the half-filled case is special. The ground state in this system largely consists of electrons spread out to minimize the Coulombic energy. Exciting this system by moving any particular electron will come at the cost of greatly increasing the electron-electron repulsion and consequently, the location of the required excitation energy corresponds roughly to U.

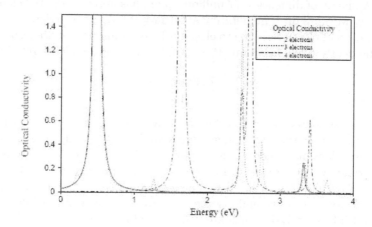

FIGURE 4.23:   Optical conductivity of the two dimer system with two, three and four electrons.

As the length of the polymer is increased, results remain qualitatively similar. In the three-dimer structure, results are shown for the conductivity in Fig. 4.24a and for the integrated conductivity in Fig. 4.24b. Once again, the spectra are quite similar for all fillings aside from the half-filled case which consists of six electrons. The shift of the conductivity to higher energies can be clearly seen as the filling fraction increases towards half filling. We now present results where the broadening parameter, ϵ, and the temperature have been varied to see the effect which they have on the results. Fig. 4.25 demonstrates that as the width is decreased, the spectrum becomes sharper. Note how the two peaks in this simple spectrum which occur near 2.5 eV are successively blended into a band as the ϵ is increased. Peaks represent transitions from one specific electron configuration to another and are weighted with the associated conductivity. The effect of changing the temperature is to excite electron configurations other than the ground state. Consequently, transitions from one excited state to another can now contribute to the conductivity. Fig. 4.26 shows how this often reduces the gap to conduction; all the major peaks in the 300 K spectrum shift slightly towards lower energies in the 3000 K spectrum but many more features develop as well as states which previously did not contribute to conductivity may make transitions to excited states. Physiological temperature of 300 K ($kT = 0.026$ eV) is small relative to the system's energy scales; consequently, lowering the temperature further causes very little change to the spectrum.

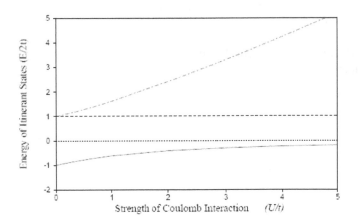

FIGURE 4.24: (a) Optical conductivity of the three-dimer system from two electrons through half filling. (b) Integrated optical conductivity of the three-dimer system shows that indeed the protofilament is an insulator. The gap for excitation of the half-filled polymer is seen easily here.

If the α-tubulin and β-tubulin sites have different energies, then the MT A and B lattices can be distinguished. The effect of this change on the optical spectrum seems to be small when individual protofilaments are considered. In addition, the sum rule shows that protofilaments remain insulators. There are a few parameters which we can consider individually and which can be compared as the size of the system changes: the threshold to conductivity, and the expectation value of the kinetic energy operator. The threshold to conductivity is simply the location of the first absorption peak in the optical

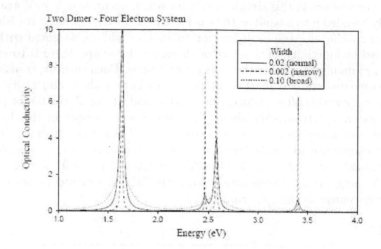

FIGURE 4.25: Optical conductivity spectrum variation with the broadening parameter.

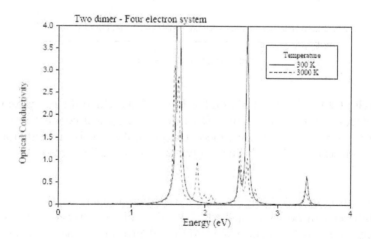

FIGURE 4.26: The variation of the optical conductivity spectrum with temperature is shown.

spectrum. It is a decreasing function of the polymer length. Starting at 0.48 eV for the two-dimer system, it seems to drop a little faster than $1/n$ for systems consisting of n dimers and containing two electrons. The fitted value in Fig. 4.27 varies $\alpha 1/n^{1.4}$. Extrapolating to larger values of n, this gap appears certain to fall into the range where these excited states could be thermally excited. The protofilament then develops a finite DC conductivity at finite temperature. It is interesting that although the first peak does move towards zero, the fraction of the conductivity it contains is gradually diminished. Some of the conduction continues to reside in a peak located close to t in the energy spectrum. In the 2D lattice, there is a phase transition in the dimensionality. DC conduction becomes possible even at zero temperature. However, we shall still be concerned about the boundary conditions since calculations are being performed on a small lattice.

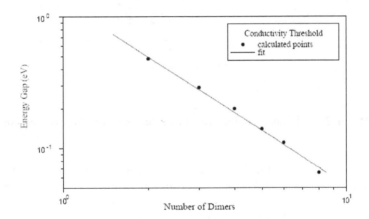

FIGURE 4.27:   Threshold to conduction as a function of protofilament length.

Consider the kinetic energy/electron which has been calculated in a Hubbard model with our parameters for t and U and with either one, two or three protofilaments. Fig. 4.28 demonstrates how the kinetic energy of each electron approaches the limit of $2t$ or 0.80 eV as the size of the system increases. We can easily see that for more than about 20 electron sites, the average kinetic energy of each electron starts to approach the infinite limit and we expect that the results obtained from our model should converge to the long protofilament limit in a similar manner. In particular, we are considering 28 electron sites in our MT model which span 3 protofilaments. While this is not the ideal situation, the boundary conditions are not expected to create huge effects in our results. We are also using open boundary conditions as they minimize the boundary effects when compared with a variety of periodic boundary conditions.

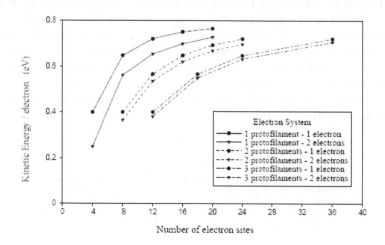

FIGURE 4.28:   Kinetic energy per electron as a function of system size.

The unit cells depicted in Fig. 4.29 are the basis for the calculations. In the A lattice used for the calculations, the α-dimers connect to β-dimers along the $\bar{3}$-start helix both within the lattice and at the seam when the MT is wrapped up. In the B lattice, the connections are $\alpha - \alpha$ or $\beta - \beta$ along the $\bar{3}$-start helix and never $\alpha - \beta$. For comparison, calculations have also been carried out with a flat tubulin sheet, specifically an A lattice MT lacking periodic boundary conditions in the lateral direction. This simulates the MT when it has been unwrapped. We recall that the MT lattice is triangular and has three distinct hopping directions. As well, since protofilaments are strongly bound together along their length but relatively weakly bound to other protofilaments, one expects the hopping parameter to be much smaller when an electron moves between protofilaments. The calculations have been carried out with the same parameters as in the one-dimensional case, $U = 2.00$ eV, $t = 0.40$ eV along the protofilaments and the site energies have been set equal to zero for all sites. The value of the hopping parameter along the $\bar{3}$-start helix is called t_ℓ (left) and that along the 8-start helix that connects diagonally from a monomer in a direction up and to the right is called t_r (right).

We consider first the results of the MT lattice which has not rolled up to form a tube but is that of a tubulin sheet where $t_r = 0$ and only the value

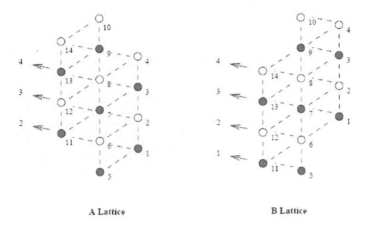

A Lattice   B Lattice

FIGURE 4.29: Lattice unit cells for conductivity calculations. The unit cell on the left corresponds to the MT A lattice while that on the right corresponds to the MT B lattice.

of t_l is varied (see Fig. 4.30). Since $t_r = 0$, the nature of the lattice remains rectangular to this point. When $t_l = 0$, this particular system is equivalent to individual protofilaments. The two peaks in the optical spectrum arise from the fact that the protofilaments have different lengths. The absorption peaks of 0.22 eV and 0.40 eV correspond roughly to those of the two-dimer and three-dimer two electron cases discussed earlier. The difference here is that the two electrons are spread over three protofilaments and consequently the effect of the electron repulsion is reduced. However, as t_l is increased from zero, the entire lattice becomes accessible and the protofilament character of the optical spectrum is lost. The two peaks of the t_l spectrum coalesce into a single peak with a large activation energy. A second but smaller absorption peak forms just above 0.50 eV. In the plot of the integrated conductivity, we can see that there is also a smaller peak at 0.16 eV when $t_l = 0.10$ eV. The source of this peak is an absorption for conduction along the direction of t_l. When t_l is increased to 0.40 eV, this peak occurs at 0.61 eV. Thus this peak seems to occur at roughly $1.5t_l$. The activation energy for conduction along the protofilament axis also increases but is not as sensitive to t_l. Finally, examination of the integrated conductivity demonstrates that the total absorption combines such that the Drude weight remains zero.

The results in Fig. 4.31 show a difference between an individual protofilament and a tubulin sheet. The behaviour here is inconsistent with that of an insulator since the Drude weight is non-zero. In the optical conductivity we find that the largest peak occurs at the same location in all directions but with different weightings. The first line traces the results for $t_r = 0.00$ eV and

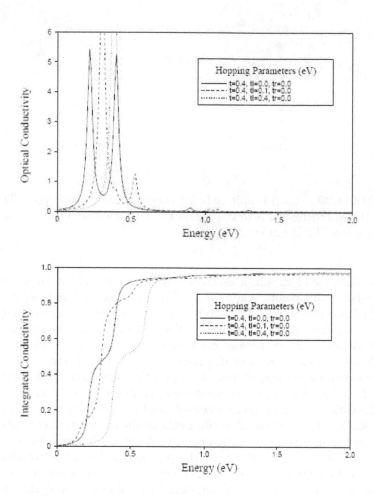

FIGURE 4.30:   (a) Optical conductivity along the protofilament axis of the tubulin sheet consisting of a 14-site lattice with two electrons while $t_r = 0$ and t_l is varied. (b) Integrated conductivity of the tubulin sheet with two electrons while $t_r = 0$ and t_l is varied.

has peaks at 0.38 eV and 0.61 eV. Once the third hopping direction is allowed, the peak absorption is raised slightly to 0.43 eV for $t_r = 0.10$ eV and a smaller peak forms at 0.74 eV. The lower peak consists mainly of hopping along the protofilament while the second peak consists largely of hopping along the t_l direction, which also corresponds to a large hopping parameter. Once t_r is raised to 0.40 eV so that it has the same magnitude as the other hopping parameters, there is only a single peak at 0.70 eV.

The effect of applying the MT A lattice boundary conditions to our system is shown in Fig. 4.32, where we see that changing only t_l to a finite quantity while still maintaining t_r as zero is sufficient to result in a non-zero Drude weight. Thus the periodic boundary conditions in the lateral direction seem to act in a manner similar to the additional hopping directions of the unwrapped lattice. When the kinetic parameters between protofilaments are set to zero, the optical conductivity spectrum shows that there are two large absorption peaks at about 0.23 eV and 0.40 eV. As t_l increased, these peaks move higher in energy and the second peak becomes smaller in integrated weight. Eventually, once $t_l = t$, there is a single large peak near 0.47 eV. Introduction of the third hopping direction serves only to raise further the absorption peak, up to 0.70 eV in the case where the hopping parameter is 0.40 eV in all directions.

From Fig. 4.32 one can see that more conduction results in the B lattice for a small t_l, such as 0.1 eV, than in the MT A lattice by comparing with Fig. 4.33. However, even the positioning of absorption peaks is similar in the two lattices. Again, it is sufficient for only one of t_l or t_r to be non-zero for the Drude weight to be finite.

Finally, we compare the MT A and B lattices with the unwrapped tubulin sheet and in addition with individual protofilaments of tubulin dimers. There is an interesting geometrical interpretation to the results, which show that the MT A and MT B lattices have identical conduction properties when $t_l = t_r$. This is because in our calculations we have considered all monomers to have the same site energy. Consequently, the B lattice can be viewed as an A lattice with the opposite helicity but the same structure. The sign of the system's helicity does not affect the conduction properties along the major axis. The unwrapped MT lattice is not as good a conductor since some of the hopping freedom has been removed from the electrons on the lattice but does remain conducting provided that both t_l and t_r are non-zero.

If we consider the case where one of t_l or t_r is quite small, however, the wrapping of the lattice is much more important. As discussed earlier, the unwrapped tubulin sheet is an insulator while the MT A and B lattices may carry electrons. This observation is especially interesting given that MTs have been observed to zip up and essentially change their structure from that of a tube to that of a sheet *in vivo*.

Our calculations based on the Hubbard model of electron hopping show that the MT is not an insulator but that its conductivity depends on the lattice geometry and whether the MT is wrapped up or not. How well or poorly does one expect the MT to conduct relative to known semi-conductors? Referring

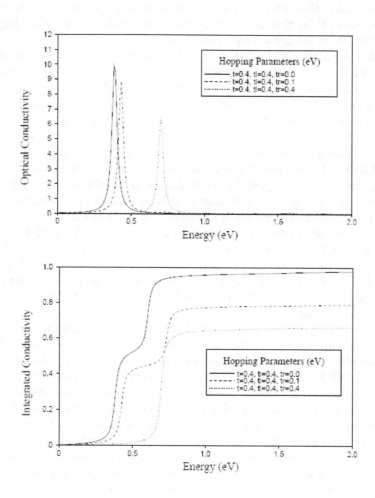

FIGURE 4.31: (a) Optical conductivity along the protofilament axis of the unwrapped 14-site lattice with two electrons while $t_l = 0.4$ and t_r is varied. (b) Integrated conductivity of the unwrapped 14-site lattice with two electrons while $t_l = 0.4$ and t_r is varied.

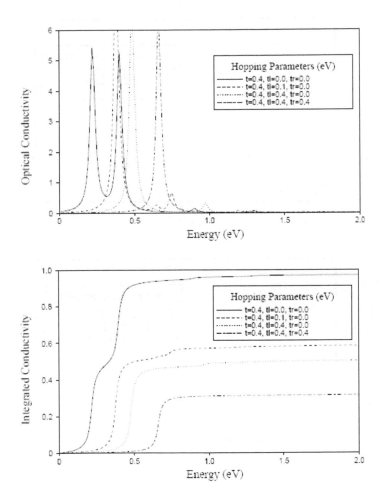

FIGURE 4.32: (a) Optical conductivity along the protofilament direction of the 14-site MT B lattice with two electrons. (b) Integrated conductivity of the 14-site MT B with two electrons shows a non-zero weight for finite t_l.

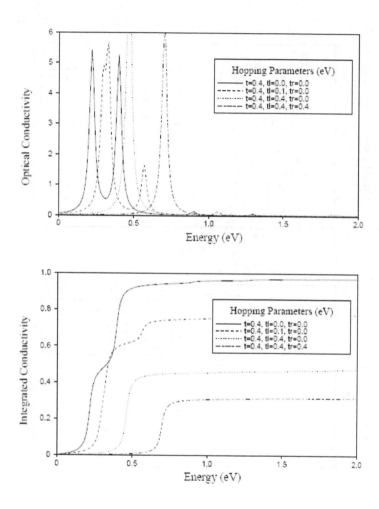

FIGURE 4.33: (a) Optical conductivity along the protofilament direction of the 14-site MT A lattice with two electrons. (b) Integrated conductivity of the 14-site MT A lattice with two electrons shows a non-zero Drude weight for finite t_1.

back to equations (4.92) and (4.93), the DC conductivity of the MT is simply given by D as

$$\frac{D}{2} = -\frac{\pi e^2}{2} a^2 < K > - \int_0^\infty \sigma_{kubo}(\omega) d\omega \qquad (4.95)$$

In this equation, the constants can be factored and what remains is the fraction that can be read from the graphs presented earlier. The Kubo fraction (KF) is the limit of the integral as the frequency tends to infinity.

$$\frac{D}{2} = -\frac{\pi e^2}{2} a^2 < K > (1 - KF). \qquad (4.96)$$

By returning constants of the system so that the dimensionality of D is correct, we find

$$D = -\frac{\pi n e^2 a^2}{\hbar} \frac{< K >}{t} (1 - KF). \qquad (4.97)$$

Finally, we are in a position to predict the conductivity of individual MTs of varying lattice types. For the following calculations, the electron density has been taken to be of two electrons within the lattice of 14 dimers and corresponds to a density of $1.8 \times 10^{18}/cm^3$. Thus the results must assume that the conductivity remains constant as the length of the polymer is extended, provided the degree of filling remains the same.

The values in Table 4.3 should be compared with the conductivity of metals, such as copper $(6 \times 10^7 \Omega m^{-1})$ and iron $(1 \times 10^7 \Omega m^{-1})$. Indeed the values can also be compared to the intrinsic semiconductors germanium $(2.5 \Omega m^{-1})$ and silicon $(4 \times 10^{-4} \Omega m^{-1})$. Given these comparisons, the MT may indeed be quite a good semi-conductor given our assumptions. What is particularly interesting is the way that the conduction properties depend on the lattice and the particular boundary conditions. The situation can be compared to carbon nanotubes which have a similar size and structure to MTs. In addition, theoretical consideration of nanotube structure predicts that the conduction properties depend on the boundary conditions. In the case of nanotubes, this means the way in which the graphene sheet is wrapped up to form the nanotube. It is quite interesting that particular sets of boundary conditions produce a semi-conducting nanotube while the appropriate choice of wrapping the nanotube gives rise to metallic conduction [67].

The ability to change from an insulator to a conductor by a simple geometrical change could be biologically relevant even without observing a single electron to be conducted along the MT since it is the conductivity property that affects the way the cell views and responds to external electromagnetic fields. Specifically, the reflectivity of MTs to electromagnetic fields is high when they are conducting and consequently they could act to direct infrared signals to the interior of the centrosome. Since the centrioles are maintained at right angles to each other, the cell would be able to determine the location in latitude and longitude of a light source if the MTs are conducting [8, 9]. Peaks in the AC conductivity are all at energies significantly above the thermal activation threshold so no conductivity is expected along a darkened MT, but photoconduction in the infrared-visible range is possible. Therefore, it is possible that this photo-activated conduction has functional repercussions. In addition to the intrinsic semi-conducting behavior of MT's, and possibly AF's, these protein filaments possess very intensely active cable properties when immersed in an ionic buffer solution. In the next section we discuss the ionic conductivity effects of AF's.

TABLE 4.3: Calculated Conductivity and Resistance of a 1-micron Polymer Composed of Tubulin Dimers

Tubulin Structure	Conductivity ($\Omega^{-1}m^{-1}$)	Resistance
Individual protofilament	non-conducting	
Tubulin sheet ($t_1 = t_r = 0.04eV$)	9.4×10^3	418 kΩ
Tubulin sheet ($t_l = t_r = 0.1eV$)	2.0×10^4	195 kΩ
Tubulin sheet ($t_1 = t_r = 0.4eV$)	6.3×10^4	62 kΩ
MT (A lattice, $t_l = t_r = 0.04eV$)	1.8×10^4	219 kΩ
MT (A lattice, $t_l = t_r = 0.1eV$)	4.5×10^4	87 kΩ
MT (A lattice, $t_l = t_r = 0.4eV$)	1.6×10^5	25 kΩ
MT (B lattice, $t_l = t_r = 0.04eV$)	1.9×10^4	211 kΩ
MT (B lattice, $t_l = t_r = 0.1eV$)	4.7×10^4	84 kΩ
MT (B lattice, $t_l = t_r = 0.4eV$)	1.6×10^5	24 kΩ

4.4.5 Actin Filaments Support Non-linear Ionic Waves

Several experiments [32; 163] indicate the possibility of ionic wave generation along AFs. As the condensed cloud of counter-ions separates the filament core from the rest of the ions in the bulk solution, we expect it to act as a dielectric medium between the two. It has both resistive and capacitive components associated with each monomer that makes up the AF. Ion flow is expected to occur at a radial distance from the surface of the filament approximately equal to the Bjerrum length. An inductive component is proposed to emerge due to the actin's double stranded helical structure that induces the ionic flow in a solenoidal manner. Due to the presence of the sheath of counter-ions around the AF, these polymers act as biological "electrical wires" [163], and have been modeled as non-linear inhomogeneous transmission lines propagating non-linear dispersive solitary waves. Recently, Lader et al. [156] applied an input voltage pulse with amplitude of approximately 200 mV and duration of 800 μs to an AF, and measured electrical signals at the opposite end of the AF indicating that AFs support ionic waves in the form of axial non-linear currents. In an earlier experiment [163], the wave patterns observed in electrically-stimulated single AFs were remarkably similar to recorded solitary waveforms for electrically-stimulated non-linear transmission lines [166]. Considering the AF's highly non-linear complex physical structure and thermal fluctuations of the counter-ionic cloud [208], the observation of soliton-like ionic waves is consistent with the idea of AFs functioning as biological transmission lines.

The electro-conductive medium is a condensed cloud of ions surrounding the polymer and separated from it due to the thermal fluctuations in the solution. The distance beyond which thermal fluctuations are stronger than the electrostatic attractions or repulsions between charges in solution is defined as the so-called Bjerrum length, λ_B. With the dielectric constant of the medium denoted by ε, the Bjerrum length is given by

$$\frac{\varepsilon^2}{4\pi\varepsilon\varepsilon_0\lambda_B} = k_B T \qquad (4.98)$$

for a given temperature T in Kelvin. Here ε is the electronic charge, ε_0 the permittivity of the vacuum and k_B is Boltzmann's constant. For a temperature of $293K$ we find that $k_B = 7.13 \times 10^{-10}m$. Counter-ion condensation occurs when the mean distance between charges, b, is such that $B/b = S > 1$. Each actin monomer carries an excess of 14 negative charges in vacuum, and accounting for events such as protonation of histidines, and assuming there to be 3 histidines per actin monomer, there exist 11 fundamental charges per actin subunit [267]. Assuming an average of 370 monomers per μm we find that there is approximately $4e/nm$. Thus we expect a linear charge spacing of $b = 2.5 \times 10^{-10}m$ so $S = 2.85$. As the effective charge, q_{eff}, or renormalized rod charge is the bare value divided by S we find $q_{eff} = 3.93e/monomer$. Consequently, it can be shown that approximately 99% of the counter-ion population is predominantly constrained within a radius of $8nm$ [216] around the polymer's radial axis [290]. Significant ionic movements within this "tightly bound" ionic cloud are therefore allowed along the length of the actin, provided that it is shielded from the bulk solution [207, 211].

The physical significance of each of the components of the electrical network and additional details are described in [275]. For actin in solution, a key feature is that the positively charged end assembles more quickly than the negatively charged end [252]. This results in an asymmetry in the charges at the ends of the filaments and F-actin's electric polarization. Actin monomers arrange themselves head to head to form actin dimers resulting in an alternating distribution of electric dipole moments along the length of the filament [147]. We assume, therefore, that there is a helical distribution of ions winding around the filament at approximately one Bjerrum length. This corresponds to a solenoid in which a fluctuating current flows as a result of voltage gradient between the two ends.

The capacitive element of the electric circuit is obtained following the observation of oppositely charged layers surrounding the filament surface. We envisage the protein surface's negative charge to be distributed homogeneously on a cylinder defining the filament surface. Furthermore, positive counter-ionic charges in the bulk are expected to form another cylinder at a radius greater than the AF itself, approximately one Bjerrum length, λ_B, away from the actin surface which includes the condensed ions. The permittivity, ε_0, is given by $\varepsilon = \varepsilon_0 \varepsilon_r$ where ε_r is the relative permittivity which we take to be that of water, i.e., $\varepsilon_r = 80$. We take the length of an actin monomer typically as $a = 5.4nm$ and the radius of the actin filament, r_{actin}, to be $r_{actin} = 2.5nm$ [44]. The next step is to consider a cylindrical Gaussian surface of length a whose radius is r such that $r_{actin} < r < r_{actin} + \lambda_B$.

Applying Gauss's law for the total charge enclosed in the cylinder we have the following expression for the capacitance

$$C_0 = \frac{2\pi\varepsilon a}{ln(\frac{r_{actin}+\lambda_B}{r_{actin}})} \tag{4.99}$$

With the parameters given above we estimate that the capacitance per monomer is $C_0 = 96 \times 10^{-6} pF$. The resistive part is obtained from Ohm's law. Taking into account the potential difference and the current I, the magnitude of the resistance, $R = V/I$, for an actin filament is given by

$$R = \frac{\rho ln(\frac{r_{actin}+\lambda_B}{r_{actin}})}{2\pi l} \tag{4.100}$$

where ρ is the resistivity. Typically, for K^+ and Na^+ intracellular ionic con-centrations are $0.15M$ and $0.02M$, respectively [274]. Kohlrausch's law states that the molar conductance of a salt solution is the sum of the conductivities of the ions comprising the salt solution. Thus

$$\sigma = \Lambda_0^{K^+} c_{K^+} + \Lambda_0^{Na^+} c_{Na^+} = 1.21(\Omega m)^{-1} \tag{4.101}$$

Using this with $\rho = \sigma^{-1}$, the resistance estimate becomes $R = 6.11M$ which is much lower than pure water since $R_{water} = 1.8x10^6 M\Omega$.

To describe the properties of the whole filament, we simply connect n sub-circuits as described above to obtain an effective resistance, inductance, and capacitance, respectively, such that:

$$R_{eff} = (\sum_{i=1}^{n} \frac{1}{R_{2,i}})^{-1} + \sum_{i=1}^{n} R_{1,i} \tag{4.102}$$

$$L_{eff} = \sum_{i=1}^{n} L_i \tag{4.103}$$

and

$$C_{eff} = \sum_{i=1}^{n} C_{0,i} \tag{4.104}$$

where $R_{1,i} = 6.11 \times 10^6 \Omega$, $R_{2,i} = 0.9x10^6 \Omega$ such that $R_{1,i} = 7R_{2,i}$. Note that we have used $R_{1,i} = R_1$, $R_{2,i} = R_2$, $L_i = L$ and $C_{0,i} = C_0$. For a $1\mu m$ of the AF we find therefore

$$R_{eff} = 1.2 \times 10^9 \Omega \tag{4.105}$$

$$L_{eff} = 340 \times 10^{-12} H \tag{4.106}$$

$$C_{eff} = 0.02 \times 10^{-12} F \tag{4.107}$$

This solenoidal flow geometry leads to an equivalent electrical element possessing self inductance. From Faraday's law we can derive an effective inductance for the actin filament in solution by

$$L = \frac{\mu N^2 A}{l} \tag{4.108}$$

where l is the length of the F-actin and A is the cross-sectional area of the effective coil given by

$$A = \pi(r_{actin} + \lambda_B)^2 \tag{4.109}$$

The number of turns is approximated by simply working out how many ions could be lined up along the length of a monomer. We would then be approximating the helical turns as circular rings lined up along the axis of the F-actin. We also take the hydration shell of the ions into account in our calculation. The hydration shell is then the group of water molecules oriented around an ion. It can be shown that $L = 1.7pH$ for the length of the monomer.

The electrical model of the AF is an application of Kirchhoff's laws to one section of the effective electrical circuit that is coupled to neighboring monomers. Taking the continuum limit [275], for a large number of monomers along an AF we derive the following equation which describes the spatio-temporal behavior of the potential along the AF:

$$LC_0 \frac{\partial^2 V}{\partial t^2} = a^2(\partial_{xx}V) + R_2 C_0 \frac{\partial(a^2(\partial_{xx}V))}{\partial t} - R_1 C_0 \frac{\partial V}{\partial t} + R_1 C_0 2bV \frac{\partial V}{\partial t} \tag{4.110}$$

Motivated by this picture, physical properties of the ionic distribution along a short stretch of the polymer (the average pitch 35-40 nm) have been modeled by Tuszyński et al. [275] as an electrical circuit with non-linear components (see Fig. 4.34). The main elements of the circuit are: (a) a non-linear capacitor associated with the spatial charge distribution between the ions located in the outer and inner regions of the polymer, (b) an inductance due to the helical geometry of the filament, and (c) a resistor due to the viscosity imposed by the solution.

From Kirchhoff's laws, one can derive an equation governing the propagation of voltage along the filament (see [275]). One of the key aspects in this model is the non-linearity of the associated capacitance [169, 280] that

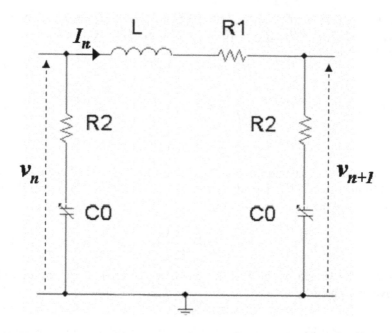

FIGURE 4.34:   An equivalent electrical circuit for a segment of an actin filament.

eventually gives rise to the self-focusing of the ionic waves. The equations developed for the model originate from the application of Kirchhoff's laws to the RLC resonant circuit of a model actin monomer in a filament. Perhaps the most important finding is the existence of the traveling wave which describes a moving transition region between a high- and low- ionic-concentration due to the corresponding inter-monomeric voltage gradient. The velocity of propagation was estimated to range between 1 and $100m.s^{-1}$ depending on the characteristic properties of the electrical circuit model. It is noteworthy that these values overlap with action potential velocities in excitable tissues [109].

Analysis of the model described above reveals the possibility of stationary waves in time that may lead to the establishment of spatial periodic patterns of ionic concentration. As mentioned above, Lader et al. [156] applied an input voltage pulse with amplitude of approximately $200mV$ and duration of $800\mu s$ to an AF, and measured electrical signals at the opposite end of the AF indicating that AFs support ionic waves in the form of axial non-linear currents. In an earlier experiment [163], the wave patterns observed in electrically-stimulated single AFs were similar to recorded solitary waveforms for electrically-stimulated non-linear transmission lines [166]. Considering the AF's highly non-linear physical structure and thermal fluctuations of the counter-ionic cloud [208], the observation of soliton-like ionic waves is consistent with the idea of AFs functioning as biological transmission lines.

This section only provides an indication as to a realistic model of actin that can support soliton-like ionic traveling waves. Modeling relies on data constrained by experimental conditions, and/or assumptions made, including the charge density, which is calculated based on the net surface charges of actin. It should also be considered that soliton velocity is directly proportional to the magnitude of the stimulus, which in a biological setting has not been formally described. Actin interacts with a number of ion channels, of different ionic permeability and conductance. Thus, it is expected that channel opening, single channel currents and other channel properties, including the resting potential of the cell, may significantly modify the amplitude and velocity of the soliton-supported waves. This should correlate with the velocity of the traveling waves along channel-coupled filaments. Other parameters that may play a role in this type of electrodynamic interaction are the local ionic gradients and the regulatory role of actin binding proteins, which can help "focus" the conductive medium or otherwise impair wave velocity. The physical significance of each of the components of the electrical network (see Fig. 4.34) and additional details are described in [275]. MT's differ from AF's in several respects. One of the key differences is the presence of peptide chains called C-termini that protrude from their surface. In the following section we outline their role in ionic wave conductivity along MT's.

4.4.6 Long-range Spatio-temporal Ionic Waves along Microtubules

MTs are long hollow cylinders made of $\alpha\beta$-tubulin dimers [69], as described in Section 4.4.4. Recently, it has become apparent that neurons utilize MTs in cognitive processing. Both kinesin and MAP2 that associate with MTs have been implicated in learning and memory [140, 285]. Dendritic MTs are implicated in particular, and it is highly probable that the precisely coordinated transport of critical proteins and mRNAs to the post-synaptic density via kinesin along MT tracks in dendrites is necessary for learning, as well as for LTP [141].

The following molecular dynamics (MD) simulation results focus on the C-termini of neighboring tubulins, whose biophysical properties have a significant influence on the transport of material to activated synapses. This affects cytoskeletal signal transduction and processing as well as synaptic functioning related to LTP. Using MD modeling we calculated conformation states of the C-termini protruding from the outer surfaces of MTs and strongly interacting with other proteins, such as MAP2 and kinesin [236]. To elucidate the biophysical properties of C-termini and gain insight into the role they play in the functioning of dendrites, we developed a quantitative computational model based on the currently available biophysical and biochemical data regarding the key macromolecular structures involved, including tubulin, their

C-termini, and associated MAP2. In the proposed model of the C-termini microtubular network, the tubulin dimer is considered to be the basic unit. Each dimer is decorated with two C-termini that may either extend outwardly from the surface of the protofilament or bind to it in one of few possible configurations.

The most dynamic structural elements of the system (i.e., its elastic and electric degrees of freedom) are envisaged as conformational states of the C-termini. Each state of the unbound C-terminus evolves so as to minimize the overall interaction energy of the system. The negatively charged C-termini interact with (a) the dimer's surface, (b) neighboring C-termini and (c) adjacent MAPs. While the surface of the dimer is highly negatively charged overall, it has positive charge regions that attract the C-termini causing them to bend and bind in a 'downward' state. The energy difference between the two major metastable states is relatively small, on the order of a few $k_B T$ [219].

To simplify our calculations, we first used a bead-spring model representing the C-terminus as a sequence of beads with flexible connections. We have accounted for the electric field exerted by the dimer, the external field generated by the environment and various short-range interactions within the simulated C-terminus, including interactions between the beads (i.e., Lennard-Jones potential, angular forces etc.).

A simplified model of the interaction between MAP2 and its ionic environment via counter-ions was used to investigate the ability of MAP2 to function as a "wave-guide" that transfers the conformational change in a C-terminus state to an adjacent MT (see Fig. 4.35). A perturbation applied to the counter-ions at one end of the MAP2 drives them out of equilibrium and initiates a wave that travels along the MAP2 (for details see [219]). Fig. 4.35 (right) depicts the main result for a localized perturbation applied for a few picoseconds to the counter-ions near the binding site of MAP2. Wave propagation along the chain of $N = 50$ counter-ions is shown as a counter-ion displacement parallel to MAP2, u_i, where i denotes the ith counter-ion. This perturbation along MAP2 propagates almost as a "kink" whose phase velocity has been found to be $v_{ph} \approx 2nm \cdot ps^{-1}$.

Our simulations indicate the ability of an ionic wave to trigger a coupled wave of C-termini state changes from their upright to downward orientations. Four examples of 'up' and 'down' states of the C-termini obtained through bead-spring model simulations are shown in Fig. 4.35 (left) and results from molecular dynamics simulations for an actual tubulin dimer are shown in Fig. 4.35 (right). Calculations of the energy minimized positions of the individual beads representing the amino acids of the C-terminus in two equivalent forms reveal that the probability of the down position which includes all cases of full or partial attachment is 15%. This means that the system has two major states with a strong bias towards the stretched-up state.

In the remainder of this section we present an MD simulation study of the interaction between MTs and MAP2, followed by preliminary experimental

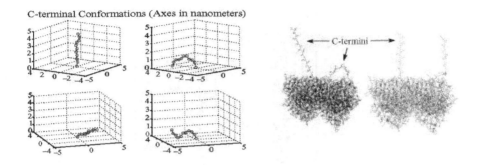

FIGURE 4.35: Examples of the conformational states of the C-termini in a tubulin dimer (right) and results from bead-spring model simulations (left).

results of ionic wave conduction by MTs.

A simplified model of the interaction between MAP2 and its ionic environment via counter-ions was used to investigate the ability of MAP2 to function as a "biological wire" that transfers the conformational changein a C-terminus state to an adjacent MT (see Fig. 4.36).

A perturbation applied to the counter-ions at one end of the MAP2 drives them out of equilibrium and initiates a wave that travels along the MAP2 [219]. Fig. 4.37 depicts the main result for a localized perturbation applied for a few picoseconds to the counter-ions near the binding site of a MAP2. Wave propagation along the chain of counter-ions is demonstrated as counterions displacement parallel to MAP2.

To validate the proposed model we have conducted experiments on electro-stimulation of MTs in a solution. A detailed description of the experiment and its results has been published elsewhere [220]. Here we only describe the main aspects of the work on this phenomenon. Fig. 4.38 depicts the experimental set-up used. Isolated taxol-stabilized MTs have been shown to be able to amplify an electric signal applied to them through a micropipette. The input signal was of 5-10 msec duration with amplitude in the range of ± 200 mV. The signal that arrived at the other end of the MT was more than twice higher than the signal recorded in a control experiment where the same two pipettes were immersed in a solution with no MT making contact to them. The calculated conductivity of MTs was found to be on the order of 10 nS, indicating a high level of ionic conductivity along an MT. By comparison, for a typical ion channel the corresponding value ranges between 5 and 200 pS.

The results of the MD modeling described above raise the possibility of transmitting electrostatic perturbations collectively among neighboring C-termini and from C-termini on one MT to those C-termini on another MT via MAP2. The importance of collective conformational states of C-termini

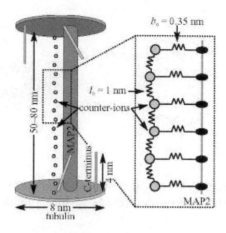

FIGURE 4.36: An illustration of a MT C-terminus conformational change transfer process. Ionic waves travel along an adjacent MAP via local perturbations to the bounded counter-ions.

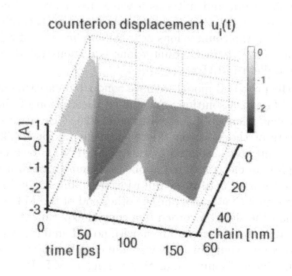

FIGURE 4.37: Propagation of a localized perturbation applied to counterions near the binding site of a MAP2.

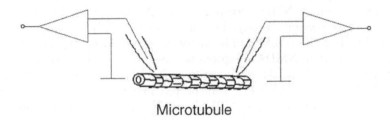

Microtubule

FIGURE 4.38: Schematic representation of the experimental set-up used for the measurements of MT's ionic conductivity. The MT is attached to two micropipettes that are connected to signal amplifiers.

and of transmitting perturbations among the C-termini as a novel information processing mechanism operating at a sub-neuronal level is clear. MAP2 and kinesin bind near to the C-termini on tubulin and the electrostatic properties of C-termini affect this binding [4, 268]. Hence the conformational states of the C-termini must at some point be taken into account in order to understand neural processing that depends on transport of synaptic proteins inside of neurons. Kim and Lisman [142] have shown that inhibition of MT motor proteins reduces an AMPA receptor-mediated response in hippocampal slice. This means that a labile pool of AMPA-receptors depends on MT dynamics, and MT-bound motors determine the amplitudes of excitatory postsynaptic currents (EPSC). A walking kinesin carries with it a protein or an mRNA molecule. Since kinesin binds to a MT on a C-terminus, as it steps on it, it brings the C-terminus to the MT surface and makes it ineffective in binding for the next kinesin over a period of time that it takes the C-terminus to unbind and protrude outside. From this we can deduce that long stretches of C-termini in the upright position are going to be most efficient at transporting kinesin and kinesin cargo while C-termini that lie flat are expected to be most efficient at detaching kinesin. Thus, in considering the trafficking of many kinesins, collective electrostatic effects of C-termini become crucially important. Collective states that correspond to transport strategies that will send optimal numbers of kinesin molecules to synaptic zones are likely to occur when synapses are activated (e.g., when synapses are generating EPSCs). One type of kinesin cargo associated with learning and memory is the NMDA receptor, which is well known to be associated with LTP. The PSD might be expected to require replenishment of NMDA receptors at a later time than AMPA receptors are replenished (see [142]). Kinesin protein (specifically KIF17) actively transports NMDA receptor 2B subunits (NR2B) to the region of the PSD [103]. Once NR2B is in the vicinity of an active synapse it dissociates from kinesin and then becomes associated with the PSD, presumably using actin transport as an intermediary step. Hence kinesin-mediated transport of NMDA receptors along MTs has a built-in negative feedback mechanism: whenever too many NMDA receptors are transported to the synaptic site (often located on a spine head), then those NMDA receptors can initiate proteolytic breakdown of MAP2. The latter event would be expected to reduce further transport of NMDA receptors to the synaptic site. The signal transduction molecule CaMKII is also critical for learning and LTP. Similar to the NMDA receptor, CaMKII is transported to active synapses via a kinesin-mediated transport mechanism. However, it appears to be the mRNA for CaMKII to a larger extent than the protein that is transported to spines, since dendrites are enriched with polyribosomes and CaMKII mRNA [261]. Ribosomes are redistributed from dendrites to spines with LTP [210]. Local translation of CaMKIIα in dendrites appears to be necessary for the late phase of LTP, fear conditioning and spatial memory [191]. Moreover, changes in synaptic efficacy are often accompanied by changes in morphology; reorganization of underlying MTs are fundamental factors for these morphological changes. Having discussed the individual properties and roles played by AF's and MT's in subcellular conduction processes, we now turn to an integrated view of their functioning within a cell. The next several sections discuss this issue in depth.

4.4.7   Dendritic Cytoskeleton Information Processing Model

Our current hypothesis states that the cytoskeletal biopolymers constitute the backbone for ionic wave propagation that interacts with, and regulates dendritic membrane components, such as ion channels to effectively control synaptic connections. Fig. 4.39 depicts a portion of the dendritic shaft where MTs are decorated by C-termini and interconnected by MAP2 (thick line). Connections between MTs and AFs are shown as well as two types of synaptic bindings. On the upper left side, actin bundles bind to the post-synaptic density (PSD) of a spineless synapse. On the lower right hand a spiny synapse is shown, where actin bundles enter the spine neck and bind to the PSD, which, at the other end, is connected to the MTN.

We envision a mechanism in which a direct regulation of ion channels and thus synaptic strength by AFs and associated cytoskeletal structures controls and modifies the electrical response of the neuron. In this picture, MTs arranged in networks of mixed polarity receive signals in the form of electric perturbations, from synapses via AFs connected to MTs by MAP2 [233], or via direct MT connections to postsynaptic density proteins by molecules such as CRIPT [212]. As discussed below, the MTN may be viewed as a high-dimensional dynamic system where the main degrees of freedom are related to the conformational state of the C- termini. The input signals perturb the current state of the system that continues to evolve. Hypothetical integration of the above ideas is outlined as follows. Electrical signals arrive at the PSD via synaptic transmission, which in turn elicits ion waves along the associated AFs at the synaptic spine. These dendritic input signals propagate in the form

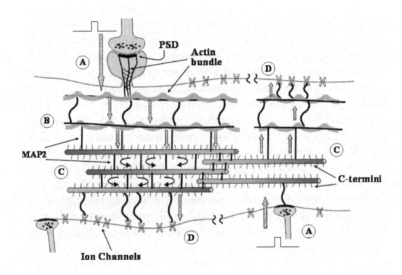

FIGURE 4.39: A scheme of the dendritic shaft with MTs arranged in networks of mixed polarity which receive signals in the form of electric perturbations, from synapses via AFs connected to MTs by MAP2, or via direct MT connections to postsynaptic density proteins by molecules such as CRIPT. A spiny synapse is depicted in the bottom right where an actin bundle enters the spine neck.

of ionic waves through AFs to the MTN where they serve as input signals. The MTN, operating as a large high-dimensional state machine, evolves these input states, e.g., by dynamically changing C-termini conformation. The output from the MTN is the state of the system that may propagate via AFs to remote ion-channels. These output functions are assumed to regulate the temporal gating state of voltage-sensitive channels. This process subsequently regulates the membrane conductive properties and controls the axon hillock behavior by changing the rate, distribution and topology of open/close channels. The overall functions of the dendrite and neuron can thus be regulated in this manner.

The attractiveness of the concept that the cytosol, with its cytoskeletal structures, may behave as a large dynamical system is clear as it provides a means for real-time computation without the need for stable attractor states. Moreover, the output is relatively insensitive to small variations in either the MTN (cytoskeletal networks) or the input patterns. It should be noted that the temporal system's state evolves continuously, even without external inputs. Recent perturbations, however, have a long-term effect on the MTN trajectories, i.e., there is a memory effect inherent to this system (not to be mistaken with synaptic LTP, which has a much longer time scale). The output from the MTN may converge at or near ion channels to regulate their temporal behavior. The issue of adaptation requires a feedback mechanism that will, at least locally, enable the change of the output function. In the context of neuronal function, with focus on processivity, synaptic strengthening, LTP, and memory enhancement, the output function may simply reflect an effect of the MTN on synaptic channel function, such that the desired state of the channel appears to have a higher probability of being open/close upon the presentation of the associated input pattern. One possibility is a Hebbian-based response where a more frequent activity of certain sub-domains of the MTN output states gives rise to higher/lower density of AFs connecting to corresponding channels.

We now explain how this integrative view may serve as a regulatory (adaptive) mechanism. The input level denoted by A in Fig. 4.38 in the previous section is associated with external electric perturbations passing through membranes, mainly synaptic inputs arriving from other neurons. These signals arrive at the cytoskeleton and affect directly MTNs, and/or actin filaments (bundles), in spiny synapses. However, the actin cytoskeleton is responsible for the propagation of signals to the MTN (see level B). These propagated signals (level C) are in turn used as inputs to the MTN, viewed as a dynamic system. We further propose that the MTN generates diverse phase space trajectories in response to different input vectors. The requirements from such a system are not too restrictive since the output-state is not an attractor of the system. In other words, information processing at this level is not necessarily based on attractor dynamics but rather on real time computations. This proposition relies on the observed ability of the MTN to propagate signals and on the specific topological features of MTNs in dendrites, in which shorter MTs of mixed polarity are interconnected by MAP2s. The output from the MTN would be a function of the evolved state vector in certain areas accessed by actin filaments and/or directly linked to ion channels (see both possibilities at level D). These output signals may modify the temporal channel activity, either by directly arriving from the MTN to the channels, or mediated by actin filaments. Hence, the channel-based synaptic membrane conductance is regulated, in particular in the axon hillock region, which is, in most cases, responsible for the generation of action potentials.

The idea that a non-specific high dimensional dynamical system may serve as a reservoir of trajectories in the context of liquid state machines (LSM) has recently been suggested as an explanation for the existence of microcircuits in the brain [170]. The basic structure of an LSM is an excitable medium (hence "liquid") and an output function that maps the current liquid state. The liquid must be sufficiently complex and dynamic to guarantee a universal computational power. This is sufficient to ensure that different input vectors will lead to separate trajectories. A network of spiking neurons and a recurrent neural network has been used as a 'liquid', whereas the output (readout) function has been implemented by simple perceptrons, threshold functions or even linear regression functions. Clearly, simpler readout functions restrict the ability of the whole system to capture complex non-linear dependencies.

4.4.8 The Inter-relation Between the Neural Cytoskeleton and the Membrane

Usually, F-actin and microtubular cytoskeletal networks are thought to fulfill separate, independent cellular roles. Highly dynamic actin networks are known for their role in cell motility, in particular the spreading of the leading edge and contraction. The more stable MT cytoskeleton is best known for its importance in cell division and organelle trafficking. However, recent studies provide a more unified role, ascribing important roles to the actin cytoskeleton in cell division and trafficking and important roles for MTs in the generation and plasticity of cellular morphology. Coordination between the actin- and MT- based cytoskeletons has been recently observed during cellular migration and morphogenesis, processes that share some similarities with neurite initiation [63]. A direct physical association between both cytoskeletons has been suggested, because MTs often preferentially grow along actin bundles and transiently target actin-rich adhesion complexes. In neurons certain plakins and neuron-specific MAPs, like MAP1B and MAP2, may play a role in linking MTs and AFs, helping in the transition from an undifferentiated state to neurite-bearing morphology.

MAP1B and MAP2 are both known to interact with actin in vitro [214, 270, 242, 249, 53]. It is likely that by crosslinking, MAP2 and/or MAP1B associated with both cytoskeletons could be involved in guidance of MTs along AF bundles. Alternatively, MAPs could shuttle from MTs to actin and could alter F-actin behavior by actively crosslinking AFs.

Little is known, however, about interactions between neuronal ion channels and the cytoskeleton. Recently, whole-cell and single channel recordings showed that acute disruption of endogenous AFs with cytochalasin D activated voltage-gated K^+ currents in these cells, which was largely prevented by intracellular perfusion with the AF-stabilizer agent, phalloidin. Interestingly, direct addition of actin to excised, inside-out patches activated and/or increased single K^+ channels. Thus, acute changes in actin-based cytoskeleton dynamics regulate voltage-gated ion channel activity in bipolar neurons. This may be indicative of a more general and quite appealing mechanism by which cytoskeletal structures control feedback mechanisms in neuronal channels.

Recent theoretical and experimental studies of the electrical properties of AFs and MTs in solution revealed their capability to act as biopolymer wires [163, 275]. This means that these protein polymer filaments are capable of conducting non-linear ionic waves and even amplifying the signal with respect to the conducting solution [219]. We conjecture a mechanism in which a direct regulation of ion channels and thus synaptic strength by AFs and associated cytoskeletal structures controls and modifies the electrical response of the neuron. According to this scenario (see Fig. 4.39 in the previous section), MTs arranged in networks of mixed polarity receive signals in the form of electric perturbations, from synapses via AFs connected to MTs by MAP2 [233], or via direct MT connections to postsynaptic density proteins by molecules such as CRIPT [212]. These signals propagate in the form of ionic waves. Specific physical properties of the propagating ionic waves will be discussed in sections that follow.

4.4.9 Relationship to Cognitive Functions

Transport of proteins and receptors in neurons is likely to have an electromagnetic basis to the extent that this function is possibly a result of MT computation. The perspective advanced in this section is that a specific fingerprint defined by a particular electromagnetic state of a microtubular array potentially corresponds to a unique unit of cognition (e.g., a basic visual parameter). Recently, it has been shown that visual components can be represented musically [57]; hence the idea that there is one common type of energy underlying divergent perceptual and cognitive processes appears likely. Activation of one electromagnetic fingerprint could, in turn, activate another electromagnetic fingerprint, irrespective of sensory input. Moreover, the subjective feels of this widespread pattern of electromagnetic energy can be specified according to those key physical properties of MTs that influence the transport of proteins to synapses. Factors influencing kinesin-mediated transport include the protein conformation of tubulin and the nature of the C-termini (see [255, 219]).

Not only is the conformational state of tubulin critical to effective transport, motor proteins appear to alter, albeit temporarily, the conformation of tubulin. Kinesin binding and that of MAP, tau, significantly alter the direction of the protruding protofilament ridges along MTs, which in turn influences their further binding abilities [241]. More than mere local adaptation to binding, MTs may alter their conformation ahead of kinesin processivity [150], supporting the notion of long-range cooperative effects between tubulin dimers located along longitudinal protofilaments of MTs. These biochemical relationships have consequences for electromagnetic fields. The dipole moment of tubulin depends on its configuration in the MT [106, 189]. Thus, electromagnetic fields among MTs could, in theory, be induced or inhibited by synaptic inputs that affect the protein conformation of tubulin directly or through alterations in kinesin or MAP binding. Synaptic effects upon MTs could be mediated through ionic currents, by propagation via actin filaments [275] or by signal transduction cascades resulting in the phosphorylation of MAPs [239].

Due to lengthwise electric dipoles of tubulin dimers, information in the form of traveling waves propagated along microtubular tracks can, in principle, be transmitted between synapses with high fidelity (see [272]). MAP2 bridges keep microtubular arrays within the dendritic core parallel and antiparallel by aligning portions of polarized MTs. The antiparallel alignment of MTs, which specifically occurs in dendrites, would severely attenuate any electromagnetic field generated by MTs. However, during enhanced kinesin-mediated transport, as is likely to occur with heightened synaptic activity, MAP2 bound to the MT would be perturbed and may even temporarily detach from the MT. A similar phenomenon might also occur due to dynein-mediated transport, which occurs largely in the opposite direction to that of kinesin-mediated transport. Assuming that at least some MAP2 stays attached to the antiparallel MTs, keeping the dendritic array intact, any net unidirectional transport along the MT array should increase the strength of the electromagnetic field associated with the fingerprint and should further result in the spread of that electromagnetic field to adjacent MTs. Once a sufficient number of MTs were engaged in dynamically sending and receiving complementary electromagnetic energies, whole neuronal compartments (e.g., dendrites) might be expected to interact. Due to the parallel/antiparallel arrangement of MTs in cortical dendrites and the ability of electromagnetic fields to pass from one dendrite to adjacent dendrites, information could, in principle, pass between neurons when such electromagnetic fields were sufficiently amplified as a result of changes in the binding of MAPs or kinesin.

While it is true that there are some 100 billion neurons and some 100 trillion synapses in the human cerebral cortex, so too does each sensory field afford the luxury of extremely high resolution. The fovea of the retina, for example, contains roughly 1 million receptor cells that relay information to the cerebral cortex, diverging to eventually drive the synaptic activity of at least 1 billion cortical cells. Although the mammalian cortex has many neurons, and even more synapses, its connectivity can be readily grasped by one recurrent theme-repetitive high-resolution topographic representation. Despite this simple organizational plan, cells in cortical areas, in particular cells in higher sensory or association areas, inexplicably show correlated responses during cognitive tasks. Experience alters the basic structure of dendrites, and as a consequence, it should also alter any electromagnetic fields generated by MTs as a result of transport proteins moving along them. As described earlier, a rearrangement of the cytoskeleton occurs during early development, with learning, and with neurodegeneration underlying dementia.

Although models of changes in synaptic efficacy, e.g., long-term potentiation or depression (LTP or LTD), offer great potential as memory mechanisms (see [175]), one often overlooked problem is that large numbers of synapses are affected in concert. If only a few synapses are changed with each memory, then the entire cortical system would have a near unlimited capacity, but if many synapses participate in each memory, as seems to be the case for LTP, then there could be a serious saturation effect. This relates back to the topographic organization of the cerebral cortex. Strictly synaptic models (electrochemical) suggest that complex neural networks increase synaptic weights (i.e., neurons that fire together wire together). Nonetheless, visual cortical regions, for example, do not have massively random interconnections among all parts of the visuotopic map, thereby making many of the changes in synaptic efficacy necessary for encoding complex perceptual features impossible.

The biophysical properties of MTs are just beginning to be understood at a molecular and atomic level and recent empirical evidence suggests interesting effects occur between MTs under certain experimentally induced conditions. Two groups, one led by Watt Webb at Cornell and another led by Paul Campagnola and William Mohler at the University of Connecticut, observed that MTs give rise to intense second-harmonic generation-a frequency doubling upon exposure to a sapphire laser in the 880 nm range. (Other frequencies were partially effective.) MTs were one of the few biological materials having electric dipoles that constructively interfered with the dipoles of neighboring MTs [31; 65]. This occurred for parallel MTs in axons, but not for antiparallel MTs in dendrites. Could such a phenomenon be expected to occur with natural learning or upon exposure to oscillatory input, an LTP-inducing tetanus or various pharmacological agents? To the extent that these induce electromagnetic energy, it is conceivable. In addition to being sensitive to electromagnetic radiation, MTs may themselves produce this kind of energy. Second-harmonic generation by MTs is consistent with their ferroelectric properties. In second-harmonic generation, the most strongly enhanced wavelength is 880 nm, supporting the frequently overlooked reports of electromagnetic signaling by cells. Guenther Albrecht-Buehler [9] observed electromagnetic energy in the near infrared region overlapping the 880 nm value that was generated by centrioles for the purpose of cell-to-cell communication, leading him to suggest that: "...one of the functions of MTs may be to play the role of cellular 'nerves'." Infrared light in this range of wavelengths also induces cell aggregation [6]. Since frequency (and hence wavelength) determines the long-range effect of dielectric polarization of a given MT, one would expect this effect to be length-dependent. The distance between MTs was shown to be critical in second-harmonic generation by MTs.

4.4.10 The Potential for Bioelectronic Applications and Neuromorphic Computing

Traditionally, biologists viewed the cell as a "fluid mosaic bag" containing enzymes, organelles, etc. Recent work has focused on finding the proteins that come together in this cellular "soup" and proceed to effect a change corresponding to a particular biological function. Through the molecular-level analysis of these protein-protein interactions, recent work in various labs has proceeded to build a model of "systems biology", in which these biochemical reactions that have been individually characterized can be brought together to form a cohesive whole, a synergistic dynamics network. From this knowledge, we hope to extract useful information that will affect fields as diverse as nanotechnology, biotechnology and materials science. Inside each cell, there is an elaborate network that serves a multitude of functions. It is an intracellular "highway" that provides a roadway for motor proteins to carry cargo and travel along in their journey to extend axonal length, for example. It controls synapses, and dendrite outgrowth. It is responsible for pulling the chromosomes apart during mitosis, for cell motility, etc.

As man-made engineered systems become more and more miniaturized, the push to build "nano-machinery" is driving physicists and engineers towards biominetics, i.e., the branch of engineering that mimics biological materials. This is highly feasible and much progress has been made in this area. However, due to the limitations inherent in trying to build mechanical devices on such a small scale, there is still much to learn. It is conceivable that through our better understanding of the cell's individual functional and structural components we will be able to advance medicine and technology beyond even our wildest dreams.

As we enter into the 21st century the dawn of the exciting new world of computational biotechnology is upon us and we are poised for a new generation of computers that do not have to be told what to do, but have the power to learn on their own. The development of a biological computer that is fast, small and evolvable is no longer considered science fiction. Scientists are combining biological materials with the latest silicon-based technology in an effort to give us not only electronic devices that are smaller, but are also more flexible in terms of structure and function.

Ironically, our brain can serve as proof of concept. Many scientists now believe that individual neurons display wiring patterns and communication powers that resemble a human computer, and therefore may provide the framework for the design of a biological computer. The brain is composed of 10 billion neurons, each of which may be communicating with 1000 others. The neuron transmits electric signals along its arms, or 'axons'. Inside each axon a parallel architecture of microtubules is interconnected with other proteins, not unlike the parallel computer's wiring. In fact, the structure of microtubules may have evolved towards optimal computational efficiency. The piezoelectric properties of these protein filaments allow them to bend as a result of electric fields or currents which makes them ideal candidates as regulators of synaptic plasticity, a mechanism that could explain the popular principle of "use it or lose it". Thus the synaptic connections which are seldom used would be switched off in favour of more active ones. Biologist Guenter Albrecht-Buehler [7] demonstrated that cells perceive their environment through a tiny organelle called the centriole, composed of microtubules that were shown to interact with electromagnetic radiation. To further show that microtubules themselves can be conductive the German biotechnology group led by Eberhard Unger [85, 86] succesfully conducted required experiments and is now working on building nano-electronic components using these and other proteins. Simultaneously, ideas of combining biological and silicon-based materials in hybrid arrangements are gaining more and more support for future bio-electronic applications. Work is underway at various labs around the world to develop a biological chip. The main objective is to design nano-electronic components using hybrid protein-silicon structures with arrays of proteins as biological oscillators that can be stimulated by electrodes or acoustic couplers. This will provide an analog circuit with rich dynamics.

DNA computers are unlikely to become stand-alone competitors for electronic computers. But digital memory in the form of DNA and proteins is a real possibility with exquisitely efficient editing machines that navigate through the cell, cutting and pasting molecular data into the stuff of life. Beyond that, the innate intelligence built into DNA molecules could help fabricate tiny, complex structures, in essence using computer logic not to crunch numbers but to build things, an idea conceived by Caltech's Winfree and Rothemund [234]. A single test tube of DNA tiles could perform about 10 trillion additions per second; about a million times faster than an electronic computer. Interfacing DNA or protein-based hybrid structures with living cells will enable a host of medical and technological applications including non-invasive diagnostic and therapeutic medical applications and computational devices.

One major component of this framework is a protein-based subcellular structure called a microtubule to which we have devoted a large portion of this section. Microtubules are composed of 13 protofilaments, arranged in a spiral pattern and each protofilament is composed of subunits of tubulin. Each tubulin possesses one of two conformational states: either both monomers can align, or one monomer can shift 30° relative to the vertical axis. Based on the conformation an individual tubulin exists in at a given moment, the conformation of its nearest neighbors can be altered as well. This conformational change can be transmitted throughout the microtubule in a similar manner as patterns are transmitted through cellular automata programs. In addition, as we argued earlier, tubulin can have several electronic states as well as dipolar states, some of which can be conceivably linked to the conformations of C-termini and the counter-ion clouds around them. The idea that we could input an electric, chemical or mechanical signal and generate another one at the output end of the system brings computing or information processing to mind. We believe that ordered protein arrays can function as a programmable processor or even a computer either inside cells such as neurons or as nano-scale devices.

Nanotechnology could and should take advantage of biology in a completely novel manner, i.e., instead of building materials that mimic biological components, we could use the components themselves. In the case of microtubules and tubulin, rather than purify it from traditional sources such as cow or pig brains, tubulin can be readily cloned into an expression vector, and purified from bacteria in mass quantities. Then purified tubulin of a predetermined sequence (and thus with desirable biophysical and biochemical characteristics) can then be routinely polymerized to form microtubules with specific structural and functional properties. Following that, the microtubules produced could be stabilized in a polymer gel capable of undergoing a phase transition at a particular value of the control parameter such as temperature, applied pressure, pH or ionic concentration. This would then trigger a signal that would initiate a computational process carried out by individual microtubules or their assemblies interconnected by MAP's or even connected to actin filaments in architecures that could mimic subneuronal designs. This technology could also be applied to sensors or in artifical intelligence systems in building an evolvable computer. The only limitations in this endeavor are those imposed by our imagination and our ability to dream up new applications.

4.4.11   Discussion

Considering the abundance of MTNs and AFs in axons and dendritic trees, the findings and theoretical models described above may have important consequences for our understanding of the signaling and ionic transport at an intracellular level. Extensive new information (see [130]) indicates that AFs are both directly [44] and indirectly linked to ion channels in both excitable and non-excitable tissues, providing a potentially relevant electrical coupling between these current generators (i.e., channels), and intracellular transmission lines (i.e., AFs, MTs). Furthermore, both filaments are crucially involved in cell motility and, in this context, they are known to be able to rearrange their spatial configuration. In nerve cells AFs are mainly located in the synaptic bouton region, whereas MTs are located in both dendrites and the axon. Again, it would make sense for electrical signals supported by these filaments to help trigger neurotransmitter release through a voltage-modulated membrane deformation leading to exocytosis [247]. Actin is also prominent in post-synaptic dendritic spines, and its dynamics within dendritic spines has been implicated in the post-synaptic response to synaptic transmission. Kaech et al. [135] have shown that general anesthetics inhibit this actin mediated response. Among the functional roles of actin in neurons, we mention in passing glutamate receptor channels, which are implicated in long term potentiation.

It is therefore reasonable to expect ionic wave propagation along AFs and MTs to lead to a broad range of physiological effects.

In summary, this section broadly described the physical conditions that enable cytoskeletal polymers such as AFs and MTs to act as electrical transmission lines for ion flows along their lengths and along MAP-mediated interconnections between adjacent filaments. In the case of AFs we propose a model in which each protein subunit is equivalent to an electric circuit oscillator. The physical parameters used in the model were evaluated based on the molecular properties of the polymer. Using the general conductivity rules that apply to electrical circuits we analyzed the properties of ionic waves that propagate along AFs and compared these values to those observed in earlier experiments. In the context of the role played by MTs in neurons we described the dynamics of C- termini states. We discussed both individual MTs and their networks including the interactions with ions and signal transmission via MAPs. Recent experiments on ionic conductivity along AFs and MTs show the validity of the basic assumptions postulated in our models. In light of these results we conjectured a new dendritic signaling mechanism that involves ion waves along protein filaments which may travel without significant decay over tens of microns, potentially affecting the function of synapses and ion channels.

We proposed a new model for information processing in dendrites based on electrical signaling involving the cytoskeleton. This model predicts that the dendritic cytoskeleton, including MTs and AFs, plays an active role in computations affecting neuronal function. These cytoskeletal filaments are affected by, and regulate ion channel activity. A molecular dynamics description of the C-termini protruding from the surface of a MT has revealed the existence of several conformational states, which lead to collective dynamical properties of the neuronal cytoskeleton. These collective states of the C- termini on MTs have been shown to have a significant effect on the ionic condensation and ion cloud propagation with physical similarities to those recently found in actin-filaments. We have been able to then provide an integrated view of these phenomena in a bottom-up scheme that demonstrates how ionic wave propagation along cytoskeletal structures impact channel functions, and thus neuronal computational capabilities.

The possibility of an evolvable, dynamic and responsive electrical circuitry within the cell provided by actin and MT filaments could be of enormous consequences to our understanding of the way cells operate internally and interact with their environment. In particular, it would cast an entirely new light on cell differentiation, cell division and cell-cell communication. While an integrated theory of this type of behavior is far from being constructed, its individual elements are gradually taking shape.

4.5 Mechanisms of Exciton Energy Transfer in Scheibe Aggregates - from ref [273]

An extremely interesting class of Langmuir-Blodgett thin films that represent compact aggregates of dye molecules composed of chromophores and fatty acids was discovered independently in the late 1930's by G. Scheibe in Germany [235] and E. E. Jelly in England [127]. These designed molecular monolayers are, therefore, commonly referred to as Scheibe or J-aggregates.

What was particularly unusual about these aggregates was the appearance of a new, very narrow absorption band of a longer wavelength than the monomer absorption band. Scheibe interpreted this as reversible polymerization effects.

Subsequently, in the late 1960's and early 1970's, it was observed by Kuhn, Möbius and their associates [41,42,43,152,168] that when irradiated with uv or visible light, donor fluorescence in these monolayers was strongly quenched. Simultaneously, an acceptor fluorescence line appeared whose amplitude was almost equal to that of the primary donor spectral line, but its peak was slightly red shifted. This observation was interpreted as giving evidence that the Scheibe aggregate acts as a cooperative molecular array that, after absorbing a photon, channels the energy laterally over distances of up to 1000 Å to a particular energy-accepting molecule (an acceptor dye).

Interestingly, the efficient energy capture and transfer phenomena that characterize the Scheibe aggregates disappear with the aggregates' loss of rigidity or regular order. Somewhat paradoxically, energy capture by acceptor molecules becomes more efficient with increasing temperature. Furthermore, optimal efficiency properties are achieved for very low acceptor-to-donor concentrations. For example, at ratios on the order of $1 : 10^4$, over 50% of the light energy is transferred to acceptor molecules. Because of the above properties, Scheibe aggregates have sometimes been called photon energy funnels.

The importance of Scheibe aggregates lies both in fundamental and applied aspects of their functioning. Their primary application is in photographic and photo-detection processes. However, potential applications in the design of solar energy cells as well as in light-operated devices such as photoresistors and photomemory have also been considered [123]. Furthermore, an understanding of the mechanism of energy transfer in Scheibe aggregates could lead to a better theory of photon absorption in living cells.

The primary motivation for this section is a critical reassessment of the physical mechanisms at play in the energy capture and transfer processess in Scheibe aggregates. Our objective is to use simple physical arguments in establishing relative importance of a number of mechanisms involved. We wish to present estimates of energy and time scales for excitons, phonons, their mutual interactions, and the heat bath (thermal effects). Feasibility arguments will be given for various possible models of transport of the exciton energy in the monolayer.

As can be seen from the experimental absorption intensity profile [168], the monomer absorption line is centred at $\lambda_{\text{mono}} = 384$ nm, which corresponds to the energy $\hbar\Omega = 3.23$ eV. Thus, we represent the monomer as a two-level system with the second-quantized Hamiltonian of the form

$$H_{\text{mono}} = \hbar\Omega A^\dagger A \qquad (4.111)$$

On the other hand, the absorption line of the dimer is centred at $\lambda_{\text{dimer}} = 368$ nm corresponding to the energy of $E_{\text{dimer}} = 3.37$ eV. Assuming that the exciton energy can be transferred between two neighbouring molecules via a hopping mechanism, the simplest second-quantized Hamiltonian for the dimer can be postulated as

$$H_{\text{dimer}} = (\hbar\Omega)(A_1^\dagger A_1 + A_2^\dagger A_2) - J(A_1^\dagger A_2 + A_2^\dagger A_1) \qquad (4.112)$$

where J is the hopping constant and it should approximately correspond to the dipole-dipole interaction energy calculated earlier in the extended dipole model. The indices 1 and 2 label the two neighbouring molecules. It is easy to diagonalize the Hamiltonian matrix for the dimer and we obtain its eigenvalues as

$$E_{\text{dimer}} = \hbar\Omega \pm J \qquad (4.113)$$

We, therefore, infer that $J \approx 140$ meV, which compares favourably with the extended dipole moment estimate we present next. Unfortunately, at this stage we are unable to deduce the sign of J and consequently predict whether the ground state is symmetric or antisymmetric. Finally, Grad et al. [92] used an estimate of $J \approx 75$ meV, which is close to the range obtained in the extended dipole model.

4.5.1   The Exciton Model

A typical Scheibe aggregate structure that will be considered in the present paper is a thin film with a bricklayer arrangement of constituent dye molecules. Two types of structurally similar molecules are being used in the aggregate: (a) an acceptor molecule (commonly thiacyanine), and (b) a donor molecule, most often oxacyanine. Their similarity is also seen in only slight differences in their molecular weights, viz. $M_a = 1.25 \cdot 10^{-24}$ kg and $M_d = 1.21 \cdot 10^{-24}$ kg for the two cases given above. The dimensions of the bricklayer structure are given by $a = 16$ Å, $b = 3.0$ Å and $c = 8.0$ Å.

In a purely excitonic model with nearest neighbour coupling, we have the dispersion relation

$$E(\vec{k}) = \hbar\Omega - 2\sum_{i=1}^{4} J_i \cos(\vec{k}\cdot\vec{d}_i) \qquad (4.114)$$

$$= \hbar\Omega - 2J_1 \cos(ak_x) - 4J_2 \cos(ak_x/2)\cos(bk_y) - 2J_3 \cos(2bk_y)$$

where $\vec{d}_1 = (a, 0)$, $\vec{d}_2 = (a/2, b)$, $\vec{d}_3 = (0, 2b)$ and $\vec{d}_4 = (-a/2, b)$. Here J_1 and J_3 are the horizontal and vertical coupling constants, respectively, while J_2 and J_4 are the two diagonal coupling constants. Due to symmetry we have $J_2 = J_4$. The constant term $\hbar\Omega$ is the on-site exciton energy.

Of utmost importance to our understanding of the underlying physical mechanisms is a reasonably accurate estimate of the coupling constants. This, of course, depends on the mutual orientation of the adjacent molecules. This problem has been dealt with in this context by Czikkely et al. [42,43] who proposed an extended dipole model. With an arrangement given in Figure 4.40 they calculated the interaction constant semiclassically via

$$ J = \frac{q^2}{\epsilon} \left(\frac{1}{a_1} + \frac{1}{a_2} - \frac{1}{a_3} - \frac{1}{a_4} \right) \tag{4.115} $$

where q is an elementary charge, ϵ the dielectric constant of the medium and the distances $a_1, \ldots, a_4$ are explained in Figure 4.40. The values of q and l were estimated to be $q = 0.22\,e$ and $l = 8.9\,\text{Å}$ [42]. For $\epsilon \approx 2.5\epsilon_0$ we then find

$$ J_1 = -0.20\,\text{meV}, \quad J_2 = -0.50\,\text{meV}, \quad J_3 = 0.52\,\text{meV} \tag{4.116} $$

For these values the dispersion relation (4.114) is shown in Figure 4.41.

At $\vec{k} = (0, 0)$ the energy exhibits a saddle point with a negative mass along one axis and a positive along the other. It is worth noting that the possibility of a negative effective mass of a Scheibe aggregate has been recently raised by Kirstein and Möhwald [145].

The ground state (lowest energy) is located at $\vec{k} = (0, \pi/b)$ and has a value of

$$ E_0 = \hbar\Omega - 2J_1 + 4J_2 - 2J_3 = \hbar\Omega - 2.65\,\text{eV} \tag{4.117} $$

which is given relative to the on-site and deformation energies. Expanding in a power series around the ground state we obtain the following

$$ E(\vec{k}) = E_0 + J_x a^2 k_x^2 + J_y b^2 (k_y - \pi/b)^2 + \cdots \tag{4.118} $$

where the effective coupling constants J_x and J_y are given by

$$ J_x = J_1 - J_2/2 = 0.06\,\text{eV}, \quad J_y = 4J_3 - 2J_2 = 3.06\,\text{eV} \tag{4.119} $$

They are both much larger than thermal energy, which at room temperature is approximately 26 meV. The effective mass (around the equilibrium state) may now be found as

$$ m_x = \frac{\hbar^2}{2a^2 J_x} = 2.2 \cdot 10^{-31}\,\text{kg}, \quad m_y = \frac{\hbar^2}{2b^2 J_y} = 1.3 \cdot 10^{-31}\,\text{kg} \tag{4.120} $$

In other words, the effective mass is approximately the same in both directions, which may be linked to a nearly circular and fairly isotropic domain of coherence.

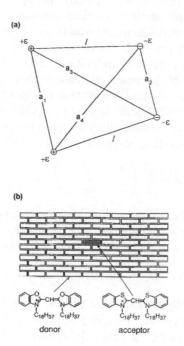

FIGURE 4.40: An arrangement of molecules (a) and extended dipoles (b) in the extended dipole calculation.

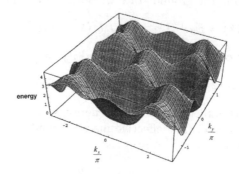

FIGURE 4.41: Dispersion relations for the exciton energy in the Scheibe aggregate.

4.5.2 Exciton Domain Size

We now attempt to estimate theoretically the exciton domain size. For simplicity the result will be derived for a one-dimensional model. In the continuum approximation the energy is of the form

$$E(k) = Jd^2k^2 \tag{4.121}$$

where d is the lattice spacing. The two-site correlation function is defined as

$$\Gamma(\xi) = \int_{-\infty}^{\infty} A^\dagger(x)A(x+\xi)\,\mathrm{d}x \tag{4.122}$$

where $A(x)$ and $A^\dagger(x)$ are the annihilation and creation operators in the site representation. In the Fourier (momentum) representation the corresponding expression is

$$\Gamma(\xi) = \int_{-\infty}^{\infty} a^\dagger(k)a(k)e^{-ik\xi}\,\mathrm{d}k \tag{4.123}$$

The mean occupation number of the k'th mode is

$$\langle a^\dagger(k)a(k)\rangle = K \exp\left[-\frac{E(k)}{k_\mathrm{B}T}\right] \tag{4.124}$$

The normalization constant K is found to be

$$K = d\sqrt{\frac{J}{\pi k_\mathrm{B}T}}. \tag{4.125}$$

We insert these equations into the correlation function, which in turn evaluates to

$$\Gamma(\xi) = \exp\left[-\frac{\xi^2}{2\xi_\mathrm{cor}^2}\right] \tag{4.126}$$

with

$$\xi_\mathrm{cor} = d\sqrt{\frac{2J}{k_\mathrm{B}T}} \tag{4.127}$$

Thus the two-site correlation function is Gaussian and we interpret the width ξ_cor as the correlation length (or the size of the coherent domain). The two-dimensional analogue, domain size A, is then approximately given by the product of the correlation length in the x and y directions, i.e.,

$$A = 2ab\frac{\sqrt{J_xJ_y}}{k_\mathrm{B}T} \approx 33ab$$

This result shows that the exciton size is inversely proportional to absolute temperature, in agreement with the analysis by Möbius and Kuhn [197], and at room temperature covers approximately 33 lattice sites.

The correlation lengths may be expressed in terms of effective mass as follows

$$\xi_{\text{cor}} = \frac{\hbar}{\sqrt{mk_{\text{B}}T}}$$

As mentioned earlier, since the effective mass is almost isotropic, it follows that the exciton domain is approximately circular.

We shall now discuss the launching of an exciton domain by a photon as shown in Figure 4.42. The first estimate we make in this context is the maximum initial velocity of the exciton domain. This occurs for $\Theta \approx 90°$, where Θ is the angle of incidence measured with respect to the normal to the film, and a completely inelastic scattering (absorption) of electromagnetic radiation. Then $p_\lambda = p_{\text{ex}}$, which means that the maximum velocity of the exciton is

$$v_0^{\text{max}} = \frac{p_\lambda}{m_{\text{eff}}} = \frac{h}{\lambda_\nu m_{\text{eff}}} \approx 1 \times 10^4 \, \text{m/s} \tag{4.128}$$

for $\lambda_\nu = 366\,\text{nm}$ [197]. This is very interesting since it is very close to the average propagation velocity of an exciton inferred from the experiments of Kuhn and Möbius [197]. As a consequence, the exciton in all likelihood is accelerated, probably by the presence of impurity acceptors.

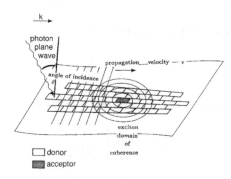

FIGURE 4.42: An illustration of the process of the creation of a coherent exciton domain by a photon; following [27].

The second estimate concerns the size of the coherent exciton domain. The Heisenberg uncertainty principle $\Delta x \Delta p \geq \hbar$ should give us an idea of the minimal spatial extent of the domain. Since $\Delta p(0) \approx \hbar \omega_\nu / c$ we easily find that $\Delta x(0) \geq 500\,\text{Å}$, which corresponds to approximately 120 lattice spacings along the brick width and 30 lattice spacings along its length.

Furthermore, based on Gaussian wavepacket spreading phenomena, one expects this domain of coherence to grow in size over time. The question we wish to pose is how large will the domain become when the centre reaches an acceptor molecule. Taking the acceptor-to-donor ratio of $1 : 10^4$ leads to an average distance of 100 lattice spacings or approximately 10^{-7} m. Taking the average speed of exciton propagation of $v_{ex} = 2 \cdot 10^4$ m/s gives the flight time of $\tau_{flight} = 5$ ps. A long-time approximation for the Gaussian wavepacket spreading can be taken as

$$\Delta x(\tau) \approx \frac{\hbar}{\Delta x(0)m_{eff}}\tau \tag{4.129}$$

This expression assumes the coupling with lattice vibrations to be negligible. Substituting the numbers discussed above gives $\Delta x(\tau_{flight}) \approx 10^{-7}$ m, i.e., the size of the domain has doubled in each direction and it now covers an area approximately equal to the entire space available between neighbouring acceptors! What we are tempted to conclude from this simple calculation is that a mechanism is needed to focus this energy and transfer it eventually to a well-localized acceptor site. One possible way of causing localization is through an inclusion of non-linear effects. Another eventuality is related to the presence of impurity sites as shallow energy potentials leading to a binding process. Both cases have been studied in the past and we briefly summarize the latter below. A discussion concerning the role of non-linearity in the formation and maintenance of a coherent domain will be given in a later section.

Bartnik et al. [19] have suggested that the acceptor eigenenergy lies slightly below that of the donor (approximately 10%). Simultaneously they assumed the acceptor hopping constant to be $10 - 20\%$ lower than that of the donor. In their simulations a one-dimensional rigid exciton lattice was adopted and the results indicated the possibility of efficient energy transfer to acceptor sites at optimal parameter values. Subsequent simulations included an additive white noise in the coupling constant J that was intended to mimic the role of thermal disorder in the Scheibe lattice. Encouragingly, increasing the noise level to a value corresponding to the standard deviation of approximately 1.5% resulted in an improved capture efficiency. A further increase in the noise level caused an eventual destruction of the exciton energy transfer to an acceptor site. This would indicate that shallow energy levels of acceptor molecules may indeed cause the observed effects.

4.5.3 Random Walk Model

Having established a basis for the existence of a *domain of coherence* we now turn to the question of its motion towards an acceptor molecule. The first input into this discussion comes from an estimate of the kinetic energy of the centre-of-mass motion of an exciton domain. We find that

$$E_{kin} \approx \frac{1}{2}m_{eff}v_{ex}^2 \approx 0.1 \, \text{meV} \tag{4.130}$$

assuming that $v_{ex} \approx 2 \cdot 10^4$ m/s. This is an exceedingly small amount of energy. For example, thermal energy (for two degrees of freedom) ranges from approximately 3 meV to 30 meV for temperatures between 30 K and 300 K, respectively. Hence, unless there is significant effective friction opposing the domain's motion, it will execute random motion similar to a quasi-free gas molecule. However, this line of thinking poses further problems. While an estimate of rms velocity of such motion is very high, namely

$$v_{rms} = \sqrt{2k_B T/m_{eff}} \qquad (4.131)$$

which ranges between $4 \cdot 10^4$ m/s and $2 \cdot 10^5$ m/s for $T = 30$ K and $T = 300$ K, respectively, its interpretation puts this picture in question. If the random walk process is unbiased, the domain's most probable position is still the original starting point, i.e., as far from the acceptor molecule as in the beginning. In order to arrive at a feasible mechanism we suggest a biased random walk picture instead. This would imply the probability of a step towards the closest acceptor site given by q and away from it by p ($q > p$). Of course, the same argument would apply to both directions and motion in the two perpendicular directions would be statistically independent. One then finds a net drift towards a more probable acceptor given by

$$\langle x \rangle = N(q - p)d \qquad (4.132)$$

where N is the number of steps, i.e., $N = t/\tau$, with τ denoting the average hopping time. Thus, the average propagation velocity is given by

$$v_{ex} = \frac{\langle x \rangle}{t} = \frac{d}{\tau}(q - p) \qquad (4.133)$$

which, with $d_\perp = 3$ Å, $d_\parallel = 16$ Å and $\tau_0 = 3 \cdot 10^{-14}$ s, indicates a need for a very strong bias along the width of the bricklayer structure in order to reach speeds exceeding 10^4 m/s (i.e., $q \approx 1$ and $p \approx 0$). In the opposite case, propagating along the long axis would require a reduction by a factor of 2, which translates into $q = 3/4$ and $p = 1/4$, so that the net velocity would be $2 \cdot 10^4$ m/s.

The biased random walk framework brings us closer to another possible point of view, namely a uni-directional motion towards an acceptor, which could be caused by the attractive presence of a shallow energy level at the acceptor site. Since we established earlier on that the initial exciton velocity cannot exceed $1 \cdot 10^4$ m/s (average propagation speed), this means that the motion should be accelerated by a force. Taking the depth of the acceptor level as $\Delta E \approx 0.3$ eV and assuming that it is removed by 100 lattice sites from the centre of the coherence domain yields an average force $\bar{F}$ of between 0.02 pN and 0.08 pN. Assuming further that the domain starts from rest we arrive at an estimate of an average propagation velocity as

$$\bar{v} \approx \sqrt{\frac{\bar{F}\Delta l}{2m_{eff}}} \approx 5 \cdot 10^5 \text{ m/s} \qquad (4.134)$$

which is off by an order of magnitude. However, this force will be distributed inhomogeneously and friction will slow the motion down.

We conclude that far away from an impurity site, a small domain (i.e., at high temperatures) is likely to execute a random walk with a bias that becomes more pronounced close to an acceptor site. However, high concentrations of acceptor sites and low temperature ranges are likely to lead to a behaviour that is far more unidirectional and far less random. However, radiative losses are greater at low temperatures, since the domain size is larger.

4.5.4 Phonons

An important aspect that so far has been ignored in our discussion is the effect of lattice vibrations on both the formation of a coherent exciton domain (giving it lifetime) and on its propagation (providing effective friction).

Following standard texts on the solid state theory [104], the energy of the lattice can be written as

$$H_{\text{ph}} = \int \hbar \omega_k \left(b_k^\dagger b_k + \frac{1}{2} \right) dk \qquad (4.135)$$

where the dispersion relation ω_k depends on the type of phonons considered (optical or acoustic, etc.) and $b_k^\dagger$, b_k are creation and annihilation operators.

Inoue [123] fitted the tail of the absorption spectrum for merocyanine using the Urbach rule [201] and obtained an estimate of the effective phonon energy as $\hbar \omega = 30.3 \, \text{meV}$. Spano et al. [257] obtained a similar value of $\hbar \omega = 29.8 \, \text{meV}$ for optical phonons in pseudocyanine. The above estimates show that the phonon energy is roughly an order of magnitude smaller than the exciton hopping energy and two orders of magnitude smaller than the on-site exciton energy, i.e.,

$$\frac{\hbar \omega}{J} \approx 0.2, \qquad \frac{\hbar \omega}{\hbar \Omega} \approx 10^{-2} \qquad (4.136)$$

The characteristic phonon temperature (the Debye temperature) is

$$T_{\text{ph}} = \frac{\hbar \omega}{k_{\text{B}}} \approx 350 \, \text{K} \qquad (4.137)$$

The characteristic phonon time scale is

$$\tau_{\text{ph}} \approx \frac{1}{\omega} \approx 2 \cdot 10^{-14} \, \text{s} \qquad (4.138)$$

which is very close to the hopping time τ_0, indicating a strong possibility of coupling between these two types of excitations. The on-site lattice fluctuations are found to be

$$u_{\text{rms}} \approx 0.1 \, \text{Å} \cdot \sqrt{\frac{T}{T_{\text{ph}}}} \qquad (4.139)$$

which is consistent with an analogous discussion presented in [20].

4.5.5 Exciton-Phonon Coupling

The root cause of exciton-phonon coupling is the distance dependence of the excitonic parameters Ω and J. As a consequence, the simplest way an interaction energy of this type can be written is

$$H_{\text{int}} = \tilde{\chi}_1 \sum_n (u_{n+1} - u_{n-1}) A_n^\dagger A_n$$

$$+ \tilde{\chi}_2 \sum_n (u_{n+1} - u_n) \left(A_{n+1}^\dagger A_n + A_n^\dagger A_{n+1} \right) \qquad (4.140)$$

It has recently been shown [133] that these two coupling terms (proportional to $\tilde{\chi}_1$ and $\tilde{\chi}_2$) tend to inhibit each other and their combined effect may be conveniently represented by the single dimensionless parameter

$$\alpha = \frac{\tilde{\chi}_1 - \tilde{\chi}_2}{2J\sqrt{Mk_B T_{\text{ph}}}} \qquad (4.141)$$

A preliminary estimate of the magnitudes of $\tilde{\chi}_1$ and $\tilde{\chi}_2$ indicates that $\tilde{\chi}_1 \approx 100\,\text{pN}$ [99] and $\tilde{\chi}_2 \approx J/(3d) \approx 40\,\text{pN}$. It will be useful to convert these values into energy units so we define

$$\chi_{\text{eff}} = (\tilde{\chi}_1 - \tilde{\chi}_2)d \qquad (4.142)$$

and obtain an estimate of χ_{eff} as $(16 - 60)\,\text{meV}$ depending on the direction.

Inoue [123] concluded that the exciton-phonon coupling applies to a two-dimensional case and inferred from the experiments for merocyanine that its strength is approximately given by $\chi_{\text{eff}} \approx 29\,\text{meV}$. Spano et al. [257] estimated $\chi_{\text{eff}} \approx 26\,\text{meV}$, which is quite consistent.

The total Hamiltonian for the aggregate can, therefore, be written as

$$\begin{aligned}
H &= H_{\text{ex}} + H_{\text{ph}} + H_{\text{ex-ph}} \\
&= \sum_n \left[\hbar\Omega A_n^\dagger A_n - J \left(A_n^\dagger A_{n+1} + A_{n+1}^\dagger A_n \right) \right] \\
&\quad + \sum_n \left[\frac{p_n^2}{2M} + \frac{1}{2}K(u_n - u_{n+1})^2 \right] \\
&\quad + \tilde{\chi}_1 \sum_n \left[A_n^\dagger A_n (u_{n+1} - u_{n-1}) \right] \\
&\quad + \tilde{\chi}_2 \sum_n \left[\left(A_n^\dagger A_{n+1} + A_{n+1}^\dagger A_n \right) (u_{n+1} - u_n) \right] \qquad (4.143)
\end{aligned}$$

which can be compactly recast into a momentum representation as

$$\begin{aligned}
H &= \sum_k \left[\hbar\Omega_k a_k^\dagger a_k + \hbar\omega_k \left(b_k^\dagger b_k + \frac{1}{2} \right) \right] \\
&\quad + \sum_{k,l} \chi_{\text{eff}}(k) \left(b_k^\dagger + b_{-k} \right) a_{k+l}^\dagger a_l \qquad (4.144)
\end{aligned}$$

The dynamics of the combined system is determined by just two parameters: α, which was introduced earlier and $\beta = T_{ex}/T_{ph}$ [133]. Depending on the values of α and β one finds four possible regimes: (a) $\alpha < 1$ and $\beta < 1$, (b) $\alpha < 1$ and $\beta > 1$, (c) $\alpha > 1$ and $\beta < 1$ and (d) $\alpha > 1$ and $\beta > 1$. It appears that we are dealing here with $\beta \gg 1$ and $\alpha \ll 1$, which leads to the breakdown of the adiabatic approximation. The maximum velocity of localized excitonic soliton wavepackets is bounded by the sound velocity. This causes a coherence dephasing of the exciton whose rate is proportional to J [133].

The presence of exciton-phonon coupling introduces non-linearity into the problem. There exist two standard approaches to treat this difficulty: (a) for weak non-linearities adopt a perturbative scheme, which can be done fully quantum mechanically, and (b) for sufficiently strong non-linearities the above method fails and one solves the problem semi-classically, but incorporating non-linear terms in the equations of motion. Below, we discuss the importance of non-linearity in the dynamics of Scheibe aggregates.

4.5.6 The Role of Non-linearity

A recent paper [22] discussed three possible regimes of behaviour in the one-dimensional approach to the exciton-phonon coupled systems. Depending on the values of the following two parameters $\gamma = \hbar\omega_{ph}/J$ and $g = |\chi|^2/(2\hbar\omega_{ph}J)$, the three regimes found correspond to: (a) delocalized, almost free exciton, (b) the small polaron limit and (c) the self-trapped exciton state. Figure 4.43 demonstrates the location of these regimes on the parameter plane.

Admittedly, the Scheibe agregate is a two-dimensional system and a straight-forward application of these results may be misleading. Nevertheless, we have evaluated the range of parameters g and γ based on the estimates arrived at earlier, i.e.,

$$\hbar\omega_{ph} \approx 30\,\text{meV} \tag{4.145}$$

$$50\,\text{meV} \leq J \leq 150\,\text{meV} \tag{4.146}$$

$$25\,\text{meV} \leq \chi_{\text{eff}} \leq 100\,\text{meV} \tag{4.147}$$

This results in $0.2 \leq \gamma \leq 0.6$ and $0.07 \leq g \leq 3.3$, which is still inconclusive but the largest area convered favours either a free exciton or a small polaron picture, at least in the quasi-one-dimensional picture. This is somewhat reassuring in light of what was argued earlier in the paper. The rather large size of the domain of coherence and a relatively weak exciton-phonon coupling emphasized in our paper is suggestive of a nearly free exciton case with a possible role of non-linearity in preventing the wavepacket from spreading.

There have been a number of studies that focussed on the role of non-linearity in more realistic two-dimensional models (but still in the continuum approximation). The first such non-linear model was proposed by Huth et al. [108]. The model was derived by treating the phonons classically and subsequently eliminating them via an adiabatic approximation which, in view of

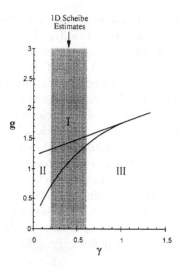

FIGURE 4.43: The three regimes of behaviour for one-dimensional exciton-phonon systems; following [22].

our findings, is not entirely justified. In the continuum limit, Huth et al. [108] derive a two-dimensional cubic non-linear Schrödinger equation [289,134,232], where the energy transfer takes place through soliton-like ring waves. The dynamics is described by collapsing ring waves and the collapse time was associated with the effective exciton lifetime [54,40].

Bang [16,18] further showed that the effect of temperature is to increase the collapse time and decrease the coherence time which seems to be in contradiction with experimental data. More recent investigations [102,98,99] involving the formation and propagation of non-linear localized coherent structures in two dimensions included the presence of Gaussian impurity potentials due to acceptor molecules. Their results are much better correlated with experimental observations.

However, what appears to be needed in a future study of this type is the introduction of a negative effective mass along one of the two directions of propagation, which might be how an exciton domain is originally created. It appears that in a hyperbolic version of the two-dimensional non-linear Schrödinger equation, the effect of a negative signature in the two-dimensional Laplacian is to increase the stability of a solitary wave, which is a bright soliton along one direction and a dark soliton along the other direction [29].

4.5.7 Conclusions

In this section we have been concerned with a number of physical mechanisms that are at play in the processes of energy capture and transfer in

Scheibe aggregates.

It was argued based on fundamental principles of quantum mechanics that the photon energy harvested by a monolayer must necessarily be highly delocalized when it is initially absorbed by donor molecules. The thus created domain of exciton coherence is then subjected to a number of influences. The main effects can be summarized as follows: (a) quantum decoherence due to wavepacket spreading – this may possibly be arrested by non-linearity, (b) thermalization which affects the size of the domain – the exciton coupling with phonons appears to conform to the semi-empirical formula $N_{\text{eff}} = 3000\,\text{K}/T$, (c) the attractive influence of acceptor molecules. The latter effect is perhaps the least studied and merits careful examination.

One effect that has not been discussed in the present work is the role of radiative losses. It is believed that an exciton domain of coherence radiates its energy at a rate that is proportional to its size, i.e.,

$$k_{\text{rad}} = N_{\text{eff}}\,(\text{ns})^{-1} \tag{4.148}$$

Mukamel and his collaborators [92,257] proposed a model to incorporate radiative losses via the incorporation of an imaginary term in the exciton-phonon Hamiltonian. As a result, these authors found that the effective size of a coherent exciton domain, N_{eff}, is proportional to this imaginary part and hence increasing the absolute temperature T causes the size of the domain, N_{eff}, to decrease in proportion to T. This, in turn, decreases the radiative decay causing an elongation of the lifetime, in agreement with experimental results.

We may, therefore, conclude that there exist several competing processes with their own time scales and it is imperative to determine which of these processes are dominant in the various concentrations. We will attempt a summary by including the following processes.

(a) uni-directional propagation of a domain of coherence: It is governed by centre-of-mass motion and its characteristic time scale is given by the time of flight

$$\tau_{\text{flight}} = \frac{\Delta l}{v_{\text{ex}}} \tag{4.149}$$

where Δl is the mean donor-to-acceptor distance and v_{ex} will be assumed on the order of $2 \cdot 10^4\,\text{m/s}$.

(b) radiative decay: Its characteristic time scale is given by the size of the domain of coherence and we adopt the formula

$$\tau_{\text{rad}} \approx (1\,\text{ns})\frac{T}{3000\,\text{K}} \tag{4.150}$$

(c) exciton-phonon interactions: assuming a weak coupling, and hence the absence of soliton formation, leads to the lifetime formula of an exciton given by [102]

$$\tau_{\text{ex-ph}} = \frac{2\hbar J}{\chi^2(2\bar{n}+1)} \qquad (4.151)$$

where

$$\bar{n} = \frac{1}{\exp(T_{\text{ph}}/T) - 1} \qquad (4.152)$$

is the mean occupation number for phonons. It is also well known that $\tau_{\text{ex-ph}}$ is inversely proportional to the width of an absorption line.

(d) diffusion processes: For high concentrations of acceptors and high temperature ranges, diffusive propagation of an exciton domain may compete with the above processes. The characteristic time scale is determined by the mean diffusion time

$$\tau_{\text{diff}} = \frac{(\Delta l)^2}{D} \qquad (4.153)$$

where $D = v_{\text{ex}}^2 \tau_{\text{ex-ph}}$.

In Table 4.4 below we have summarized the results of our estimates for the four different time scales at two temperatures $T = 30\,\text{K}$ and $T = 300\,\text{K}$ as well as two concentration values $N_{\text{d}} : N_{\text{a}} = 10^2$ and 10^4. The idea is to find out which mechanisms dominate the picture in each of the four regimes of control parameter values (high/low temperature and high/low acceptor concentrations).

TABLE 4.4:   A Summary of Characteristic Time Estimates

$N_{\text{d}}/N_{\text{a}}$	T	τ_{flight}	τ_{rad}	$\tau_{\text{ex-ph}}$	τ_{diff}
10^2	30 K	$10^{-12}\,\text{s}$	$10^{-11}\,\text{s}$	$10^{-13}\,\text{s}$	$0.6\times 10^{-11}\,\text{s}$
10^2	300 K	$2\times 10^{-12}\,\text{s}$	$10^{-10}\,\text{s}$	$5\times 10^{-14}\,\text{s}$	$4\times 10^{-11}\,\text{s}$
10^4	30 K	$10^{-10}\,\text{s}$	$10^{-11}\,\text{s}$	$10^{-13}\,\text{s}$	$4\times 10^{-8}\,\text{s}$
10^4	300 K	$2\times 10^{-10}\,\text{s}$	$10^{-10}\,\text{s}$	$5\times 10^{-14}\,\text{s}$	$4\times 10^{-7}\,\text{s}$

Based on this table the following can be concluded. Firstly, exciton-phonon interactions lead to a rapid thermalization process of the exciton domain. Depending on the strength of the coupling we then deal with a "dressed-up" collective exciton structure. Phonons are not expected to determine the capture process at an acceptor site. Secondly, diffusive propagation of the exciton domain is very unlikely to play a major role, except possibly at very high concentrations of acceptor molecules. Depending on the actual parameter values, either radiative losses or the speed of uni-directional propagation towards an acceptor will determine the fate of an exciton domain. It would now appear that at low acceptor concentrations the most important effect is the radiative loss process, which limits the probability of exciton energy capture. However, at high acceptor concentrations exciton propagation velocity is sufficiently high to reach an acceptor molecule before any excessive energy loss has occurred.

In summary, we wish to emphasize that the processes of exciton energy capture and transfer via Scheibe aggregates may be more complex and multifaceted than any hitherto proposed model has emulated.

4.6 Conformational Transitions in Proteins

Biochemical reactions are controlled by the intra-molecular transitions between a multitude of conformational sub-states of proteins. This conclusion follows from the observations of non-exponential initial reaction stages. The first (already of historical importance) experiment performed a quarter century ago by Frauenfelder and coworkers [15, 82] concerned the kinetics of ligand binding to myoglobin. Myoglobin, like its more complex relative, hemoglobin, which is made up of 4 myoglobin-like units, is a protein that stores molecular oxygen. It is well known that a replacement of oxygen with carbon monoxide results in poisoning the organism. This is related to the fact that the CO binding reaction, as opposed to the $O2$ binding one, is irreversible (see Fig. 4.44a). The experimentalists broke up the heme-CO bond in a non-thermal way using a laser flash and observed the process of ligand rebinding to heme under various conditions after the photolysis (see Fig. 4.44b). At 300 K, only the bimolecular reaction of binding from the solution was observed, at usual exponential time courses. The novelty of the experiment was the study of the process at low non-physiological temperatures. Under such conditions the curve of the time course of the bimolecular reaction reveals non-exponential time dependence of the uni-molecular reaction of ligand binding from the protein matrix.

A second type of experiment was performed due to the powerful technique called patch clamp [237]. It enabled observations of ionic current fluctuations through single protein channels (Fig. 4.45a). Soon, it appeared that most of the channels occur in two discrete states: "open" and "closed" and that the statistics of opening and closing times very often shows a non-exponential distribution density (Fig. 4.45b).

In standard kinetic experiments with an ensemble of molecules, the initial distribution of microstates is not specifically prepared and usually not much different from the local equilibrium distribution which occurs in the absence of the pre-exponential stage of the reaction, even if the reaction rate is controlled by the intra-molecular dynamics. However, in experiments using the patch clamp technique, a single protein channel molecule can be observed changing

(a)   (b)

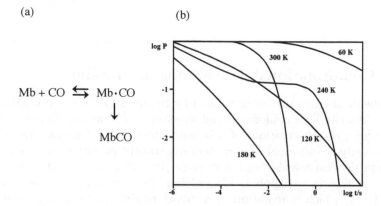

FIGURE 4.44:   (a) An irreversible process of carbon monoxide binding to myoglobin consists of two steps: a reversible bimolecular reaction of the ligand adsorption from the solution and an irreversible uni-molecular reaction of the ligand covalent binding to heme from the protein interior. (b) Sketch of the time dependence of the rebinding of CO molecules after the photo-dissociation of CO-bound sperm whale myoglobin in various temperatures, after Frauen-felder and coworkers [15, 82]. $P(t)$ represents the fraction of the myoglobin molecules that have not rebound CO at time t after the laser flash. At low temperatures, only the uni-molecular reaction of CO rebinding from the protein interior is observed. Its time course is evidently non-exponential. The exponential stage observed at 240 K and at higher temperatures is attributed to the bimolecular reaction of CO rebinding from the solution. The latter process masks the exponential stage of the uni-molecular reaction of CO rebinding from the protein interior (incomplete masking has been observed for the horse myoglobin by Post et al. [218]).

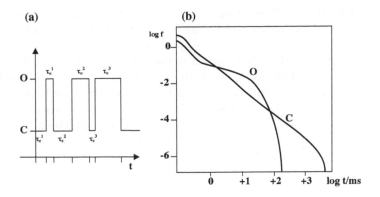

FIGURE 4.45: (a) A schematic "telegraphic noise" recorded with the help of the patch clamp technique. The ionic current flowing through a single protein channel fluctuates between the two values corresponding to two "chemical" states of the protein: open (O) and closed (C). (b) Sketch of time dependence of the closed time, C, and open time, O, distribution density $f(t)$, observed with the help of the patch clamp technique for the protein K^+ channel of NG 108-15 cells, after Sansom et al. [240]. Both curves show a short-time non-exponential behavior.

its state stochastically between an open and a closed state. As a result, experiments with single molecules bring first-passage time distribution densities $f(t)$, separately for the forward and backward reaction, each formally treated as irreversible. Each time after the reactive transition, the molecule starts its microscopic evolution from a conformational substate within the transition state of the return reaction. The initial distribution of conformational substates confined only to the transition state is attained also in the first experiment mentioned, where an ensemble of molecules being initially in some thermodynamically stable state is non-thermally excited to the unstable state P. The presence of non-exponential initial stages in the time courses discussed above implies that some, if not all, biochemical processes are controlled by the intra-molecular dynamics of the proteins involved.

Because the experiments at hand cannot elucidate the nature of the conformational transition dynamics within the protein native state in detail, the problem of modeling this dynamics is to some extent left open to speculation. In two classes of models provided hitherto in the literature, the speculative element seems to be kept within reasonable limits. We refer to them symbolically as the protein-glass and protein-machine models [155]. In essence, the question concerns the form of the reciprocal relaxation time spectrum above the gap (Fig. 4.46). The non-exponential time course of the processes discussed indicates this spectrum to be quasi-continuous, at least in the range from 10^{-11} to 10^{-7} s. The simplest way to tackle problems without a well-defined time scale separation is to assume that the dynamics of a system looks alike on all time scales, i.e., the spectrum of relaxation times has self-similarity symmetry. This assumption is the core of any protein-glass model (Fig. 4.46a). An alternative leads to the protein-machine class of models in which the variety of conformations composing the native state is supposed to be labeled with only a few "mechanical" variables. The reciprocal relaxation time spectrum is then a sum of several more or less equidistant sub-spectra (Fig. 4.46b). We now elaborate on the characteristics of each class of these models.

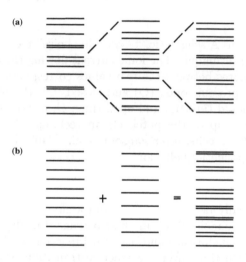

FIGURE 4.46: Schematic spectra of the reciprocal relaxation times of conformational transition dynamics within the native state of protein. (a) The protein-glass model: the spectrum looks approximately alike on several successive time scales. (b) The protein-machine model: the spectrum consists of a few more or less equidistant sub-spectra.

4.6.1   The Protein-glass Model

Time scaling, considered to be a generic property of glassy materials, can originate either from a hierarchy of barrier heights in the potential energy landscape, or from a hierarchy of bottlenecks in the network joining conformations between which direct transitions take place. A hierarchy of inter-conformational barrier heights (Fig. 4.47) was proposed more than ten years ago by Frauenfelder and co-workers in order to combine the results of various experiments concerning the process of ligand binding to myoglobin (see the review by Frauenfelder et al. [82]). A reasonable mathematical realization of such a hierarchy in the context of applications to proteins is spin glasses [260, 26]. The mathematical realization, on the other hand, of hierarchical networks is lattices with the effective dimension between 1 and 2, e.g., geometrical fractals or percolation lattices [154]. The process of diffusion on fractal lattices can (but does not have to) be interpreted as simulating structural defect motions in the liquid-like regions between solid-like fragments of the secondary structure. It will be discussed in more detail in the next section.

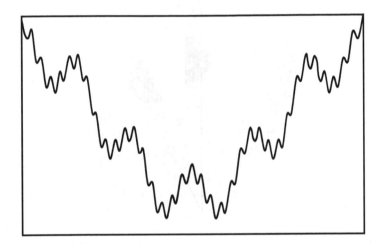

FIGURE 4.47: An example of a one-dimensional potential with a hierarchy of barrier heights. The curve represents a superposition of three sinusoids of appropriately scaled periods and amplitudes. The addition of subsequent, more and more subtle components leads, in the asymptotic limit, to the Weierstrass function, which is everywhere continuous but nowhere differentiable.

4.6.2   The Protein-machine Model

The protein-machine model was proposed for the first time thirty years ago by Chernavsky, Khurgin and Shnol [46]. Speculative when first proposed, the model has recently become increasingly justified experimentally [153]. In the simplest case, the mechanical variables can be identified with the angles describing the mutual orientation of rigid fragments of the secondary structure or larger structural elements (see Fig. 4.48). The mechanical co-ordinate may be also identified with a "reaction coordinate", if this can be determined. There is strong evidence that the mechanical coordinates are related to the "essential modes" of motion, studied extensively in recent years with the help of molecular dynamics simulations by the Garcia group [119, 110, 205]. In the continuum limit, successive conformational transitions along a given mechanical coordinate are to be approximated by diffusion in an effective potential, the simplest being parabolic [2, 153]. The reciprocal relaxation time spectrum for such a model is exactly equidistant (see Fig. 4.46b).

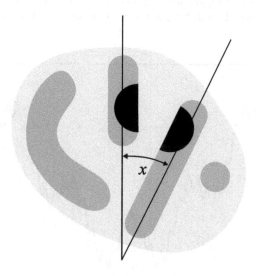

FIGURE 4.48:   A schematic cross-section of the fundamental structural unit of each protein, a domain.   Heavily shaded are solid-like fragments of secondary structures and lightly shaded are surrounding liquid-like regions. Black is the catalytic center usually localized at two neighboring solid-like elements. In models of the protein-machine type, the dynamics of conformational transitions is treated as a quasi-continuous relative diffusion of solid-like elements along a mechanical co-ordinate, identified here with the angle x. The picture can be reinterpreted on a higher structural level: solid-like elements then represent whole domains moving in a multi-domain protein complex. After Kurzynski [153].

Each of the models of conformational transition dynamics mentioned above may be true to some extent but one class of models of the protein-glass type, namely that of the random walk on fractal lattices, seems to offer the greatest theoretical power. We devote the next Section to this class of important models.

4.7 Vesicle Transport and Molecular Motors

4.7.1 Chemo-Chemical Machines

Consider two chemical reactions: ATP hydrolysis:

$$ATP \rightarrow ADP + P_i \tag{4.154}$$

and phosphorylation of a certain substrate Sub according to:

$$Sub + P_i \rightarrow SubP \tag{4.155}$$

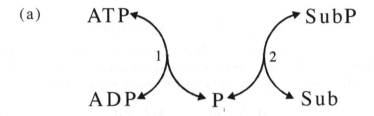

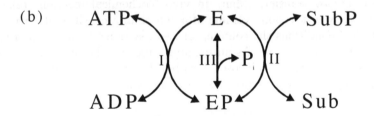

FIGURE 4.49: Coupling of two phosphorylation reactions through a common reagent P_i (a) and through a kinase enzyme E (b).

In both reactions we ignore the participation of a molecule of water, of which there is an excess. The equilibrium constants K_1 and K_2 are determined by the following quotients:

$$K_1 = \frac{[ADP]^{eq}[P_i]^{eq}}{[ATP]^{eq}}, \qquad K_2 = \frac{[SubP]^{eq}}{[Sub]^{eq}[P_i]^{eq}} \tag{4.156}$$

The first reaction is exoergic from left to right ($K_1 M^{-1} > 1, \Delta G_0 < 0$) while the second is assumed to be endoergic in this direction ($K_2 M^{-1} < 1$,

$\Delta G_0 > 0$). The second reaction can, however, take place from left to right if both reactions take place in the same reactor and if the first reaction ensures a sufficiently high concentration of the common reagent P_i (see Fig. 4.49a). It at first appears that this system acts as a chemo-chemical machine where the first reaction transfers free energy to the second. But is it really so?

The thermodynamic forces acting in the two reactions are given respectively by:

$$A_1 = k_B T \ln K_1 \frac{[ATP]}{[ADP][P_i]}, \qquad A_2 = k_B T \ln K_2 \frac{[Sub][P_i]}{[SubP]} \qquad (4.157)$$

Under stationary conditions, the corresponding fluxes are

$$J_1 = \dot{X}_1, \qquad J_2 = \dot{X}_2 \qquad (4.158)$$

where

$$X_1 = [ADP], \qquad X_2 = [SubP] \qquad (4.159)$$

and they are equal. The forces in (Eq. 4.157) are not, however, independent from these fluxes. The flux and force for each reaction taken separately are always of the same sign and hence we have simultaneously that:

$$A_1 J_1 \geq 0, \qquad A_2 J_2 \geq 0 \qquad (4.160)$$

Therefore, there cannot possibly be any free energy transformation but at most an entropy transfer. Many in vivo biochemical reactions take place simultaneously in the same volume of a cell or in a consitutent part. Under physiological conditions the coupling of two reactions via a common reagent is not easily implemented. As an example consider the first reaction in the glycolysis chain, i.e., the glucose phosphorylation (Sub = Glu) into Glu6P which is coupled to ATP hydrolysis. For this reaction we have:

$$K_2 = \frac{[Glu6P]^{eq}}{[Glu]^{eq}[P_i]^{eq}} = 6.7 \times 10^{-3} M^{-1} \qquad (4.161)$$

Under stationary but non-equilibrium physiological conditions, $[Pi] = 10^{-2} M$ and $[Glu6P] = 10^{-4} M$, hence for the reaction to move forwards the force A2 must be positive and the concentration of glucose [Glu] must exceed $1.6M = 300g/dm^3$ which is an unrealistic value.

Nature has found a different way to achieve the coupling between ATP hydrolysis and the reaction of phosphorylation on the substrate, namely through an enzyme that catalyzes the two reactions simultaneously (see Fig. 4.49b). According to the terminology presented in Section 4.4.4.2 enzymes that catalyze the transfer of a phosphate group are called kinases. If one of the coupled reactions is ATP hydrolysis, then we refer to the enzyme as having ATPase activity. The sketch presented in Fig. 4.49b includes not two but three reactions labeled by the Roman numerals I, II and III. Note that besides the transfer of the phosphate group from ATP to Sub it is also possible to detach this group from the enzyme unproductively. This is characterized by the equilibrium constants given below:

$$K_I = \frac{[ADP]^{eq}[EP]^{eq}}{[ATP]^{eq}[E]^{eq}}, \quad K_{II} = \frac{[E]^{eq}[SubP]^{eq}}{[EP]^{eq}[Sub]^{eq}}, \quad K_{III} = \frac{[E]^{eq}[P_i]^{eq}}{[EP]^{eq}}$$
$$(4.162)$$

Since an enzyme cannot affect chemical equilibrium conditions, the constants in (Eq. 4.163) are independent of the constants in (Eq. 4.162) and the following relations are satisfied:

$$K_I K_{III} = K_1, \quad \frac{K_{II}}{K_{III}} = K_2 \qquad (4.163)$$

Similarly to the relationships between equilibrium constants one finds corresponding relationships between thermodynamic forces:

$$A_1 = A_I + A_{III}, \quad A_2 = A_{II} - A_{III} \qquad (4.164)$$

where the forces acting reactions I, II and III are defined analogously to those in (Eq. 4.157). Using the definitions in (Eq. 4.159) one can identify reaction fluxes in the two schemes shown in Fig. 4.49

$$\dot{X}_1 \equiv J_1 = J_I \qquad \dot{X}_2 \equiv J_2 = J_{II} \qquad (4.165)$$

Under the stationary conditions [E]=const and [EP]=const, there is also the relation:

$$J_{III} = J_I - J_{II} \qquad (4.166)$$

It can be concluded from (Eq. 4.166) and (Eq. 4.164) that the dissipation function for the system of the three reactions illustrated in Fig. 4.49b can be represented as:

$$\Phi = A_I J_I + A_{II} J_{II} + A_{III} J_{III} = (A_I + A_{II})J_I + (A_{II} - A_{III})J_{II} = A_1 J_1 + A_2 J_2$$
$$(4.167)$$

Each of the three terms in (Eq. 4.165) are non-negative but this does not mean that each of the two terms in (Eq. 4.167) must also be non-negative. It is sufficient that for $J_1, J_2 > 0$ when $A_I, A_{II} > 0$, we have $A_{III} > A_{II}$. The other condition: $A_{III} < -A_I$ is impossible to satisfy under physiological P_i concentrations. However, for $J_1, J_2 < 0$, when $A_I, A_{II} < 0$, the inequality $A_{III} > -A_I$ can take place. In both cases we deal with a real transformation of the free energy. The first case, namely the transfer of the free energy from subsystem 1 to subsystem 2, takes place in the already discussed process of glucose phosphorylation at the expense of ATP hydrolysis. The second case, namely the transfer of free energy from subsystem 2 to subsystem 1, occurs, for example, in the next two stages of the glycolysis chain where ADP phosphorylation to ATP proceeds at the expense of even higher energy substrates.

The kinetic scheme shown in Fig. 4.49b can be generalized to the case of two arbitrary coupled chemical reactions:

$$R_1 \rightleftharpoons P_1 \quad \text{and} \quad R_2 \rightleftharpoons P_2 \tag{4.168}$$

It is worth emphasizing that biochemical reactions can be coupled through extensive enzymatic pathways, e.g., cascades. We have assumed for simplicity that both reactions are mono-molecular. One reaction is a donor of free energy and the other is a free energy acceptor. Each biochemical reaction must be catalyzed by a protein enzyme. Separately, each reaction takes place in the direction determined by the second law of thermodynamics such that the amount of chemical energy dissipated is positive (see Fig. 4.50a and b). Only when both reactions occur simultaneously using the same enzyme according to the second law of thermodynamics can the second reaction be forced to take place against the second law. In this case, the latter would transfer a part of its free energy recovered from dissipation doing work on it (see Fig. 4.50.c). The mechanism of energy transfer is very simple: if both reactions occur in a common cycle, they must proceed in the same direction.

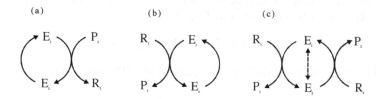

FIGURE 4.50: (a, b) Two different chemical reactions proceeding independently from each other. We assume that they are catalyzed by the same enzyme but the reagent for the first reaction R_1 is coupled to the state E_1 of the enzyme, and the reagent for the second reaction, R_2, is coupled to the state E_2. Reactions take place under stationary conditions as a result of keeping the values of concentrations of the reagents and products fixed but different from those at equilibrium. These values are selected such that the first reaction proceeds from R_1 to P_1 and the second from P_2 to R_2. (c) If both reactions take place simultaneously using the same enzyme, the direction of the first reaction can force a change of direction of the second reaction. This new direction would be opposite to the one dictated by the stationary values of the respective concentrations. The broken line denotes a possible unproductive transition between the states E_1 and E_2 of the enzyme.

To emphasize the similarity between a chemo-mechanical machine and a wheel and axle device, we note that just as turning a wheel requires the input of work acting against the force of gravity due to the weight attached through a rope to a wheel, the first reaction performs work against the chemical force acting on the second reaction forcing it to proceed in the opposite direction. Friction associated with the motion of the wheel and the axle corresponds to the energy dissipation in the common reaction cycle. Slippage of the axle with respect to the wheel is mirrored by the possible direct reaction between the states E_1 and E_2 of the enzyme. However, there is an essential difference between the wheel-and-axle device and a chemo-chemical machine. The former is a machine characterized by macroscopic spatial organization while the enzymes enabling the operation of a chemo-chemical machine are micro- or at worst mesoscopic entities. Viewed macroscopically, the chemo-chemical machine is a more or less spatially homogeneous solution of enzymes with a typical concentration of 10^{-6} M, i.e., close to 10^{15} molecules per cubic centimeter or 10^3 molecules per cubic micrometer (the typical size of a bacterial cell or an organellum of a eukaryotic cell).

Similarly to the suspension molecules in the solutions observed by Brown, macromolecular enzymes playing the role of biological machines move about and, in particular, change their chemical state due to thermal fluctuations. On a short time scale, the energy is borrowed from and returned to the heat bath. The stochastic motion of biological machines is not purely random, and results from their highly organized structure and the constant input of free energy, mainly due to the hydrolysis of ATP. This is clearly seen in experiments with single biomolecules that employ an ever more precise and powerful arsenal of techniques [125, 126].

4.7.2 Biological Machines as Biased Maxwell's Demons

In 1871 James Clerk Maxwell, pondering the foundations of thermodynamics, contemplated the functioning of a hypothetical being that would be able to observe the velocities of individual gas molecules moving about in a container. The special feature of the container would be a partition with an opening that can be covered by a latch (see Fig. 4.51a). This 'Gedenken experiment' has been referred to in the literature as Maxwell's demon since it was imagined that a demon, 'Maxwell's demon', would be in charge of closing and opening the hole in the partition allowing only sufficiently fast particles to move from right to left and only sufficiently slow ones to move from left to right across the partition. This would, of course, over time result in a temperature increase in the left part of the container and a temperature decrease in the right part of the container. The thus created temperature gradient is clearly in contradiction with the second law of thermodynamics due to the work performed in the process by thermal fluctuations alone in a gas at a thermodynamic equilibrium.

Less than 100 years later another great physicist, Richard Feynman [77], presented this problem in a more provocative manner shown in Fig. 4.51b where a mechanical wheel with a ratchet and a pawl is sketched that can rotate around its axis and which has wing-like plates attached to it. The

(a)

(b)

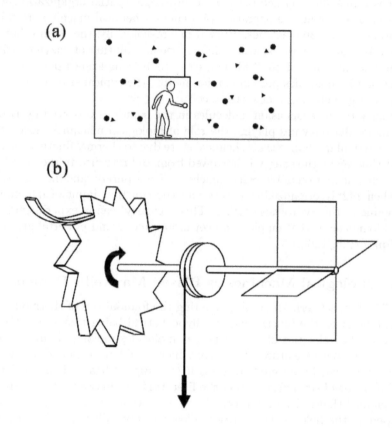

FIGURE 4.51:   (a) Maxwell's demon. (b) Feynman's version of Maxwell's demon.

surfaces of the plates are bombarded with gas molecules exerting pressure on both sides of each wing. However, the presence of the pawl prevents the ratchet from rotating in one of the directions. As a result the kinetic energy of gas fluctuations is being transformed into the rotational kinetic energy of the ratchet's one directional motion around the axis. This can, in principle, be utilized to raise a weight against the force of gravity.

The logical error in both these arguments is made when we treat microscopic systems (gas molecules) the same way as macroscopic ones (hole in the partition, pawl in the wheel). In particular, the pawl device cannot possibly react to the collision of a single molecule unless it, too, is a microscopic object subject to the same types of thermal fluctuations. Random bending of the pawl assists with the rotation of the wheel in the opposite direction. Similarly, with the latch controlled by Maxwell's demon, to measure the speed of individual molecules, they must be in contact via a physical interaction. To react to such an interaction, the observing device must be microscopic itself but then, of course, it will be rapidly brought to thermal equilibrium as a result of interactions with chaotic moving molecules around it. Hence, it will behave chaotically the same way as an average gas molecule does and, consequently, no net macroscopic force will be generated. These unfavorable fluctuations can be reduced by lowering the temperature of the pawl or by freezing the head of Maxwell's demon. In general, this can be done by reducing the entropy or supplying some amount of free energy. However, when this is done the contradiction with the second law of thermodynamics is automatically removed and these mesoscopic machines will work according to normal rules of behavior discussed earlier. In fact, biological motors that are the centerpiece of this section provide an interesting example of such biased Maxwell's demons.

4.7.3 Pumps and Motors as Chemo-chemical Machines

From a theoretical point of view, it would be very convenient to treat all molecular biological machines as chemo-chemical machines. In this section we will address this issue in the case of molecular pumps and motors. In general, a chemo-chemical machine can be viewed as a 'black box' to which enter and from which exit molecules that take part in the coupled chemical reactions shown in Fig. 4.52. The kinetic scheme of the reactions can be arbitrary as long as it involves at least one cycle so that the enzyme cannot be used up in the course of the reaction. In particular, four possible kinetic schemes can be devised assuming one substrate-enzyme intermediate for each catalyzed reaction. Both reactions can be coupled through the free enzyme E (see Fig. 4.53a), both can be coupled through the intermediate complex M (see Fig. 4.53b), the intermediate complex for one reaction can be the free enzyme for the second reaction (see Fig. 4.53c) and both reactions can proceed as alternating half-reactions (see Fig. 4.53d).

The scheme in Fig. 4.53a applies to substrate phosphorylation, that in Fig.

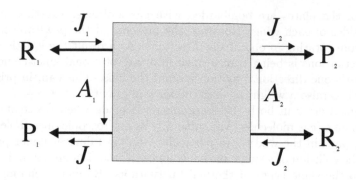

FIGURE 4.52:   The general scheme of a chemo-chemical machine coupling two reactions: (a) $R_1 \Longleftrightarrow P_1$ that produces free energy and (b) $R_2 \Longleftrightarrow P_2$ that consumes free energy. Each of the two reactions can take place in either direction which is determined by the sign of the flux J_i. The forces A_i are given by the values of the relevant concentrations kept stationary.

4.53b applies to molecular motors and those in Figs. 4.53c and 4.53d apply to molecular pumps. Treating molecular pumps as chemo-chemical machines poses no great problems. The molecules that are transported from one side of a biological membrane to the other, inside or outside the compartment considered, can be considered to be in different chemical states, while the transport process across the membrane can be regarded as an ordinary chemical reaction (see Section 4.4.4.4). In Figs. 4.54a and 4.54b, simplified kinetic cycles of the calcium and the sodium-potassium pumps are presented, respectively [262, Ch. 12]. It is clear that both schemes are indeed identical to the one presented in Fig. 4.53d.

Molecular motors present a slightly more complicated case. In Fig. 4.55, a simplified version of the Lymn-Taylor-Eisenberg kinetic scheme of the mechano-chemical cycle of the acto-myosin motor is presented [117]. The scheme indicates how the ATPase cycle of myosin is related to a detached, weakly-attached and strongly-attached state of the myosin head to the actin filament. Both the substrate and the products of the catalyzed reaction bind to and rebind from the myosin in its strongly attached state, whereas the very reaction takes place either in the weakly-attached or in the detached state. Only completion of the whole cycle with the ATP hydrolysis achieved in the detached state results in the directed motion of the myosin head along the actin track. ATP hydrolysis in the weakly bound state alone is ineffective, and corresponds to slippage. However, the question remains how should a load acting on the motor be represented in terms of concentrations. In Fig. 4.55 the letter A denotes the actin filament either before or after translation of the myosin head by a unit step. There is serious experimental evidence that an

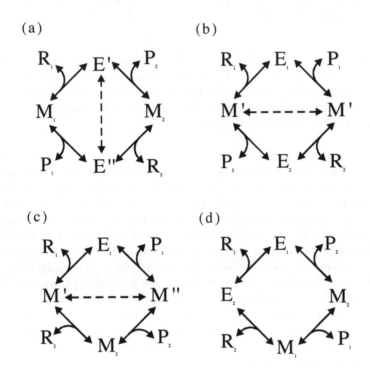

FIGURE 4.53: The four possible kinetic schemes for the system of two coupled reactions assuming that each of them proceeds through one intermediate state of the enzyme. The broken lines represent unproductive (without binding to or unbinding from substrates) direct transitions between different states of the enzyme. They cause a possible mutual slippage of the two component cycles.

(a)   (b)

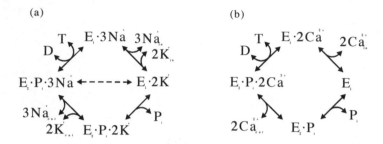

FIGURE 4.54: Simplified kinetic cycles of the calcium pump (a) and the sodium-potassium pump (b). E_1 and E_2 denote two states of an enzyme that represents a pump with a reaction center oriented to the interior and exterior compartment, respectively. Here, T, D and P_i stand for ATP, ADP and inorganic phosphate, respectively.

external load attached to the statistical ensemble of myosin heads (organized, in the case of myofibrils, into a system of thick filaments) influences the free energy of binding of the myosin heads to thin actin filaments [17]. The associated changes of the binding free energy can be expressed as the changes of the effective rather than actual concentrations of the actin filament A before and after translation. As a consequence, the acto-myosin motor can indeed be effectively treated as a usual chemo-chemical machine. The output flux J_2 is proportional to the mean velocity of the myosin head along the actin filament and the force A_2 is proportional to the load.

At a macroscopic level, the action of molecular pumps and motors manifests itself as a directed transport of a substance. The possible functioning of mesoscopic machines on a macroscopic scale is due to an appropriate organization of the statistical ensemble. Namely, molecular pumps are embedded in the two-dimensional structure of the membrane (see Fig. 4.56a), while molecular motors move along a structurally organized system of tracks: microfilaments or microtubules (see Fig. 4.56b). More will be said about these protein filaments and their self-assembly later.

However, not all biological molecular machines perform work on a macroscopic scale. Examples of such behavior are molecular turbines, e.g., the F_0 portion of the ATP synthase. Since there is no mechanism coordinating the rotational motion of individual turbines, there is no macroscopic thermodynamic variable characterizing this motion. Both portions F_0 and F_1 must, therefore, be treated jointly from the macroscopic point of view, together giving rise to a reversible molecular pump - the H+ATPase.

A large number of ingenious techniques have been designed for molecular pumps and motors that enable precise observation of the behavior of these single molecules. For molecular pumps and ion channels such a technique is the patch-clamp method [237]. In the case of molecular motors these are var-

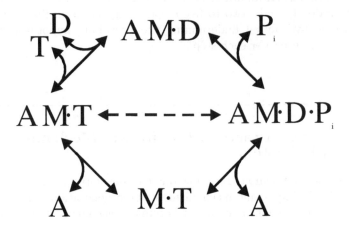

FIGURE 4.55: A simplified version of the Lymn-Taylor-Eisenberg kinetic scheme of the mechano-mechanical cycle of the acto-myosin motor. Here, M denotes the myosin head, A the actin filament, T, D and P_i stand for ATP, ADP and an inorganic phosphate, respectively.

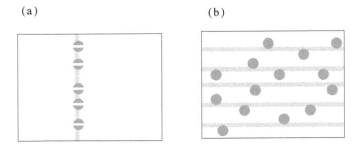

FIGURE 4.56: The functioning of biological molecular machines on a thermodynamic scale at a macroscopic level is possible due to an appropriate organization of the system. Molecular pumps are embedded into the two-dimensional structure of the membrane (a), while molecular motors move along an organized system of tracks - microfilaments or microtubules (b).

ious motility assays [186, 125, 126]. All these observations reveal a stochastic nature of the behavior exhibited by molecular motors. In a particularly impressive experiment Kitamura et al. [144] demonstrated that a single myosin head can randomly make two to five steps along the actin filament per one ATP molecule hydrolyzed undermining the prevailing hypothesis that one ATP molecule is used per each step.

4.8   Muscle Contraction; Biophysical Mechanisms, Contractile Proteins

In this section we demonstrate how one can develop a connection between a physiologically important mechanism of force generation through muscle contraction and its molecular basis involving molecular motors and protein filaments.

4.8.1   Biophysics of Muscles

In general a muscle is attached, via tendons, to at least two different bones; see Fig. 4.57 that shows muscles in the human arm and Fig. 4.58 showing the muscles in the forefinger.

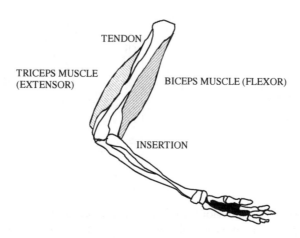

FIGURE 4.57:   The triceps and biceps muscles in the human arm.

At a joint, two or more bones are flexibly connected, e.g., the elbow, knee

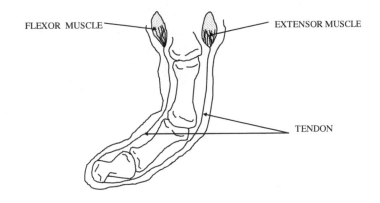

FLEXOR MUSCLE

EXTENSOR MUSCLE

TENDON

FIGURE 4.58: The forefinger. Note that the tendons carrying the forces exerted by the muscles go over joints that change the direction of the force.

and hip. A pull is exerted by a muscle when its fibers contract under stimulation. Muscles that bring two limbs closer together, like the biceps muscle in the human arm, are called flexor muscles. On the other hand, those that extend a limb outwards, such as the triceps, are called extensor muscles.

Flexor muscles are used, for example, when the upper arm is used to lift an object in the hand whereas the extensor muscle is used when throwing a ball. The human skeleton is a very sophisticated device that transmits forces to and from various parts of the body. It is the muscles which move the parts and generate forces, use up chemical energy and hence perform work. Thus they provide the power for movement in most many-cell organisms [88, 25].

Muscles generate forces by contracting after they have been stimulated electrically. The tendons or rope-like attachments experience a net tension, after a series of these contractions, which increase with the number of electrical stimuli per second. Muscles try to shorten the distance between the attachment points of the tendons and it is noteworthy that they cannot act to push these points apart. Thus pairs of muscles are necessary to operate a limb, e.g., when the knee is bent the hamstring muscle, at the base of the thigh, shortens. To straighten the knee the quadriceps muscle at the front of the thigh shortens. In Fig. 4.59 we give examples of counter-acting muscle pairs in the human arm and leg.

Other types of muscles may join back on themselves and cause the constriction of an opening when they contract. Such muscles, called sphincters, serve several functions. The sphincter at the lower end of the esophagus prevents the backflow of stomach fluids. Another sphincter muscle in the eye changes the curvature of the lens of the eye to allow clear vision of near and distant objects.

Several muscles act simultaneously in concert in the shoulder to produce

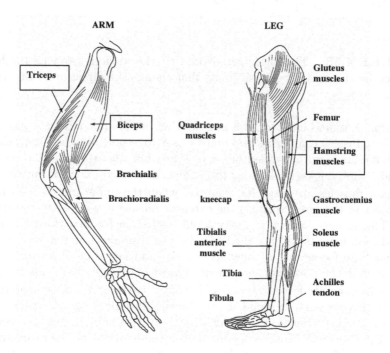

FIGURE 4.59:   Counteracting muscle pairs in the arm and the leg.

the total force exerted on the arm. The graphical method of adding forces yields the expected result: The upward and downward components of the forces cancel leaving a large horizontal force (see Fig. 4.60).

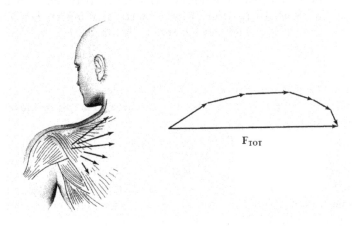

FIGURE 4.60: Principle of simultaneous action of muscles.

Single muscles may be extracted from any organism, such as a frog, and put into salt solution. These may be electrically stimulated by applying a voltage pulse to the solution. The length of the muscle may be kept fixed by clamping the ends of the muscle. We refer to measurements of the tension of a muscle of fixed length as an isometric force measurement. On the other hand, an isotonic force measurement is performed at a fixed load. We find that the maximum isometric force, F_{max}, which a muscle can generate is proportional to its cross-sectional area, A. Thus

$$F_{max} = \sigma A \tag{4.169}$$

If F_{max} is in newtons, then the constant of proportionality has units of N/m^2 (or pascals, Pa) if A is in m^2. A remarkable result is that σ is approximately the same for all vertebrate muscles and its value is $\sigma \approx 0.3 \times 10^6 Pa$. Muscles consist of fiber bundles where each fiber can contract with a given force. This provides a physiological basis behind the linear relationship between the contractile force, and the muscle cross sectional area..

4.8.2 Biophysical Mechanisms, Contractile Proteins

Animal and human posture and motion are controlled by forces generated by muscles. A muscle is composed of a large number of fibers which can be contracted under direct stimulation by nerves. A muscle is connected to two bones across a joint and is attached by tendons at each end. The contraction of a muscle produces a pair of action-reaction forces between each bone and the muscle. The work W performed by a contracting muscle is the product of the force of contraction and the shortening distance $\Delta\ell$, i.e.,

$$W = \sigma A \Delta \ell \tag{4.170}$$

where σ is the tensile stress produced by the muscle, i.e., the force of contraction per unit area and A is the cross-sectional area. The power developed by the muscle is then

$$P = \sigma A \frac{\Delta \ell}{\Delta t} = \sigma A v \tag{4.171}$$

and the speed of shortening $v = \frac{\Delta \ell}{\Delta t}$ appears to be a constant across species, since as the length of a muscle fiber increases, the time of contraction increases by the same factor. Thus, it has been found that the power generated is regulated by the value of the cross-sectional area A.

Muscles in the body are not constantly generating the maximum force F_{max} they can exert. It is a matter of common experience that the smaller a weight, W, you hold the faster you can lift it. Defining the contraction velocity, v, of a muscle to be the shortening length of a muscle divided by the shortening time, one can fit the results of an isotonic measurement (tension fixed) to the relation

$$v = b \frac{(F_{max} - W)}{(W + a)} \tag{4.172}$$

where a and b are constants, the former with the dimensions of force and the latter of velocity. This relation is known as Hill's Law. According to equation (Eq. 4.172) v = 0 when the load W is equal to the maximum force F_{max}. In Figure 4.61 we plot the contraction velocity, v, against the load W and it can be seen that as the load is reduced the muscle contracts more rapidly. In terms of the parameters in equation (Eq. 4.172) the maximum contraction velocity for zero load ($W = 0$) is given by

$$V_{max} = \left(\frac{b}{a}\right) F_{max} \tag{4.173}$$

The plot in Figure 4.61 is called a force-velocity curve. When force velocity curves are measured for car engines automotive engineers call F_{max} the stall force that will stop the engine.

In terms of their molecular structure, muscles are made up of fiber-like bundles called fascicles. Bundles such as these are made up of very long cells with a diameter of approximately 0.4 mm and lengths of about 40 mm. The

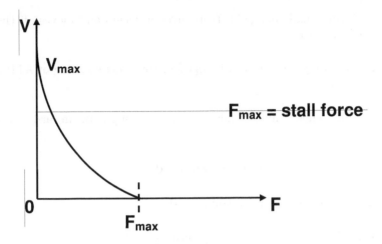

FIGURE 4.61:   Force-velocity curve.

forces which the fibers can exert are additive so that the total force generated is proportional to the number of fibers and hence the cross sectional area as was pointed out in earlier sections. Fiber cells come in two types. One type are thin filaments called actin protein fibers and the others are thick filaments known as myosin protein fibers (see Figure 4.62). There are 200 myosin molecules per thick filament.

Fibers are arranged in an alternating pattern of thick and thin filaments which may slide over each other. When we contract our muscles, motor proteins (see Figure 4.63) or molecular cross-bridges pull the thin filaments over the thick ones and when the muscles are extended filaments do not overlap. Motor proteins are fixed to myosin fibers and move over the actin fiber, very much like a collapsing telescope. The muscle force generated is larger the more activated motor proteins there are. Force-velocity curves have been measured for individual motor proteins and are quite similar to Hill's Law in equation (Eq. 4.172), so there is a molecular basis for this law.

The physiology of skeletal muscle contraction can be summarized as a sequence of the following molecular-level events:

- Nerve impulse reaches neuromuscular junction (acetycholine)

- It causes a depolarization wave due to a large influx of positive ions

- Action potential spreads along plasma membrane, down T tubules

- Calcium is released from sarcoplasmic reticulum

- Calcium binds to troponin

- Conformational change in Tropomyosin occurs and moves off the "hot spot" on actin

- The actin myosin complex is together following splitting of ATP on the myosin head

- The high energy myosin head swivels, pulling actin inward in a power stroke

- ADP is released from the myosin head

- Myosin and actin remain complexed

- ATP binds to myosin releasing actin

- These events repeat themselves until all calcium returns to sarcoplasmic reticulum

- Rigor Mortis takes place where rigidity is due to the lack of ATP upon death

Rigor Mortis is a state when neither ATP nor Ca^{2+} are present, and the myosin heads are bound to the actin. Relaxation State occurs when ATP is present, but Ca^{2+} is not, so the myosin heads are not bound to the actin. Contraction State is when both ATP and Ca^{2+} are present, and the myosin heads are bound to the actin. The most likely result is that when the muscle changes from rigor to relaxation the distance between the myosin heads will increase and as the muscle changes from rigor to contraction the distance between the heads increases.

The so-called Sliding Filament Theory maintains that muscle contraction is due to the sliding of myofilaments with no associated change in their length. A reduced width of the sarcomere which is the basic functional unit is the end result of contraction. This motion requires an energy input in the form of ATP that is cleaved by myosin ATPase. Thus during muscle contraction, the myosin and actin filaments slide past each other and cause the muscle to shorten. When broken down, myosin filaments are made up of many myosin heads which bind to the actin and then rotate causing the myosin and actin filaments to slide past each other. Each myosin head is made up of three main helical protein strands: the regulatory light chain (RLC), the essential light chain (ELC) and the heavy chain . These myosin heads are situated in pairs, known as dimer pairs, and they are joined together at their base. Most of the biophysical studies of muscle dynamics focused mainly on the action of the two filaments, and of the myosin heads individually, not as a pair. The purpose of the pairing of the myosin heads is still unknown.

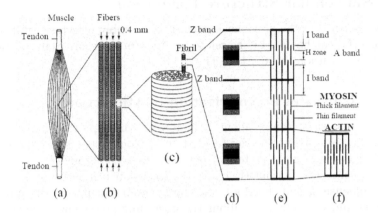

FIGURE 4.62: The composition of muscle fibers. Fibers (b) consist of fibrils which have light and dark bands (d). Myosin and actin filaments are designated (e) and (f).

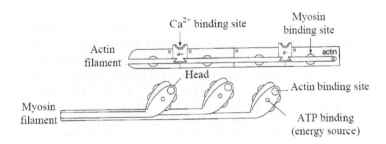

FIGURE 4.63: The principal components of the motor protein complex.

4.9 Subcellular Structure Formation

As we showed in the previous section, the assembly of cytoskeletal protein filaments [11] can be used by the cell to produce forces resulting in cell motility and internal tension [5, 167, 111, 83].

4.9.1 Aspects of Polymerization of Microtubules

When a physicist thinks about polymers he/she typically thinks about a system with unusual geometries (e.g., stripes, blocks, spaghetti-like, micelles, etc.). Furthermore, for a system of polymers as seen through its many physical properties, the normal (Euclidean) dimension has to be replaced by a fractal dimension, and therefore, the challenge in calculating the properties of such systems is to find out about the governing power laws, which are the consequence of the new spatial fractal dimension.

What we wish to show in this section is something quite different although it concerns a special type of biopolymer-microtubules. Microtubules are polymers consisting of protein called tubulin. They are found in almost every living system inside its cells. In various cells they play a number of important roles. Indeed every movement of the cell or material movement inside the cell (cell division, for example) is done with the help of microtubules. At present we are still far away from a complete understanding of all these mechanisms. On the other hand, we can also do experiments with MT's outside the cell (in vitro experiments). We hope that we can show here a first step towards the quantitative understanding of these effects. However, the challenge is to understand the time dependence of polymerization processes under a variety of in vitro conditions. Let us begin by posing a general question: "What are the main differences between the living systems and the inanimate systems we consider in physics?" A first trivial answer could be: "Living organisms cannot be in thermal equilibrium (even if they have a defined temperature)." Indeed a living organism has to consume energy (food) and we have a continuous energy flow through the system. Only this fact gives the possibility that such systems develop cyclical time dependence. Also, we always have energy dissipation in living systems. If a steady state exists in such a system, it must be a state far from thermal equilibrium. Following [221], such systems are called dissipative structures. Haken [105] calls them synergetic systems. Dissipative structures have the property of time dependence, but also the possibility of coherence (a well-known physical example is the laser). It should be clear that coherence is of crucial importance for biological systems.

To emphasize, living systems are driven complex systems with some degrees of freedom far from thermal equilibrium. For microtubules it is the pumping of chemical energy into the free tubulin, through the conversion from GDP to GTP tubulin, which itself makes all the subsequent time dependent effects possible.

Microtubules are formed both in vivo and in vitro from the dimers of tubulin which is a protein existing in two main homologous varieties, namely: α-tubulin and β-tubulin while a third variety, γ-tubulin, is involved in the nucleation of microtubules (MT's). An individual MT is a hollow 25-nm radius cylinder built usually from 13 linear protofilaments each of which is composed of alternating α- and β- subunits. Furthermore, different surface lattice types have been observed (the so-called A and B lattices) but the geometry of the MT lattice is a separate issue since in this paper our interest is limited to one-dimensional approximations for polymerization processes. The reader is referred to a recent review article for a more in-depth discussion [12] on the structure of MT's. MT polymerization and depolymerization processes due to their peculiar nature have been termed the dynamic instability phenomena [195, 194]. In spite of over a decade of analyses of these effects, a complete quantitative understanding of the microscopic nature of dynamic instability is still elusive. The rate of MT polymerization significantly depends on the concentrations of tubulin, GTP and ionic specifications of the solution. Although tubulin dimers form a regular array, each microtubule is characterized by its polarity, i.e., its distal (plus) end grows faster than the proximal (minus) end. Only tubulin dimers to which GTP is bound are capable of binding to a MT. Upon binding, GTP rapidly hydrolyses into GDP except perhaps for the top layer or two which is commonly referred to as the GTP-cap. It is this chemical energy supply, transfer and dissipation which is at the centre of the dynamic instability issue. It is also noteworthy that the geometry of growing MT's (mainly straight protofilaments) differs from that of shrinking ones (curved protofilaments). The above tends to imply that polymerization involves the formation of axial bonds while depolymerization consists in breaking the already weakened lateral bonds. It is, therefore, conceivable that mechanical stress which may be associated with the presence of unhydrolyzed GTP in different places on the surface of a MT may explain the stochastic nature of the dynamic instability phenomenon.

Experimental observations indicate the existence of several stages in the development of a single MT. An initial nucleation stage from seed oligomers slowly reaches an asymptotic density of MT ends after a period of approximately 3000s. This stage is followed by an almost continuous growth process which is randomly interrupted by catastrophic disassembly which is immediately followed by a so-called rescue event leading to a subsequent growth phase. While the pattern of catastrophes and rescues is repeated over and over again in the life history of a single MT, it exhibits little regularity. The rates at which the assembly and disassembly processes occur are quite different and significant fluctuations in growth and shrinkage velocities have been measured. The rate of disassembly is typically 10 - 20 times higher than the rate of growth, both in vivo and in vitro. The rate of growth for both the plus and minus ends increases almost linearly with tubulin-GTP concentration [113, 279]. The exact dependence of the frequency of catastrophes on both the tubulin concentration and the growth velocity is not known but it seems to exhibit a decreasing tendency as a function of the tubulin concentration. The frequency of rescues has been found to be almost linearly proportional to tubulin concentration for both ends. Kinetic model parameters show dependence on both the GTP and GDP concentrations.

In summary, we wish to point to the importance of two main effects in MT polymerization:

- The cycle of growth and catastrophic decay, followed by a rescue event. Using a graphic description: "the individual MT behaves like the antenna of a snail". This, at first, very strange behaviour may have biological relevance. MT's which grow inside the cell may be probing to find specific sites in the available space to which they can attach.

- Coherent oscillations emerge spontaneously in high density ensembles of microtubules. Besides coherence inside the cell, outside the cell (in vitro experiments) coherence has also been seen. At a higher concentration of tubulin, where we have many MT's, the condensed mass can show characteristic damped oscillations. Oscillatory nature of MT assembly may be coupled to other cycles taking place in the cell, for example the calcium waves or division cycles.

4.9.2 Simple Models of Microtubule Assembly

The first type of protein filament whose polymerization process is discussed here is the microtubule. Observations of individual MTs reveal periods of almost steady growth interrupted by brief periods of very rapid shortening as illustrated in Figure 4.64. The transition from a growth period to a shortening period is known as a catastrophe event while the reverse is known as a rescue. The assembly dynamics at each end of the MT differ. The "plus" end is between three and five times as dynamic as the "minus" end. That is to say that both the growth and shortening of the positive end of the MT occur at rates at least three times of those at the negative end but both ends are dynamic in free MTs. Through the individual MT assembly dynamics [195, 113] one can see the property of their assembly known as *dynamic instability*. What is intriguing is that the stochastic individual behavior [38] results in smooth collective oscillations observed to take place at high tubulin concentrations [36, 178]. The answer is given simply by the application of statistical mechanics to ensemble averaging. A complete explanation of the above requires an application of the master equation formalism [81] or, alternatively, the use of chemical kinetics [184, 116, 254].

The assembly dynamics for an individual MT are stochastic where the rate of shortening is about ten times the rate of growth [178, 271]. The probability of a single MT nucleating, growing and shortening depends on the local concentration of tubulin with bound GTP, tubulin with bound GDP and other molecules relevant to the assembly process which are found within the cytoplasm. Many theoretical models have successfully reproduced the single MT growth behavior such as that illustrated in Fig. 4.64.

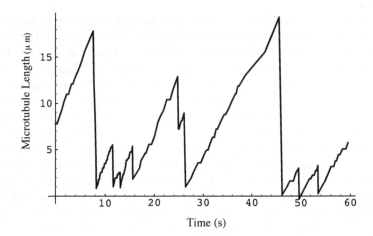

FIGURE 4.64: The growth of a single microtubule is erratic. Slow, steady growth is interrupted by large catastrophe events at about 8 s and 46 s while a rescue event occurs between two catastrophe events at about 26 s.

The challenge is to use the same model to explain ensemble dynamics and dynamic instability; and in addition, to explain changes in the dynamics when reaction conditions are altered. Suppose the length of our MT is given by the value of the variable, x, at an instant of time, t. We descretize the time variable in the interest of writing a recursion formula for the length of the MT. In the simplest model, there are only two possibilities, the addition of a subunit or the complete collapse of the structure. Addition of a subunit may be modeled as follows [254]:

$$x_{t+1} = x_t + a \tag{4.174}$$

where a is the subunit length. Complete collapse of the microtubule is simply modeled using:

$$x_{t+1} = 0 \tag{4.175}$$

Now suppose that the probability of the addition of a tubulin subunit is p, then

$$x_{t+1} = r(x_t + a) \tag{4.176}$$

where

$$r = \begin{cases} 0, & \text{if } s > p \\ 1, & \text{if } s \le p \end{cases} \tag{4.177}$$

and s is a random number between 0 and 1. Even a crude model such as that just described may successfully capture the essential dynamics of MT growth when compared by Hurst analysis or a recursive map technique [251]. However, it lacks any predictive power given that individual collapse events are random.

When an aggregation of MTs is studied, some collective properties develop. The most obvious is the phase transition behavior of the system in that certain conditions lead to overall disassembly and largely a pool of tubulin dimers, while other conditions promote self-assembly of tubulin in MTs. Phases of MT polymerization are determined largely by the temperature and concentration of tubulin. For a given temperature, some critical concentration of tubulin is required in order to keep MTs from disassembling. D.K. Fygenson, E. Braun, and A. Libchaber [90] have discussed this phase transition and Sept et al. [251] have further classified the non-disassembly phase into one where MTs assemble but are not nucleated and one where MTs are both nucleated and assembled. For a collection of MTs in an assembly phase, dynamic instability is the term given to the observation that MTs are growing in the immediate vicinity of other MTs which are shortening. Dynamic instability has been observed *in vivo* and *in vitro* and highlights both the non-equilibrium nature of the problem that crucially depends on the energy supply via GTP hydrolysis [121] that affects the structure of microtubules leading to catastrophes [122] and thus creating the stochastic nature of individual MT growth. Despite these observations, ensembles of MTs show collective oscillations given suitable conditions. Specifically, when the concentration of assembled tubulin is measured, it is observed to undergo smooth oscillations which are damped out as the energy source of GTP is depleted. Sept et al. [251] have modeled the assembly dynamics from a chemical reaction-kinetics standpoint and found good agreement with the experimental data.

The principal elements in the model of Sept et al. [251] can be summarized by the equations which follow. For simplicity, the MT is considered to be a linear polymer rather than an object of 13 protofilaments. We shall denote a MT of length n subunits by MT_n. Note that in solution, the free tubulin subunits may be bound to either GTP or GDP at their exchangeable nucleotide site and shall be denoted T_{GTP} and T_{GDP}; respectively, only tubulin bound to GTP is able to polymerize. The simple reaction set consists of addition, nucleation and catastrophic collapse:

$$MT_n + T_{GTP} \underset{k_d}{\overset{k_a}{\rightleftharpoons}} MT_{n+1} \quad \text{(addition)} \tag{4.178}$$

$$nT_{GTP} \overset{k_n}{\to} MT_n \quad \text{(nucleation)} \tag{4.179}$$

and

$$MT_n \xrightarrow{k_c} nT_{GDP} \quad \text{(collapse)} \tag{4.180}$$

For simplicity, we have again assumed that all collapses are complete and to keep the model of MT assembly simple, one normally selects a specific number of dimers, n, that will be required for nucleation. The exact value of this choice does not seem to have a large impact on the dynamics as long as it is relatively small. The first of the preceding equations is reversible and a rate constant for the back reaction is considered in general. This is not required for this discussion. These equations can also be supplemented by the reactivation of tubulin, to make it assembly competent, which will occur when the concentration of GTP is high:

$$T_{GDP} + GTP \xrightarrow{k_r} T_{GTP} + GDP \tag{4.181}$$

The free energy change associated with this reaction is less than the free energy change of GTP hydrolysis in solution and the difference is attributed to a structural change in the tubulin dimer. It is this conformational change which presumably makes assembly possible. The rate constants of these reactions, k_j, should also be in agreement with the probabilities assigned to corresponding reactions in the individual MT model. Furthermore, temperature dependence can be built into the rate constants using empirical data on the free energy of reactants and products. When at least one auto-catalytic reaction is added to the system, to provide a non-linear element, the dynamics change significantly. Consider an induced catastrophe event,

$$MT_n + T_{GDP} \xrightarrow{k_i} (n+1)T_{GDP} \tag{4.182}$$

incorporated into the model. This reaction has been able to reproduce oscillations observed *in vitro* caused by either temperature jumps or the injection of GTP into the system. It also identifies domains where tubulin is incorporated into MTs steadily and regimes where disassembly is favored. In addition, the model can reproduce the spatial pattern of MT assembly which is observed within cells simply by including a diffusive term in the equations of the assembly dynamics. This allows the model to be compared to the experiments performed independently by the Tabony [265] and Mandelkow groups [179]. Agreement has been found between this model and those empirical results [252]. Generally, the statistical dynamics of MT assembly are now believed to be well understood except for the physical mechanism that triggers the catastrophes and the origin of the rescue events.

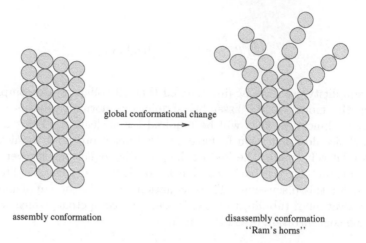

global conformational change

assembly conformation   disassembly conformation
                                         "Ram's horns"

FIGURE 4.65: The global conformational change that occurs between assembly and disassembly phases is illustrated. The once straight protofilaments become curved after the individual tubulin subunits undergo a structural change.

Again the assumption was simply that the energy difference between the GTP and GDP states is stored in the MT lattice and may be released upon disassembly or catastrophe events. More recently, Hyman et al. reported the observation of a structural change accompanying GTP hydrolysis. The change which was discovered was a length change in the monomer spacing from 4.05 nm to 4.20 nm. Thus energy may be stored locally as lattice deformation [122]. This 4% change in tubulin's length results in a new moiré pattern when the MTs are imaged by electron cryo-microscopy and different positions of equivalent peaks between the X-ray crystallographic diffraction patterns of GDP-MTs and GMPCPP-MTs. Note that GMPCPP is a non-hydrolyzable analog of GTP that keeps MT's stable. Tran et al. [271] argued that three conformational states exist with a metastable intermediary state between growing and shrinking conformations. Tran et al. [271] seem to believe that these may be tubulin with GTP bound, with $GDP \cdot P_i$ bound and finally with GDP bound at the exchangeable site. This is similar to the hypothesis of Semënov [250] in his review of MT research. It therefore appears that in addition to the multitude of α- and β-isotypes and the numerous post-translational modifications, tubulin may also exist in several conformations.

F. Gittes, E. Mickey, and J. Nettleton [93] have measured MT flexural rigidity by thermal fluctuations in the MT shape and found interestingly that it is increased when MTs are treated with MAP; the MTs are most stabilized by preventing GTP hydrolysis. Thus in addition to the previously mentioned fact that hydrolysis of GTP can result in destabilizing the MT with respect to catastrophic disassembly, it also makes the polymer more flexible [76]. It is interesting to note that the application of taxol, which reduces MT assembly dynamics, actually increases the mechanical rigidity of MTs [190].

4.9.3 Developing a Stochastic Model

How can we explain all these effects, both stochastic and coherent with essentially a single model? We can describe the processes discussed: growing, shrinking, decaying and rescues - with the help of chemical reactions involved. Mostly, they are normal chemical reactions of association and disassociation. Because of the energy pumping present, we never have thermal equilibrium and therefore we do not have a detailed balance which would mean: $A+B \rightleftharpoons C$ [14]. We instead have a reaction cycle. Before we provide the reader with a detailed description of the stochastic model proposed, we first give a brief overview of the modelling efforts to date.

Numerous experimental and theoretical studies were aimed at extracting characteristic features of life histories of MT's in order to build a predictive model [45,278,39,21,182]. The most popular statistical tools included the construction of histograms and correlation functions [95]. For example, a histogram of delays before catastrophes for both ends has a characteristic Poissonian shape which is suggestive of statistically independent, small probability events [279]. Histograms for length distributions have an exponential tail and sometimes a peak corresponding to relatively short microtubules [276, 187]. One of the main objectives of the present study is to determine precise criteria for the existence of bell-shaped as well as exponential probability distributions for MT lengths as seen in some experiments. An inverse proportionality of the frequency of catastrophes to the growth velocity of a MT has been demonstrated to be consistent with the presence of a GTP cap at the plus end [80]. A recent paper [206] investigated the validity of the standard assumption that the growth and shrinkage of MT's is governed by first-order chemical kinetics.

The main objective of this section is to develop the simplest possible stochastic model consistent with our knowledge about the chemical reactions involved with the MT growth and shrinkage. Henceforth, we will assume that the probability for a certain reaction is proportional to the probability of finding the appropriate reactants at the reaction site, which is often but not always proportional to the product of reactant concentrations and the reaction rate [14]. We will not use effective (concentration dependent) reaction rates. Furthermore, we assume that the same reaction rates should hold for all concentrations of reactants. Finally, the nucleation reaction will not be explicitly included in the model since it can be effectively incorporated via an appropriate choice of the boundary and normalization conditions. The key reactions are listed below in Table 4.5.

The first reaction is a nucleation (seeding) process and it is known to be much slower than the MT growth process. The second reaction is a normal

TABLE 4.5:   A List of Key Chemical Reactions Involved
in Microtubule Assembly

Reaction Type	Equation	Rate
Nucleation	$n_c T \rightarrow MT(n_c)$	k_n
Growth	$MT(n) + T \rightarrow MT(n+1)$	k_g
Shrinkage	$MT(n) \rightarrow MT(n-1) + D$	k_s
Catastrophe	$MT(n) \rightarrow nD$	k_c
Reactivation (pumping)	$D + GTP \rightarrow T + GDP$	k_p

growth process which may occur at either the plus or the minus end and
hence it covers in effect two processes. The same can be said about its inverse reaction, shortening (by one unit). The fourth reaction describes the
transition from the growing to the collapsing state, called a catastrophe. The
rescue reaction is the recovery from the collapsing state. Since the disassembled tubulin contains GDP, it is assumed that the hydrolysis energy may
have been stored at least partially in a MT in the form of mechanical stress
that could contribute to the causes of catastrophic disassembly. A sufficient
supply of GTP in solution ensures a reactivation of tubulin from Tu-GDP to
Tu-GTP, the latter being a prerequisite for MT assembly.

The above scheme may require an intermediate step in the reactivation of
tubulin in the pool such that starting from a collapse catastrophe, dimers of
tubulin exist in a state which is borne out by direct experimental observations
of GTP and GDP intermediates in MT assembly

$$MT(n) \xrightarrow{k_d} MT(n-1) + D_e$$
$$D_e \xrightarrow{k_t} D \tag{4.183}$$

It is worth emphasizing that the intermediate products of D_e may include
oligomers of different types making it a possibly complex process which may
even lead to induced decays in neighbouring MT's. To simplify the description
even more we will, in the next section, consider a special situation where the
minus end is fixed and the rescue events are ignored.

Although various measurements resulted in different numerical values, to
gain an intuitive feel for the orders of magnitude involved we just give a
representative sample of the parameters measured. The values in Table 4.6
have been published [278], and pertain to interphase mitotic spindles at $10\mu M$.

4.9.4   A Stochastic Model Without Rescues

The behaviour of an individual MT is stochastic, primarily due to the dynamic instability but also because growth and shrinkage events are not known
to take place with absolute certainty. In experiments with a fixed minus end
we may ignore the nucleation reaction. We also ignore the intermediate steps

TABLE 4.6: Typical Parameter Values for Microtubule Assembly Following [278]

Parameter	plus end value	minus end value
growth velocity v_g:	3.3 $\mu m \ min$	1.5 $\mu m \ min$
shortening velocity v_s:	1.6 $\mu m \ min$	0.8 $\mu m \ min$
mean growth time t_g:	33 s	50 s
mean shrinkage time t_s:	14 s	7 s
catastrophe frequency f_{cat}:	0.03 s^{-1}	0.02 s^{-1}
rescue frequency f_{res}:	0.07 s^{-1}	0.15 s^{-1}
collapse velocity v_d:	44 $\mu m \ min$	29 $\mu m \ min$

involving free tubulin assuming a virtually inexhaustible pool of Tu-GTP present and hence use the symbol T for any tubulin dimers. In the initial model without rescues we assume that the collapse is instantaneous. This adiabatic approximation is in good agreement with the fact that the collapse velocity is much higher than the growth velocity. Therefore, we do not distinguish between a growing and a shortening subpopulation of microtubules. Thus, the key reactions to be included are:

$$MT(n) + T \longrightarrow MT(n+1) \quad \text{with probability} \quad t_1$$
$$MT(n) \longrightarrow MT(n-1) + T \quad \text{with probability} \quad t_2$$
$$MT(n) \longrightarrow MT(0) + nT \quad \text{with probability} \quad t_3 \qquad (4.184)$$

and an identity reaction (a pause step) which is added for completeness.

$$MT(n) \longrightarrow MT(n) \qquad \text{with probability} \qquad 1 - t_1 - t_2 - t_3 \quad (4.185)$$

since the sum of all reaction probabilities must add up to unity.

We describe all stochastic equations with the help of random numbers r_n defined by

$$r_n = 1 \quad \text{with probability} \quad p_n$$
$$r_n = 0 \quad \text{with probabiltiy} \quad 1 - p_n \qquad (4.186)$$

A compact stochastic equation to describe all these transitions is

$$l(t+1) = (1 - r_3(1 - r_1)(1 - r_2))(l(t) + r_1 a - r_2 a - a r_1 r_3 + a r_2 r_3) \quad (4.187)$$

Here, $l(t)$ describes the length of the microtubule, dependent on the time t measured in suitable time steps (of one); a is the length of one segment which is added in the process of growing. The probabilities p_1, p_2, p_3 are connected with the probabilities t_1, t_2, t_3 for the processes by

$$t_1 = p_1(1 - p_2)(1 - p_3)$$
$$t_2 = p_2(1 - p_1)(1 - p_3)$$
$$t_3 = p_3(1 - p_1)(1 - p_2)$$

The probabilities t_n are just the transition rates t_{rn} for the process multiplied by the probability to find the reactants at the reaction region. In fact, for the simplified reaction set discussed in this section, the values of t_n are found exceedingly easily. Since we know that there is only one MT with the length n (number of layers), the corresponding probabilities are therefore given by

$$t_1 = c_T t_{r1}; \; t_2 = t_{r2}; \; t_3 = t_{r3} \qquad (4.188)$$

Here c_T denotes the concentration of Tu-GTP in the solution. Of course, one can also connect the transition rates t_{rn} with the reaction constants k_n using the following relation

$$t_{rn} = k_n \, \Delta t \qquad (4.189)$$

where Δt is the time step in the simulation. For example, a stochastic equation which describes the length evolution $\ell(t)$ as a function of time t for an individual microtubule can be descriptively written as

$$\ell(t + \Delta t) = \begin{cases} \ell(t) + a & \text{for} \quad 0 < r \le t_1 \\ \ell(t) - a & \text{for} \quad t_1 < r \le t_1 + t_2 \\ 0 & \text{for} \quad t_1 + t_2 < r \le t_1 + t_2 + t_3 \\ \ell(t) & \text{for} \quad t_1 + t_2 + t_3 < r \le 1 \end{cases} \qquad (4.190)$$

where t is the discrete time variable measured in the units corresponding to the smallest time scale for individual events Δt and a is the length of a tubulin dimer.

We now intend to generalize the results of this section to a more experimentally relevant case of a large number of microtubules present but at concentration values low enough to preclude synchronization, competition for tubulin pool or any significant interactions or correlation effects. Our goal is to derive equations which deal with ensemble averages for sets of microtubules.

4.9.5   The Averaged Picture; Master Equations

We now turn our attention to the situation where a large number of MT's is present so that an experimentalist can measure various associated averages for an ensemble of MT's with different lengths. A direct approach to this problem is to derive and solve the underlying master equation. The master equation describes the evolution of the probability distribution $P(n, t)$ which characterizes the system in a given state n (i.e., the number of segments assembled to form a microtubule) at time t. To do that we only need to know the transition probabilities for: (a) processes leading into the given state n (rate in) and (b) processes leading out of state n and into any other possible state (rate out). Furthermore, these probabilities t_{rj} are simply proportional to the reaction rates k_j introduced in the previous section.

A generic master equation always takes the form

$$\frac{d}{dt}P(n,t) = (\text{rate in}) - (\text{rate out}) \tag{4.191}$$

Since in the averaged description we do not know the exact state of the system we have

$$t_1 = P(n,t)c_T t_{r1}$$
$$t_2 = P(n,t)t_{r2}$$
$$t_3 = P(n,t)t_{r3} \tag{4.192}$$

When specified for our reaction scheme in eq. 4.184, the master equation becomes

$$\frac{d}{dt}P(n,t) = c_T t_{r1}P(n-1,t) + t_{r2}P(n+1,t) - (c_T t_{r1} + t_{r2} + t_{r3})P(n,t) \tag{4.193}$$

for $n > 0$. The master equation for $n = 0$ follows again from eq. 4.191, and has the form

$$\frac{d}{dt}P(0,t) = t_{r2}P(1,t) + t_{r3}\sum_{n=1}^{\infty}P(n,t) - c_T t_{r1}P(0,t) \tag{4.194}$$

Note that the $n = 0$ state is the nucleation state and it plays an important role in view of the boundary conditions provided through it. A stationary solution of the above equations is readily obtained as

$$c_T t_{r1}P(n-1) + t_{r2}P(n+1) - (c_T t_{r1} + t_{r2} + t_{r3})P(n) = 0 \quad (n > 0) \tag{4.195}$$

and

$$t_{r2}P(1,t) + t_{r3}\sum_{n=1}^{\infty}P(n,t) - c_T t_{r1}P(0,t) = 0 \quad (n = 0) \tag{4.196}$$

which must be solved subject to the probability normalization condition

$$\sum_{n=0}^{\infty}P(n) = 1 \tag{4.197}$$

The continuum limit of the master equation is obtained by expanding the probability $P(n \pm 1, t)$ in a series as follows:

$$P(n \pm 1, t) \longrightarrow P(x,t) \pm \frac{\partial}{\partial x}P(x,t) + \frac{1}{2}\frac{\partial}{\partial x^2}P(x,t) \pm \ldots \tag{4.198}$$

to arrive at a first order linear partial differential equation for $P(x,t)$:

$$\frac{\partial}{\partial t}P(x,t) = -(c_T t_{r1} - t_{r2})\frac{\partial}{\partial x}P(x,t) - t_{r3}P(x,t) \qquad (4.199)$$

where x is the continuum limit equivalent of n. Interestingly, the above PDE has a physically relevant exact solution of the form

$$P(x,t) = e^{-t_{r3}t}f(x - (c_T t_{r1} - t_{r2})t) \qquad (4.200)$$

where f is a prior arbitrary function which is subject to an arbitrary initial condition, and, of course, must satisfy the normalization of probability $\int_0^\infty P(x,t)dx = 1$ and a boundary condition.

4.9.6   Stochastic Models with Rescues

In vitro experiments clearly demonstrate that most of the catastrophes lead to only a partial disassembly stopping at a particular point along the MT protofilament structure. The MT then experiences a rescue event which restarts a growth phase. In order to account for rescues we modify our chemical reaction set as follows:

$$MT(n) + T \longrightarrow MT(n+1) \qquad \text{with probability } t_1 \qquad (4.201)$$
$$MT(n) \longrightarrow MT(n-1) + T \qquad \text{with probability } t_2 \qquad (4.202)$$
$$MT(n) \longrightarrow MT(n-m) + mT \qquad \text{with probability } t_3 \qquad (4.203)$$
$$MT(n) \longrightarrow MT(n) \qquad \text{with probability } 1 - t_1 - t_2 - t_3 \qquad (4.204)$$

where the third reaction describes an incomplete collapse, i.e., allows for a rescue event to follow. Here m is a random integer number in the range:

$$0 < m < \frac{\ell(t)}{a}$$

which describes the size of the assembled tubulin material in a MT following a catastrophe event.

A stochastic equation describing all these transitions is

$$l(t+1) = (1-r_3(1-r_1)(1-r_2))(l(t)+r_1 a - r_2 a - ar_1 r_3 + ar_2 r_3)) + r_3(1-r_1)(1-r_2)ma \qquad (4.205)$$

and it simply encapsulates the following possibilities (see Sec. 4.9.4 for comparison):

$$\ell(t+1) = \begin{cases} \ell(t) + a & \text{for} \quad 0 < r \le t_1 \\ \ell(t) - a & \text{for} \quad t_1 < r \le t_1 + t_2 \\ \ell(t) - ma & \text{for} \quad t_1 + t_2 < r \le t_1 + t_2 + t_3 \\ \ell(t) & \text{for} \quad t_1 + t_2 + t_3 < r \le 1 \end{cases} \qquad (4.206)$$

For the probability that a collapsing MT of length n is rescued at length $(n-m)$ we have

$$w(n,m) = w(n) = \frac{1}{n} \qquad (4.207)$$

With the same assumption about the equal probability of rescues the resulting master equation becomes

$$\frac{d}{dt}P(n,t) = c_T t_{r1} P(n-1,t) + t_{r2} P(n+1,t) + t_{r3} \sum_{n'=n+1}^{\infty} \frac{1}{n'} P(n',t)$$

$$-(c_T t_{r1} + t_{r2} + t_{r3}) P(n,t) \qquad (4.208)$$

for $n > 0$. For the purpose of obtaining a stationary solution, the exact form of the equation at $n = 0$ is of no consequence to us and we simply use $P(0) = C$ where C is found from the normalization condition. We then switch to the continuous picture by expanding P(x,t) to first order to obtain

$$\frac{\partial}{\partial t}P(x,t) = -(c_T t_{r1} - t_{r2})\frac{\partial}{\partial x}P(x,t) - t_{r3} P(x,t) + t_{r3} \int_x^{\infty} \frac{1}{x'} P(x',t) dx'$$

$$(4.209)$$

It can be demonstrated that the stationary solution of eq. (4.209) is given by the exponential function:

$$P(n) \sim n\, e^{-\frac{t_{r3}}{(c_T t_{r1} - t_{r2})}n} \qquad (4.210)$$

The assumption made about a uniform probability distribution for rescues is highly limiting and hence may result in solutions which are not appropriate for all experimentally observed situations. Several physical effects which would require a modification of this assumption are: (a) an accumulation of stress-free tubulin dimers which could occur at specific locations within the microtubule leading to a higher probability of rescues at these points. For instance, tubulin clusters close to the seed (older parts of a MT) could be more stable than those close to the growing tip, (b) the assembly of tubulin into a MT is closely related to the chemical energy pumping via GTP. GTP inside the MT hydrolyzes into GDP and releases some energy, conceivably in the form of mechanical stress. Assuming a steady flux of energy through the MT may or may not result in an equal energy concentration throughout the structure. This process could chiefly depend on the rate of assembly, the rate of GTP hydrolysis and the rate of energy diffusion inside a microtubule.

Assuming that the catastrophes start from the distal end, for every tubulin which disassembles we have a constant probability q that the collapse is <u>not</u> stopped (no rescue). The probability for rescue is then $(1 - q)$ and the probability that a rescue occurs after $n - m$ collapse events is easily summed over to give

$$w(n-m) = \frac{1-q}{1-q^n}\, q^{n-m} \qquad (4.211)$$

The limit of $q \longrightarrow 1$ brings us back to the previously discussed case of no rescues. The resultant master equation for $n > 0$ now becomes:

$$\frac{d}{dt}P(n,t) = c_T t_{r1}P(n-1,t) + t_{r2}P(n+1,t) + t_{r3}\sum_{n'=n+1}^{\infty}P(n',t)w(n'-n)$$

$$-(c_T t_{r1} + t_{r2} + t_{r3})P(n,t) \tag{4.212}$$

The continuum limit of this master equation carried out to second order is an integro-differential equation of the form

$$\frac{\partial P}{\partial t} = \frac{1}{2}(c_T t_{r1} + t_{r2})\frac{\partial^2 P}{\partial x^2} - (c_T t_{r1} - t_{r2})\frac{\partial P}{\partial x} - t_{r3}P + t_{r3}\int_x^{\infty} w(x'-x)P(x')dx'$$

$$\tag{4.213}$$

A formally similar equation has been recently derived by [81] in the context of the dynamics of the lateral cap. Fourier transforming the continuum-limit master equation directly yields the exact solution

$$P(x,t) = \int_{-\infty}^{+\infty} e^{-ik(x-At)}e^{-B(1-w(k))t+k_{r2}t}f(k)dk \tag{4.214}$$

where $w(k)$ is the Fourier transform of $w(x'-x)$ and $f(k)$ defines the initial condition (as its Fourier transform). In order for $P(x,t)$ to describe a stationary solution, $f(k)$ must select a specific value of the wavenumber k, namely k_0 such that

$$k_0 = i\frac{B}{A}[1 - w(k_0) + \frac{k_0^2}{B}] \tag{4.215}$$

under this condition, $k = k_0$, the stationary solution in real space takes the form:

$$P(x) \sim exp[-\frac{B}{A}(1 - w(k_0) + \frac{k_0^2}{B})x] \tag{4.216}$$

The function $f(k)$ contains the important information regarding the initial condition and normalization; its form poses a significant mathematical problem. As a consequence of the assumptions made about the rescue probability, we have arrived at a stationary probability distribution which is exponential. Any other solution will evolve in time from the initial condition to the asymptotic form which is exponential. The conclusion regarding the exponential form of the histogram applies to all cases when w depends on the difference $(n-m)$. Conversely, for a histogram not to be exponential (e.g., bell-shaped), the function w must depend n and m separately as we have seen in the first case (see eq. (4.207)).

4.9.7   A Model with a Finite Collapse Velocity

In order to refine the model we consider collapse catastrophes with a finite rate of shrinking such that the set of chemical reactions included in the model is now:

$$M_g(n) + T \longrightarrow M_g(n+1) \quad \text{with probability} \quad t_g \qquad (4.217)$$

$$M_g(n) \longrightarrow M_g(n-1) + T \quad \text{with probability} \quad t_s \qquad (4.218)$$

$$M_g(n) \longrightarrow M_c(n) \quad \text{with probability} \quad t_c \qquad (4.219)$$

$$M_c(n) \longrightarrow M_c(n-1) + D \quad \text{with probability} \quad t_d \qquad (4.220)$$

$$M_c(n) \longrightarrow M_g(n) \quad \text{with probability} \quad t_r \qquad (4.221)$$

where the subscript "g" refers to growing microtubules and "c" to shrinking ones. The above reactions account for catastrophes, rescues, as well as gradual growth and shrinkage events. In the continuum limit (going to the second order terms in the expansion this time) we obtain two master equations for the probability distributions of the growing state $P_g(x,t)$ and the collapse state $P_c(x,t)$ in the form of coupled second order linear PDE's:

$$\partial_t P_g(x,t) = \frac{1}{2}(c_T t_g + c_s)\partial_{xx} P_g(x,t) - (c_T t_g - c_s)\partial_x P_g(x,t) - t_c P_g(x,t) + t_r P_c(x,t)$$
$$(4.222)$$

and

$$\partial_t P_c(x,t) = \frac{1}{2} t_d \partial_{xx} P_c(x,t) t_d \partial_x P_c(x,t) - t_r P_c(x,t) + t_c P_g(x,t) \qquad (4.223)$$

which represent two coupled diffusion equations with entrainment terms (gradients). However, stationary solutions of the above system are once again exponential. The reason lies in the assumption about the rescue probability which depends only on the difference between the catastrophe coordinate and the coordinate of the rescue point. Since there is no significant qualitative difference between the results for finite and infinite collapse velocity we will henceforth keep the collapse velocity infinite for simplicity of exposition.

4.9.8 Conditions for Stationary Bell-Shaped Distributions

In many experiments the distribution of lengths of MT's in a stationary state showed a characteristic bell-shape form. Gliksman [95] pointed out that this type of behaviour can be attributed to the correlations between the free tubulin pool at a concentration c_T and the assembled MT's. The main difference between this assumption and those made by us earlier is that increasing the size of a MT results in a corresponding decrease in the amount of free tubulin dimers.

The exact mechanism for rescues is not known, but we can think of at least two possibilities: a) We may assume that not only the free tubulin, but also the MT itself, is consuming energy provided by pumping. This means that at certain segments of the MT we find GTP-tubulin units instead of GDP-tubulin units. What are the consequences? On the one hand we have a release of mechanical stress which would stabilize the MT at this point and could result in a rescue event. On the other hand, we have - after a characteristic time - the conversion of GTP tubulin to GDP tubulin again. If we assume a steady flux of energy this would end up in an equal distribution of high energy tubulin inside the MT. So if the disassembly process is initiated from the free end, we have for every segment a constant probability q (dependent on pumping) that the decay is not stopped and a probability $(1 - q)$ that the decay is stopped, which means that we have a rescue event at this point. The probability $w(n - m)$ that we have a rescue at the point m, therefore, has the form

$$w(n - m) = q^{n-m}(1 - q)$$

and $w(n - m)$ is now a decreasing function. But we already showed that a form $w(n - m)$ always results in an exponential stationary distribution of MT length. b) The accumulation of GTP-rich tubulin could possibly stabilize the MT at specific points. In this case rescue would be a memory effect. The older part of the MT must contain more stabilizing points; the probability of rescue must increase if we go from the free end to the other end. If we try this approach for a simple form of $w(m)$, the resulting solution for the stationary state is expected to show a bell shaped histogram.

Consequently, the master equation that results here is almost identical to those obtained before, i.e.,

$$\frac{d}{dt}P(n,t) = [C_T - (n - 1)]t_{r1}P(n - 1, t) + t_{r2}P(n + 1, t) + t_{r3}$$

$$\sum_{n'=n+1}^{\infty} P(n')w(n' - n) - [(C_T - n)t_{r1} + t_{r2} + t_{r3}]P(n,t) \text{ for } n > 0 \quad (4.224)$$

As before, the rescue probability is assumed to depend on the relative length of the catastrophe-to-rescue distance. Furthermore, we have replaced c_T with the quantity $(c_T - n)$ denoting the number of free tubulin dimers in solution which provides a strong correlation effect between MT's and free tubulin. The numerical results obtained by us for the stationary solution of (8.1) indicate bell-shaped behaviour. Let us now see what happens when we consider N MT's of different lengths and couple them to a finite pool of free tubulin. The state of the system would then be given by specifying the lengths of all MT's $(n_1, n_2, \ldots, n_N)$ and the number of free tubulin dimers would be: $c_T - n_1 - n_2 - \ldots - n_N$. Somewhat surprisingly, in spite of these new complications arising in the master equations, a new symmetry arises which makes the problem still solvable:

$$P(\vec{n}) = P(n_1, n_2, \ldots n_N) = P(n_1 + n_2 + \ldots + n_N) \quad (4.225)$$

i.e., the probability distribution only depends on the total amount of tubulin assembled in all MT's. Furthermore, $P(\vec{n})$ is the solution of the previously discussed master equation. Consequently, one could hope that if $P(\vec{n})$ had a bell-shape form, so would $P_1(n_1)$ defined for a single MT as the probability to find MTs with length n_1. Unfortunately, this is not the case because the sum appearing in the defining relation for $P_1(n_1)$

$$P_1(n_1) = \sum_{n_2} \sum_{n_3} \cdots \sum_{n_N} P(n_1 + n_2 + n_3 + \ldots + n_N) \qquad (4.226)$$

rapidly destroys the presence of any peak in $P(\vec{n})$ and it can be demonstrated that as $n \longrightarrow \infty$, $P_1(n_1)$ reverts back to an exponential form: $P_1(n_1) \sim e^{-cn_1}$. Since the above property is identically satisfied for a discrete case with an arbitrary number of tubulin dimers, it also holds in the continuum limit.

Finally then, we have discovered a simple assumption that yields the correct answer. Bell-shaped length distributions emerge naturally if we assume that the probability of rescues is not uniform but increases in some specific locations along a given MT, for example is significantly higher close to the nucleation points.

To summarize, we have demonstrated that a transition from exponential to bell-shaped distribution functions must be linked with a change of the internal state of a dynamic microtubule. We believe that the assumption about every process being independent of the length of the MT must inevitably lead to an exponential histogram for an ensemble of non-interacting MT's. Conversely, when we assume that, e.g., the probability of rescues depends only on the distance from the proximal end of a MT, we are able to obtain bell-shaped histograms. As we discussed earlier, the latter possibility may emerge naturally from relaxation processes following GTP hydrolysis and a redistribution of stresses in the protofilaments.

4.9.9 Coherence Effects

Now we wish to examine another type of experiment. We first assume that MT seeds have been placed in the solution to result in a nucleation process. What will happen next?

1. If the pumping rate is zero, we have no MT. Only GDP-Tublin is present which we now simply call D.

2. If we increase the pumping rate a little bit, we produce GTP-Tubulin (which we now call T), which itself can convert into seeds. The seeds will grow to be MT's. As a consequence, we have MT's in low concentration, showing the same (growing and decaying) picture as we just discussed. Every MT behaves independently.

3. If we further increase the pumping rate, the whole system of MT's suddenly shows coherent behaviour [36], [215], [157], [188], [277], [183]. If we start from a certain initial condition we see oscillations in the polymerized mass concentration and in the average length of the MT's.

To describe these effects in a quantitative manner we need the whole cycle of reactions, i.e.:

$$
\begin{aligned}
D-&>T & &\text{with transition rate } k_n & &\text{(Pumping)} \\
n_cT-&>MT(n_c) & &\text{with transition rate } k_p & &\text{(Nucleation)} \\
MT_{(n)}+T-&>MT(n+1) & &\text{with transition rate } t_g & &\text{(Growing)} \\
MT(n)-&>nD_e & &\text{with transition rate } k_c & &\text{(Decay)} \quad \text{(4.227)} \\
D_e-&>D & &\text{with transition rate } k_p & &\text{(Relaxation)}
\end{aligned}
$$

The resultant master equations are now complicated, because the state of the system is now given by the state of all the MT's plus the number of D's plus the number of T's plus the number of D_e's. After some typical approximations which amount to neglecting correlations (a chemist would say: "the experiment is conducted under continuous stirring conditions"), the equations become

$$
\begin{aligned}
\dot{c}_{M(n)} &= k_g c_T[\theta(n-n_c-1)c_{M(n-1)} - c_{M(n)}] \\
&\quad - k_c c_{M(n)} + \delta(n-n_c)k_n n_c c_T^{n_c}
\end{aligned}
$$

$$
\dot{c}_T = -k_g c_T \sum_{n=n_c}^{\infty} c_{M(n)} + k_p c_D - k_n n_c c_T^{n_c}
$$

$$
\dot{c}_D = k_c \sum_{n=n_c}^{\infty} n c_{M(n)} - k_p c_D \qquad (4.228)
$$

where $n > n_c$ and we have neglected the intermediate step. This is a set of many coupled equations. Fortunately, they can be reduced by introducing a collective variable c_N which is the overall concentration of all the MT's

$$
c_N = \sum_n c_{M(n)} \qquad (4.229)
$$

and a variable c_M which is the concentration of polymerized tubulin dimers (total polymerized mass)

$$
c_M = \sum_n n c_{M(n)} \qquad (4.230)
$$

As a consequence of the adiabatic approximation, the reduced system of differential equations becomes

$$
\begin{aligned}
\dot{c}_M &= k_g c_T c_N + k_n n_c c_T^{n_c} - (k_c^0 + k_c^1 c_{D_e})c_M \\
\dot{c}_N &= k_n c_T^{n_c} - (k_c^0 + k_c^1 c_{D_e})c_N \\
\dot{c}_T &= k_p c_D - k_g c_T c_N - k_n n_c C_T^{n_c} c_N \\
\dot{c}_D &= k_t c_{D_e} - k_p c_D \\
\dot{c}_{D_e} &= (k_c^0 + k_c^1 c_{D_e})c_M - k_t c_{D_e} \qquad (4.231)
\end{aligned}
$$

when c denotes a concentration variable corresponding to the species indicated in the subscript.

Unfortunately, having solved the resultant equations numerically we don't see any oscillations. The reason is that these equations are essentially linear. In chemistry we know about the existence of chemical oscillations which are explained by autocatalytic reactions which make the kinetic equations non-linear [221]. Unfortunately, this approach does not apply to long polymer molecules like the MT's. Our idea instead is that a collapse event in one MT can induce another collapse event in neighbouring MT's. We already showed this possibility in our first reaction cycle. The MT decays into excited tubulin products D_e which themselves can induce the decay of other MT's. We therefore have to add two more reactions to the scheme

$$MT(n) + D_e -> (n+1)D_e \qquad (4.232)$$

and

$$MT(n) + 2D_e -> (n+2)D_e \qquad (4.233)$$

We can express the effective change in the kinetic equations by making k_c the reaction rate D_e-dependent, i.e.,

$$k_c(D_e) = k_{c0} + k_{c1}D_e + k_{c2}D_e^2 \qquad (4.234)$$

It is clear that these self-induced reactions can only be in effect for a higher pumping rate which means a higher concentration of D_e's

A detailed numerical analysis of the above system of equations indicated a possible transition to synchronized oscillations in the total microtubule mass. The requirements for generating oscillations can be listed as follows:

- reactions must be far away from equilibrium (the presence of pumping) which is induced by the nucleotide exchange $GDP \longrightarrow GTP$.

- There must be autocatalytic steps.

- There must be two stationary states (at least). This is similar to a phase transition.

Oscillations in the total assembled microtubule mass for high-tubulin concentrations have been observed [36], [215], [157], [188], [277], [183] and a number of models have been proposed [177], [184], [116], [132]. Our intention in this section was to sketch a possible route to the emergence of oscillations based on the already introduced formalism.

4.9.10   Summary and Conclusions

The main questions arising in the analysis of elongation data for micro-tubules can be listed as follows:

1. What causes a transition from exponential to bell-shaped histograms?

2. Is there length dependence in the rates for catastrophes and rescues?

3. Are there time-dependent correlations between catastrophe events in a given microtubule?

4. What is the mechanism via which oscillations in the assembly process of microtubules in high-tubulin concentration experiments emerge?

We have attempted to shed light on some of the above question, especially (1), (2) and to a lesser degree (4). Some of these issues are inter-related and most of them require more detailed experimental studies. We believe, however, that the theoretical framework within which rapid progress can be made should involve the master equation formalism as presented in this work.

4.9.11   Assembly of Actin Filaments

The pioneering work of Oosawa and Asakura [209] established that spontaneous polymerization of actin monomers requires an unfavorable nucleation step followed by rapid elongation. Elongation is more accessible experimentally than nucleation, so it is much better understood, with a complete set of rate constants for association and dissociation of ATP-actin and ADP-actin subunits at both ends of the filament [216]. Nucleation has been studied by observing the complete time course of spontaneous polymerization as a function of actin monomer concentration and then finding a set of reactions and rate constants that fit these kinetic data [269]. These studies concluded that actin dimers are less stable than trimers, which form the nucleus for elongation.

In addition to nucleation and elongation, Oosawa et al. established that actin filaments can break and anneal end to end. Inclusion of a fragmentation reaction improved the fit of nucleation-elongation mechanisms to the observed time course of polymerization under some conditions [281, 51, 28]. For annealing, there has been kinetic evidence both for [143, 229] and against [35] its role in length redistribution after sonication, but the most direct evidence from electron micrographs supports rapid annealing [202].

The basis of the standard nucleation-elongation model is one or more unfavorable nucleation steps followed by more favorable elongation [209]. Earlier models for actin polymerization showed that the critical size for the nucleus is somewhere between a dimer and a trimer, and that the number of explicit nucleation steps does not affect the results of the model. It was also found that choosing a critical nucleus larger than three or four monomers did not affect the results of the model [269, 51, 84]. With these considerations in mind, a simple five-step model was proposed by Sept et al. [252]:

$$A + A \underset{k_{-1}}{\overset{k_{+1}}{\rightleftharpoons}} A_2 \qquad k_{+1} = 10 \mu M^{-1} s^{-1} \qquad k_{-1} = 10^6 s^{-1} \qquad (4.235)$$

$$A + A_2 \underset{k_{-2}}{\overset{k_{+2}}{\rightleftharpoons}} A_3 \qquad k_{+2} = 10 \mu M^{-1} s^{-1} \quad k_{-2} = 10^3 s^{-1} \qquad (4.236)$$

$$A + A_3 \underset{k_{-3}}{\overset{k_{+3}}{\rightleftharpoons}} A_4 \qquad k_{+3} = 10 \mu M^{-1} s^{-1} \quad k_{-3} = 10 s^{-1} \qquad (4.237)$$

$$A + A_4 \underset{k_{-4}}{\overset{k_{+4}}{\rightleftharpoons}} N \qquad k_{+4} = 10 \mu M^{-1} s^{-1} \quad k_{-4} = 0 s^{-1} \qquad (4.238)$$

$$A + N \underset{k_{-}}{\overset{k_{+}}{\rightleftharpoons}} N \qquad k_{+} = 10 \mu M^{-1} s^{-1} \quad k_{-} = 1 s^{-1} \qquad (4.239)$$

where A represents the concentration of actin monomers, A_i for a filament with i actin monomers and N represents the concentration of all longer filaments. The rate constants for the last reaction have been experimentally measured [216] and lead to the correct critical concentration ($C_c > 0.1 \mu M$), but the other rate constants are only approximations from kinetic simulations, chosen to reproduce the time course of polymerization over a limited range of actin monomer concentrations. Filaments longer than 4 subunits are assumed to be stable and the back-reaction rate k_{-4} is set to zero. This is appropriate, since most filaments are much longer than four monomers. The coupled first-order differential equations that arise from the set of reactions above are 'stiff-equations' due to the large differences in the forward and back reaction rates.

This set of equations produces correct polymerization curves, but the average length as a function of actin concentration is completely incorrect. The mean lengths of the observed filaments are almost independent of the initial concentration of actin monomers, while this simple model predicts a mean length with quite a different behavior, especially at high concentrations of actin (see Fig. 4.66). To solve this problem, additional processes are added to the model below.

The average length is simply given by the total amount of polymer divided by the total number of filaments formed. Since the reaction above produces the correct time course and extent of polymerization but not the correct average length, the simple model produces the incorrect number of filaments. The average length is too low, so the actual mechanism must produce fewer filaments. The number of filaments in the system is represented as the number of filaments that are formed by the addition of a monomer onto a polymer A_4. Since there is no back reaction rate for this process, the equation for the change in N, the filament number concentration, is simply

$$\dot{N} = k_{+} A A_4 \qquad (4.240)$$

Molecular and Cellular Biophysics

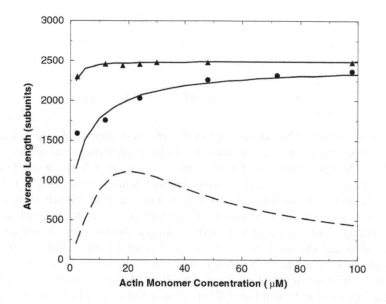

FIGURE 4.66:   Dependence of the number average length (L_n) of actin filaments on the purity and monomer concentration during polymerization. The actin monomer concentration during polymerization was varied as indicated. The filaments formed during each 30 ms interval were allowed to elongate over the succeeding course of the reaction. (•) Singly gel filtered actin. (▲) Doubly gel filtered actin. Theoretical lengths calculated by kinetic simulation [252] using the nucleation-elongation model and the rate constants in the text without annealing/fragmenting (dashed line) and with both annealing and fragmenting (solid line).

Addition of monomers to existing filaments increases the concentration of polymerized actin, but does not increase N. To include filament annealing, we must also consider the breakup of a filament or fragmentation. To include filament annealing and fragmenting in the reaction scheme, the following reaction is added:

$$N + N \underset{k_f}{\overset{k_a}{\rightleftharpoons}} N \tag{4.241}$$

where k_a represents the annealing rate and k_f the fragmentation rate of a filament. Since two filaments are joined to form a new filament the annealing rate has a quadratic dependence and the new equation for the change in N is

$$\dot{N} = k_+ A A_4 - k_a N^2 + k_f N \tag{4.242}$$

The addition of a monomer to the fast growing barbed end of a filament is a diffusion limited process, so it is reasonable to assume that annealing of two filaments is also limited by diffusion, since similar bonds are formed and filaments diffuse more slowly than monomers. Following sonification, an initial rate constant for annealing was found as $k_a = 10\mu M^{-1} s^{-1}$ for very short filaments, but as annealing progressed and the filaments became longer, the rate fell off rapidly with time [202]. Kinosian et al. [143] found the rate of annealing after sonication value of $2.2\mu M^{-1} s^{-1}$. If k_a is diffusion limited, the diffusion that we need to consider is the relative diffusion of the two filaments. In fact, actin filaments above 40 nM filament concentration have been shown to exhibit reptation motion [64, 129, 136]. The transverse diffusion is controlled by the mass of the polymer and the density of the polymer network, but the diffusion constant along the tube is inversely proportional to the length of the filament having the form [136]

$$D_{//} = \frac{k_B T}{\zeta L} \tag{4.243}$$

where ζ is the friction coefficient. The annealing rate constant k_a is therefore chosen to be proportional to $D_{//}$, namely

$$k_a = k_a' / L \tag{4.244}$$

where all of the constants are absorbed in the variable k_a' .

Erickson [74] estimated the fragmentation rate to be in the neighborhood of $k_f \approx 10^{-8} s^{-1}$. This rate should be proportional to the length of the filament and the gel network that is formed may also affect the amount of fragmentation. Within the gel, each individual filament is constrained by its neighbors. Doi [64] showed that the number of rods within a distance b of a given rod is given by $bL^2 N$ where L is the filament length and N is still the filament concentration. Choosing an additional fragmentation rate proportional to this quantity results in:

$$k_f = k_{f1}' L + k_{f2}' L^2 N \tag{4.245}$$

where once again all the constants are absorbed into the two factors k'_{f1} and k'_{f2}. Subsequently, the constant k'_{f2} is treated as a free parameter in a minimization scheme.

Replacing k_a and k_f in (Eq. 4.242) yields

$$\dot{N} = k_+ A A_4 - k'_a \frac{N^2}{L} + k'_{f1} L N + k'_{f2} L^2 N^2 \tag{4.246}$$

The amount of polymerized protein in the system is given by the expression

$$P = A_0 - A - 2A_2 - 3A_3 - 4A_4 \tag{4.247}$$

where A_0 represent the initial actin concentration. Using this expression, the average length of a filament is given simply by $L = P/N$ and we obtain

$$\dot{N} = k_+ A A_4 - k'_a \frac{N^3}{P} + k'_{f1} P + k'_{f2} P^2 \tag{4.248}$$

Note that as the actin concentration is increased and more protein polymerizes, the rate of annealing decreases while the fragmentation rate increases. The values for the rate constants depend on the actin concentration and the filament density and length.

During the polymerization of actin monomers, the rates for annealing and fragmentation will change with time, since they depend on the filament length and density. The rate of annealing immediately following sonication has been measured to be in the range of $2.2 - 10 \mu M^{-1} s^{-1}$. Since $k_a = k'_a/L$ and $L > 30$ (subunits) following sonication, we chose $k'_a = 300 \mu M^{-1} s^{-1}$ so that $k_a = 10 \mu M^{-1} s^{-1}$ for $L = 30$. The factors k'_{f1} and k'_{f2} were treated as free parameters for fitting the two curves for mean polymer length vs. actin concentration for singly and doubly filtered actin. The rate constants for the nucleation-elongation steps were fixed at the values previously used to reproduce polymerization curves. Singly filtered actin has been estimated to contain about one part in 50000 CapZ, and other proteins, such as severing proteins, could also be present at higher concentration in the singly gel filtered than doubly gel filtered actin [37]. These latter proteins cut actin filaments, and since they act with equal probability along the length of a filament, the severing rate is proportional to L. Thus, the presence of low concentrations of severing proteins results in a larger value for k'_{f1} for singly filtered actin. It is also possible that the CapZ that is present in the single filtered actin in some way increases the fragmentation rate, perhaps by changing the structure of a filament upon when it binds. With these points in mind, the values chosen were: $k'_{f2} = 1.8 \times 10^{-8} \mu M^{-1} s^{-1}$, and $k'_{f1} = 2.0 \times 10^{-7} s^{-1}$ and $1.1 \times 10^{-8} s^{-1}$ for the single and doubly gel filtered actin respectively. In agreement with the experimental observations of Murphy et al. [202], the annealing rate constant falls off rapidly with time following sonication. This model, which includes both annealing and fragmentation, agrees with the observed lengths over a wide range of starting actin monomer concentrations.

4.10   Cell Division: Mitosis and Meiosis, Stages and Checkpoints, Physical Models, Tensegrity, Cleavage Furrow

One of the most fundamental processes within the cell cycle (see Fig. 4.67) of higher organisms is the process of cell division in eukaryotic cells which is referred to as mitosis.

4.10.1   Cell Division

Mitosis insures genetic continuity since the new daughter cells have exactly the same number and kind of chromosomes as the original mother cell and the same genetic instructions are passed on. As shown in Fig. 4.68, this process can be divided into five or six basic stages (1): interphase, prophase, (late prophase/prometaphase), metaphase, anaphase and telophase which are briefly described below. (a) Cells may appear inactive during interphase, but they are quite the opposite. This is the longest period of the complete cell cycle during which DNA replicates, the centrioles divide, and proteins are actively produced [193]. During interphase, two centromeres are formed by the replication of a single centrosome.

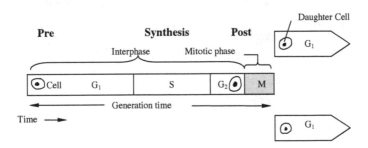

FIGURE 4.67:   The eukaryotic cell cycle.

Each centrosome has a pair of centrioles. Microtubules (MT's) extend from the centrosomes to form asters. The nucleus contains one or more nucleoli. At this stage, the chromosomes are still in the form of loosely packed chromatin fibers. (b) During prophase, the nucleoli disappear. The chromatin fibers become more tightly packed and fold into chromosomes. Each duplicated chromosome is actually two identical chromatids joined at the centromere. Also, during prophase, a network of microtubules arranged between the two centro-

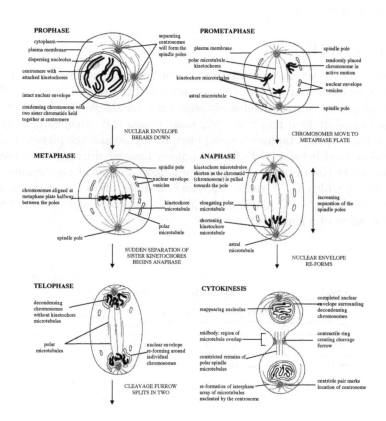

FIGURE 4.68: Stages of mitosis.

somes forms the mitotic spindle. The centromeres move apart from each other during this stage. (c) In the late prophase or prometaphase the nuclear envelope disappears. Spindle fibers (bundles of microtubules) extend from one cell pole to the other, where the two centromeres are now located. Chromosomes then begin to attach themselves to special microtubules called kinetochore microtubules. (d) During metaphase, the chromatid pairs (chromosomes) line up along the equator of the cell. The centromeres of each chromosome are aligned with one another at what is called the metaphase plate. (e) During anaphase, the paired centromeres of each chromosome divide, allowing the once paired chromatids to move away from each other. Each liberated chromatid is now called a chromosome. These newly formed chromosomes now move toward opposite poles of the cell. By the end of anaphase, the two poles of the cell each have a complete collection of chromosomes. (f) Finally, during telophase, the cell elongates and two daughter nuclei begin to form at each pole of the cell where the chromosomes have gathered. The nucleoli then reappear and the chromosomes become less tightly coiled and revert to chromatin fiber. The cell pinches in (or furrows) at the centre until two daughter cells are formed, at which point mitosis is complete. An essential component of mitosis is the formation of a mitotic spindle made up of dynamic microtubules which spatially organize and then separate the divided chromosomes. The mitotic spindle morphogenesis poses a serious puzzle due to a combination of deterministic and stochastic behavior present. Although the mitotic apparatus has to be functionally very precise, its assembly is accomplished without a detailed blueprint; the construction of the mitotic spindle is thus an example of a stochastic phenomenon where random molecular processes play a crucial role. Assembly of microtubules and the mitotic spindle is now being modelled mathematically using the principles of polymer physics and chemical kinetics.

4.10.1.1 Centrioles, Centrosomes and Aster Formation

The centrosome provides the mechanical framework that enables the division of the cell. Centrioles replicate autonomously like mitochondria and peroxisomes. They begin from centers which contain proteins needed for their formation (tubulin, etc.). Then the procentrioles form, each one growing out a single microtubule from which the triplet can form. Once a centriole is made, daughter centrioles can grow out from the tubules at right angles to each other. These then add to the daughter cell (in a dividing cell), or they move to the periphery and form the basal body for the cilium. Centrioles divide prior to the cell division. From the centrosomes originate the microtubules that pull chromosomes apart and push the centrosomes apart (spindle microtubules). The astral microtubules protrude radially from the centrosomes. The centrosome-microtubule system (CTR-MT) in which CTR acts as the nucleation centre for MTs takes a direct part in the cells' ability to maintain a structural and functional polarity, and to carry out a division cycle. This system is considered to be the major agent of cell morphogenesis. Formally analogous to the semi-conservative duplication of the chromosome, CTR duplication is likely to rest upon a different mechanism. Deciphering this mechanism is a major issue in cell biology. The centrosome duplication pathway might be so important for cell survival as to have been conserved throughout eukaryotic cell evolution. One can therefore speculate that a relatively small number of essential genes may be involved in this pathway. Thus, one could benefit from the particular features observed in widely divergent systems and from the particular experimental approaches which have been developed in each case, in order to get at general principles of centrosome inheritance. The major events that characterize CTR duplication anticipate the major cell cycle decisions. This temporal sequence would suggest that some controls on the cell division cycle might be connected with the progression of the CTR duplication cycle. The duplication of chromosomes (represented by a red bar which doubles) during S phase and the duplication of centrosome (represented by a green dot which doubles) over a period which encompasses most of the interphase are due to independent mechanisms. They both require however a coupling with the progression of the cell cycle engine. The centrosome reproduction cycle requires several interconnected structural events, involving the duplication of both centrioles (through orthogonal budding of pro-centrioles in late G1, elongation during S-G2, semi-conservative segregation in G2-M and distribution to the daughter cells as a pair of orthogonal centrioles which eventually disorient) and the maturation of the centrosome matrix (demonstrating an accumulation during interphase, a profound structural remodelling during M phase, and eventually a dispersion or degradation at the outset of mitosis). The molecular mechanisms ensuring the coupling between the sequential events of the centrosome reproduction cycle and the mitotic progression have still to be unravelled and biophysical models to be developed.

4.10.1.2 Chromosome Segregation

The equal apportionment of daughter chromosomes to each of the two cells resulting from cell division is called chromosome segregation. It occurs during mitosis and meiosis and is regulated by DNA replication. Centromeres are the attachment points of kinetochores, and are critical for segregation. Centromere structures vary among organisms. Molecular motors attached to kinetochores move chromosomes during segregation. In biological experiments chromosomes have been subjected to micromanipulation in order to learn more about their movement in mitosis. By tugging on them, the forces produced by the spindle have been measured. The spindles themselves have been chopped apart to locate the motors for chromosome movement. The current preoccupation of cell biologists like Bruce Nicklas [203, 204] is to connect cell mechanics with molecular biology. Pulling on chromosomes alters the phosphorylation of chromosomal proteins. Different phosphorylation states signal the cell either to go ahead and divide or to pause, allowing time for error correction. Now we wish to understand in more detail how mechanical tension produced by mitotic forces provides the chemical signals that regulate the cell cycle. Models for chromosome movements are being developed based on lateral interactions of spindle microtubules. The various stages of the cell cycle can be summarized as follows. Cell birth and the onset of DNA synthesis (G1); The period of DNA synthesis (S), which ends when all nuclear DNA has been replicated and hence the number of chromosomes has doubled, and the between the end of DNA synthesis and the beginning of the mitotic phase (G2). After a cell has entered S, it is committed to completing the cell cycle, even when environmental conditions are extremely adverse.

Cells have been shown to divide (a) without their membranes, (b) with all their motor proteins extracted and even (c) with no chromosomes. However, they cannot divide without microtubules or centrosomes. Similarly, insufficient ATP and calcium supplies limit the ability of cells to divide. One of the key figures in the field of cell biology, Ted Salmon stated his working hypothesis [230] that mitosis will be explained by a combination of several mechanism involving: (a) the molecular and structural properties of the centrosome which organizes and nucleates the polymerization of spindle microtubules, (b) the assembly of microtubules which orient and participate in the generation of chromosome movements and (c) the microtubule motors such as the kinesin and dynein families of proteins which appear to generate polarized forces along the lattice of microtubules, at kinetochores and within the spindle fibers. The mathematical formalism necessary for this task must encompass: (a) stochastic polymerization processes, (b) non-equilibrium chemical kinetics which follows from the structure, (c) tensegrity which reflects the mechanical properties of the networks formed and (d) electronic and dipolar characteristics that are due to the protein structure, bound water and ionic solutions.

The separation of chromosomes is accomplished by the cytoskeleton's largest constituents, the microtubules. During mitosis, microtubules connect to each of the chromosomes and align them along the cell's equatorial plate. A mysterious balance of forces prevents the microtubules from separating the chromosomes until all the chromosomes have become aligned and division may proceed in unison. Since MT's play the key role in mitosis, new cancer therapies are being developed (e.g., including the use of taxol in the treatment of ovarian and breast cancers) which directly target microtubule assembly processes. Colchicine, colcemid and nocadazol inhibit polymerization by binding to tubulin and preventing its addition to the plus ends. Vinblastine and vincristine aggregate tubulin and lead to microtubule depolymerization. Taxol stabilizes microtubules by binding to a polymer. It is, therefore, of crucial importance to our understanding of molecular level cell functioning to be able to quantify some of the most dominant processes that govern MT behavior. It is interesting to note in this context that microtubules themselves have been specialized in their functions in the dividing cells as is shown in Fig. 4.70

Cell division involves more than just chromosome segregation. In particular

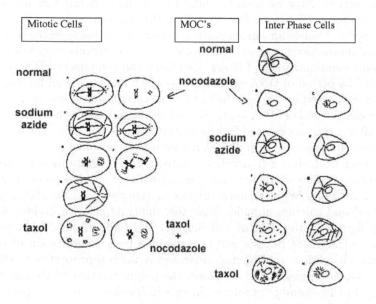

FIGURE 4.69: Disruption of the cytoskeletal organization by various chemical agents at different stages of the cell cycle.

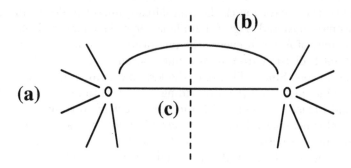

FIGURE 4.70: Specialization of MTs. (a) Astral, (b) Polar, (c) Kinetochore.

it leads to the emergence of two daughter cells via the formation of a contractile ring and a physical separation of the two halves of the dividing cell. This process is being modelled mathematically with reasonable success (see Fig. 4.71) although it does not exhibit the level of detail for the full inclusion of the cytoskeletal structures.

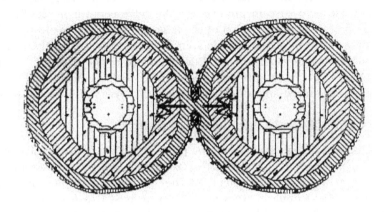

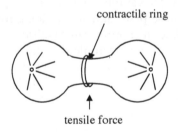

contractile ring

tensile force

FIGURE 4.71: Development of a cleavage in the simulations by Alt and Dembo [10].

Cytokinesis involves the formation of a cleavage furrow in the area near the old metaphase plate. The furrow begins a shallow groove in the cell surface. Just below this groove is where a ring of cytoplasmic microfilaments begins to contract. The cleavage furrow deepens until the parent cell is pinched in two. The basic requirement for cytokinesis is the presence of actin and MT's and a sufficient supply of ATP. The plane of cytokinesis is specified by astral microtubules and the position of the spindle. Astral MT's are sufficient to position a cleavage furrow but its activation is after mitosis. Recent evidence shows that cleavage furrow initiation requires signalling from mitotic spindle MTs. Cytokinesis is generally coupled to mitosis, but not always so. Before the start of mitosis, the cytoskeleton's actin is distributed almost uniformly except for a concentration in the cortical layer, a situation which is so stable that even exposure to an astral signal doesn't disturb it. Eventually, however, progressive stiffening takes place in the cortex area giving rise to a zone of concentrated acto/myosin presence at the cell's equator in the form of a contractile ring. The organization of the axis and the equator of the dividing cell is only a function of the mitotic spindle and asters. Recently, computer simulations [10] of the emergence and development of the cleavage due to the hydrodynamics of a cross-linked polymeric fluid with a controllable viscosity successfully reproduced spatio-temporal patterns observed in experiments.

4.10.1.3 Spindle and Chromosome Motility

The discovery of microtubule motor proteins with a clear involvement in mitosis and meiosis has attracted great interest, since motor proteins could account for many of the movements of the spindle and chromosomes in dividing cells. Microtubule motors have been proposed to generate the force required for spindle assembly, attachment of the chromosomes to the spindle and movement of chromosomes toward opposite poles. Kinesin motors have been shown to be necessary for establishing spindle bipolarity, positioning chromosomes on the metaphase plate and maintaining forces in the spindle. Evidence also exists that kinesin motors can facilitate microtubule depolymerization, possibly modulating microtubule dynamics during mitosis. The force generated by microtubule polymerization/ depolymerization is thought to contribute to spindle dynamics and movements of the chromosomes.

Many of the newly identified kinesin-related proteins (KRPs) localize to the spindle in mitotically dividing cells and are implicated in spindle function by both their cellular localization and mutant effects. Remarkably, several KRP's have been demonstrated to be chromosome-associated, providing a direct link between the chromosomes and microtubules of the spindle. The demonstration of kinesin motor protein function in spindle and chromosome motility represents a major step forward in understanding the basis of cell division.

Chapter 4 Questions and Problems

QUESTION 4.1 Discuss CA models and potential role. Are these models suitable for modeling specific systems, or do they provide conceptual frameworks? (or both?)

QUESTION 4.2 Amino acids are the basic building blocks of proteins. Describe an "average" amino acid.

QUESTION 4.3 Compare Monte Carlo and Molecular Dynamics methods for biomolecules.

QUESTION 4.4 Discuss the protein folding problem.

QUESTION 4.5 It is surmised that the spatial organization of microtubules affects their function. To investigate these properties it is sometimes necessary to polymerize tubulin in vitro. One method of polymerizing tubulin in vitro involves an initial solution of pure tubulin. Another method involves adding nucleation seeds to the initial solution. Below are two example graphs showing the amount of tubulin polymerized against the concentration of tubulin in the initial solution for the two methods. The critical concentration is the concentration at which tubulin begins to polymerize.

(a) Explain why the critical concentration at which tubulin begins to polymerize is different for the two methods.

(b) Note the average concentration of tubulin in cells. What can you conclude about microtubule formation in cells from the two graphs?

(c) From the graphs it can be seen that the critical concentrations are $15\mu M$ and $< 5\mu M$ for graphs (a) and (b) respectively. What is expected to happen to microtubules in solution over time at tubulin concentrations (i) above the critical concentration, (ii) below the critical concentration and (iii) at the critical concentration? (Ignore the process of dynamic instability.)

(d) Tubulin bound to GTP or GDP alters the critical concentration of tubulin polymerization, thus creating different critical concentrations for the (+) and (-) ends of a microtubule. What is expected to happen to microtubules in solution over time at tubulin concentrations (i) below the critical concentrations for both the (+) and (-) ends, (ii) above the critical concentration for both the (+) and

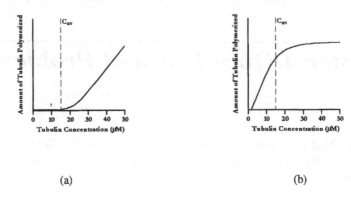

(a)   (b)

FIGURE 4.72: Graphs for tubulin polymerized in vitro for (a) an initial solution of pure tubulin; (b) an initial solution of tubulin and seeds. C_{av} is the average concentration of tubulin in cells ($10 - 20\mu M$).

(-) ends and (iii) above the critical concentrations for the (+) end and below the critical concentration at the (-) end? (Ignore the process of dynamic instability.)

PROBLEM 4.6 Use Fick's Law to estimate the rate at which water diffuses through the skin, based on the following assumptions:

- There is pure water inside the skin, and zero concentration of water outside.
- $1\ m^3$ of pure water has a mass of 1000 kg.
- The skin area of an average person is about 1.75 m^2.
- The skin is a uniform layer about 20 μm thick. ($1\mu m = 1 \times 10^{-6}$ m)
- The diffusion coefficient for water in skin is $5.0 \times 10^{-14} m^2/s$.

Express your answer in litres per day. Does your result seem reasonable?

PROBLEM 4.7 Consider an isotropic random walk in three dimensions with independent, identical displacements of length a, given by the following probability distribution function (PDF):

$$p(\vec{x}) = \frac{\delta(r-a)}{4\pi a^2} \qquad (r = |\vec{x}|)$$

Derive the formula for the PDF of the position after n steps.

PROBLEM 4.8 When we breathe, oxygen diffuses from the alveoli in our lungs into nearby capillaries. The distance from the centre of the alveoli to the capillary is about $0.1mm$. For efficiency, the oxygen should move to the capillaries in a time shorter than the time between breaths. If we take this time to be $0.5s$, what diffusion constant must oxygen have in the lungs? If we had no circulatory system and the oxygen had to diffuse to the brain, how long would it take it to travel the approximately $0.25m$?

PROBLEM 4.9 What molar concentration of solute in tree sap would be necessary to raise the sap $1m$ in the trunk of a tree? Why would this not work for a $50m$ tree? Assume this is a relatively low concentration so Equation 4.14 holds.

PROBLEM 4.10 Give an example of a non-fractal object and calculate its fractal dimension.

PROBLEM 4.11 Consider the diffusion process in the axon of a neuron. A neuron can be simplified to a spherical cell body with a long thin appendage of length L. A material, M, formed in the cell body diffuses down the appendage, all along which the substance is consumed at a constant rate Q. The concentration of the material in the cell body is C_0, and the flux of the material at the end of the appendage is 0.

(a) Show that:

$$C = - \left(\frac{Q}{D} \right) x^2 + Ax + B$$

is a quasi-stationary state solution to the diffusion equation (Eq. 4.17) where $A = \frac{QL}{D}$ and $B = C_0$. State all approximations used.

(b) Assuming C_0 and D are fixed and the material is consumed at constant rate Q along the entire length of the appendage find the upper limit of the appendage length L.

PROBLEM 4.12   *One-dimensional diffusion and filament walking.*

In this question we will consider the mean time it takes a particle released at some position in a one dimensional box to be absorbed at one end or the other. A biological application of this problem is the binding of proteins to specific sites on a strand of DNA. Such proteins may first attach non-specifically to the DNA, with an affinity that is strong enough that it doesn't come off, but weak enough that it can wander along the one dimensional strand.

(a) A particle is released at some position x $(0 < x < b)$. The mean time to capture from this position is $W(x)$. After a time step of duration τ , the particle has moved either to position $x + \delta$ or $x - \delta$, each with probability 1/2. By considering the mean time to capture from each of these two new positions, show that the function $W(x)$ satisfies the equation:

$$\frac{d^2W}{dx^2} + \frac{1}{D} = 0$$

(b) Consider a box of length b in which the $x = 0$ end is a perfect absorber, and the $x = b$ end is a perfect reflector. Solve the equation above for $W(x)$ with the appropriate boundary conditions for this box. Sketch or plot $W(x)$ for this box.

(c) Finally, find the average of this expression over all possible starting points inside the box.

PROBLEM 4.13   Rotational diffusion of bacterium.
How long, on average, would you have to wait for a $100\mu m$ diameter spherical particle sitting in water at $300K$ to have changed its orientation $60°$ due to rotational diffusion? Why doesn't your answer depend on what the particle is made of (i.e., its density?)

PROBLEM 4.14 A scientist is investigating the one-dimensional diffusion of motor proteins across a thin microtubule (a protein filament). By analyzing diffusion over a time of 30 seconds, he obtained the following table of displacements:

TABLE 4.7:   Data for Problem 4.14

Number of particles	Approximate displacement in μm
10	-50
35	-25
50	0
35	25
10	50

(a) What is the mean displacement of the particles?
(b) What is the mean square displacement of the particles?
(c) What is the diffusion coefficient of the particles?

PROBLEM 4.15 A spherical virus has a radius of $40nm$. What is the expected diffusion coefficient of such a particle in water at $20°C$?

PROBLEM 4.16 Two spherical enzymes, A and B, have respective diffusion coefficients of $7.0 \times 10^{-11} m^2 s^{-1}$ and $4.3 \times 10^{-11} m^2 s^{-1}$. Their densities are $\rho_A = 1.15 \times 10^3 kgm^{-3}$ and $\rho_B = 0.95 \times 10^3 kgm^{-3}$. What is the ratio of their molar masses?

PROBLEM 4.17 The diffusion coefficient for sodium ions crossing a biological membrane 10 nm thick is $D = 1.0 \times 10^{-18} m^2 s^{-1}$. What flow rate of sodium ions would move across an area 10 nm $\times$ 10 nm if the concentration difference across is $0.50 mol/dm^3$?

PROBLEM 4.18 The diffusion coefficient of potassium across the same membrane is 100 times greater than for sodium. If the concentration on one side is 150 millimoles per dm^3 and on the other is 5 millimoles per dm^3, what would be the potassium flow rate across $5.0 nm \times 5.0 nm$ of the membrane?

PROBLEM 4.19 Suppose we use Einstein's Formula to estimate the diffusion constant of water molecules. The viscosity of water is 1.0×10^{-3} in SI units. What radius would you have to assign? Compare with the van der Waals radius of oxygen (1.4 Å).

PROBLEM 4.20 Give an explanation of the fact that the diffusion constant of globular proteins is proportional to $M^{-1/3}$ where M is the (molar) mass. This relation does not hold for DNA. Why? How do you think the diffusion constant of short DNA strands depends on the mass of DNA?

PROBLEM 4.21 The plasma membrane of a cell has a surface charge per unit area of about $7 \times 10^{-4} Coulomb/cm^2$ on one side and an equal, but opposite, surface charge on the other side.

(i) Estimate the electrical field strength inside the plasma membrane. You will find a very large electrical field strength as compared to the electrical fields encountered in household electrical devices. The electrical field between the poles of a 12 volt car battery is, for instance, only about $10^{-2} N/C$.

(ii) Estimate the force on an ion placed inside the plasma membrane of a cell.

PROBLEM 4.22 Estimate the molar conductance of the Cl^- ion using the equation

$$\mu = \frac{1}{6\pi\eta r}$$

and

$$\Lambda_0 = e(\mu^+ + \mu^-)$$

Assume that the spherical ion is surrounded by a "hydration shell" of water molecules. Use the van der Waals radius for the radius of the Cl^- ion and the dimensions of the water molecules.

Chapter 4 References

1. Adessi, C., Walch, S., and Anantram, M.P. *Phys. Rev. B* **67**, 081405(R), 2003.

2. Agmon, N. and Hopfield, J.J. *Journal of Chemical Physics* **79**, 2042-2053, 1983.

3. Agmon, N. and Madhavi, Sastry G. *Chemical Physics* **212**(1), 207-219, 1996.

4. Al-Bassam, J., Ozer, R.S. , Safer, D., Halpain, S. , and Milligan, R.A. *J. Cell Biol.* **157**, 1187-1196, 2002.

5. Alberts B., Bray D., Lewis J., Raff M., Roberts K., and Watson J.D., *Molecular Biology of the Cell.* Garland Publishing, London, 1994.

6. Albrecht-Buehler G. *Proc. Natl. Acad. Sci. U.S.A.* **102**, 5050-5055, 2005.

7. Albrecht-Buehler, G. *Cell Motility and the Cytoskeleton* **27**, 262-271, 1994.

8. Albrecht-Buehler, G. *Cell Motil. Cytoskel.* **32**, 299-304 , 1995.

9. Albrecht-Buehler, G. *Cell Motil. Cytoskel.* **40**, 183-192, 1998.

10. Alt, W. and Dembo, M. *Math. Biosci.* 156, 207. 1999.

11. Amos, L.A. and Amos, W.B. *Molecules of the Cytoskeleton.* Macmillan Press, London, 1991.

12. Amos, L.A., *Trends in Cell Biology* **48**, 5, 1995.

13. Andersen, P. *Acta Physiol Scand.* **48**, 178-208, 1960.

14. Atkins, P.W., *Physical Chemistry*, Oxford University Press, Oxford, 1990.

15. Austin, R. H. et al., *Biochemistry* **14**, 5355, 1975.

16. Bang, O. *PhD thesis*, The Technical University of Denmark, DK-2800 Lyngby, Denmark, 1993.

17. Baker, J., LaConte, L. E. W., Brust-Mascher, I., and Thomas, D. D. *Biophys. J.* **77**, 2657-2664, 1999.

18. Bang, O., Christiansen, P.L., Rasmussen, K.Ø., and Gaididei, Y.B. In *Les Houches*, 1995. To appear.

19. Bartnik, E.A., Blinowska, K.J. and Tuszyński, J.A. *Nanobiology* **1**, 239–250, 1992.

20. Bartnik, E.A., Blinowska, K.J., and Tuszyński, J.A. *Phys. Lett. A* **169**(1, 2), 46–50, 1992.

21. Bayley, P.M., Martin, S.R., and Sharma, K.K., *American Insitute of Physics Conference Proceedings* **226**, 187-199, 1991.

22. Brizhik, L.S., Eremko, A.A. and La Magna, A. *Phys. Lett. A* **200**, 213–218, 1995.

23. Bliss, T. V. P., and Collingridge, G. L. *Nature* **361**, 31-39, 1993.

24. Bockris, J. O'M. and Khan, S.U.M. *Applied Physics Letters* **42**(1), 124-125, 1983.

25. Bruinsma, R. *Physics, 6A and 6B*, International Thomson Publishing, 1998.

26. Bryngelson, J. D., Onuchic, J. N., Socci, N. D., and Wolynes, P. G. *Function and Genetics* **21**, 167, 1995.

27. Bartnik, E.A. and Tuszyński, J.A. *Phys. Rev. E* **48**, 1516–28, 1993.

28. Buzan, J. M. and C. Frieden. *Proc. Natl. Acad. Sci. U.S.A.* **93**, 91-95, 1996.

29. Christiansen, P.L. private communication, 1996.

30. Caceres, A., Mautino, J., and Kosik, K.S. *Neuron* **9**(4), 607-18, 1992.

31. Campagnola, P.J., Millard, A.C., Terasaki, M., Hoppe, P.E., Malone, C.J., and Mohler, W.A. *Biophys J.* **82**, 493-508, 2002.

32. Cantiello, H., Patenande, C., and Zaner, K. *Biophys. J.* **59**, 1284-1289, 1991.

33. Caplow, M. Ruhlen, R. and Shanks, J. *J. Cell. Biol.* **127**, 779, 1994.

34. Careri, G., Buontempo, U., Galluzzi, F., Scott, A.C., Gratton, E. and Shyamsunder, E. *Phys. Rev. B* **30**, 4689-4702, 1984.

35. Carlier, M. F., Pantaloni, D., and Korn, E. D. *J. Biol. Chem.* **259**, 9987-9991, 1984.

36. Carlier, M.F., Pantaloni, D., Hill, T.L., and Chen, Y. *Proc. Natl. Acad. Sci. U.S.A.* **84**, 5257-5261, 1987.

37. Casella, J. F., Barron-Casella, E. A., and Torres, M.A. *Cell Motil. Cytoskel.* **30**, 164-170, 1995.

38. Cassimeris L. *Cell. Motil. Cyto.* **26**, 275-281, 1993.

39. Cassimeris, L., Pryer, N.K., and Salmon, E.D., *Journal of Cell Biology* **107**, 2223-2231, 1988.

40. Christiansen, P.L. Bang, O., Pagano, S. and Vitiello, G. *Nanobiology* **1**, 229–237, 1992.

41. Czikkely, V., Dreizler, G., Försterling, H.D., Kuhn, H., Sondermann, J., Tillmann, P. and Wiegand, J. . *Z. Naturforsch.* **24a**, 1823, 1969.

42. Czikkely, V., Försterling, H.D., and Kuhn, H.. *Chem. Phys. Lett.* **6(1)**, 11–14, 1970.

43. Czikkely, V., Försterling, H.D., and Kuhn, H. *Chem. Phys. Lett.* **6**, 207–210, 1970.

44. Chasan, B., Geisse, N., Pedatella, K., Wooster, D., Teintze, M., Carattino, M., Goldmann, W., and Canteillo, H. *Eur. Biophys. J.* **30**, 617-624, 2002.

45. Chen, Y., and T.L., *Proceedings of the National Academy of Sciences of the U.S.A.* **82**, 1131-1135, 1985.

46. Chernavsky, D.S., Khurgin, I.U.I. and Shnol', S.E. *Biofizika.* Sep-Oct **32**(5), 775-81. 1987.

47. Christie, B.R., and Abraham, W.C. *Neuroreport* **5(4)**, 385-8, 1994.

48. Christie, B. R., Eliot, L. S., Ito, K., Miyakawa, H. and Johnston, D. *Journal of Neurophysiology* **73(6)**, 2553-2557, 1995.

49. Consta, S. and Kapral, R. *The Journal of Chemical Physics* **101(12)**, 10908-10914, 1994.

50. Coombs, J.S., Eccles, J.C., and Fatt, P. *J Physiol.* **130(2)**, 326-373, 1955.

51. Cooper, J. A., Buhle, E. L. Jr., Walker, S. B., Tsong, T. Y. , and Pollard, T.D. *Biochemistry* **22**, 2193-2202, 1983.

52. Cope, F.W. *Journal of Biological Physics* **3(1)**, 1-41, 1975.

53. Correas et al., *Biochem J.* **269(1)**, 61-64, 1990.

54. Christiansen, P.L., Pagano, S., and Vitiello, G. *Phys. Lett. A* **154**(7, 8), 381–384, 1991.

55. Cragg, B. G. and Hamlyn, L. H. *J Physiol.* **129(3)**, 608-627, 1955.

56. Craig, A.M., Wyborski, R.J. and Banker, G. *Nature* **375**, 592-594, 1995.

57. Cronly-Dillon, J., Persaud, K., and Gregory, R.P. *Proc Biol Sci.* 1999 Dec 7, **266**(1436), 2427-33, 1999.

58. Dai, J. and Sheetz, M.P. *Biophysical Journal* **68**, 988-996, 1995.

59. Davies, O. R. and Inglesfield, J. E. *Phys. Rev. B* **69**, 195110, 2004.

60. Davydov, A.S. *Biology and Quantum Mechanics.* Pergamon Press, Oxford, 1982.

61. de Pablo, P. J., Moreno-Herrero, F., Colchero, J., Gómez Herrero, J., Herrero, P., Baró, A. M., Ordejón, P., Soler, J.M., and Artacho, E. *Phys. Rev. Lett.* **85**, 4992-4995, 2000.

62. Dehmelt, L., Smart, F.M., Ozer, R.S., and Halpain, S. *The Journal of Neuroscience* **23(29)**, 9479-9490, 2003.

63. Dehmelt, L. and Halpain, S. *Journal of Neurobiology* **58**(1), 18-33, 2004

64. Doi, M. *J. Physiol., Paris.* **36**, 607-617, 1975.

65. Dombeck, D.A.; Kasischke, K.A.; Vishwasrao, H.D.; Ingelsson, M.; Hyman, B.T., and Webb, W.W. *Proc. Natl. Acad. Sci. U.S.A.* **100**, 7081-7086, 2003.

66. Doniach, S. and Sondheimer, E. *Green's Functions for Solid State Physicists*, W.A. Benjamin Inc., Don Mills, 1974.

67. Dresselhaus, M.S. , Dreselhaus, G., and Eklund, P.C. *Science of Fullerenes and Carbon Nanotubes*, Academic Press, San Diego, 1996.

68. Drewes, G., Ebneth, A., and Mandelkow, E-M. *TIBS* **23**, 307-311, 1998.

69. Dustin, P. *Microtubules.* Springer-Verlag, Berlin, 1984.

70. Eccles, J. C. *The Physiology of Synapses.* Springer, Berlin, 1964.

71. Eley, D. D. *Mol. Cryst. Liq. Cryst.* **171**, 1-21, 1989.

72. Eliot, L. S. and Johnston, D. *Journal of Neurophysiology* **72(2)**, 762-777, 1994.

73. Endres, R. G., Cox, D. L., and Singh, R. R. P. *Rev. Mod. Phys.* **76**, 195, 2004.

74. Erickson, H. P. *J. Molec. Biol.* **206**, 465-474, 1989.

75. Eyzaguirre, C. and Kuffler, S.W. *The Journal of General Physiology* **39**, 121-153, 1955.

76. Felgner, H., Frank, R., and Schliwa, M. *J. Cell. Sci.* **109**, 509-516, 1996.

77. Feynman, R. P., Leighton, R. B., and Sands, M. *The Feynman Lectures on Physics* Vol. I. Ch. 46, Addison-Wesley, Reading, 1963.

78. Fink, H.-W. *Cellular and Molecular Life Sciences (CMLS)* **58(1)**, 2001.

79. Fisher, R. E., Gray, R. and Johnston, D. *Journal of Neurophysiology* **64(1)**, 91-104, 1990.

80. Flyvbjerg, H., Holy, T.E., and Leibler, S., *Physical Review Letters* **73**, 2372, 1994.

81. Flyvbjerg H., Holy T.E., and Leibler S. *Phys. Rev. E* **54**, 5538-5560, 1996.

82. Frauenfelder, H., Sligar, S. G., and Wolynes, P. G. *Science* **254**, 1998-1603, 1991.

83. Frey E., Kroy K., and Wilhelm J. *Adv. Struct. Biol.* **5**, 135-168, 1998.

84. Frieden, C., and Goddette, D. *Biochemistry* **22**, 5836-5843, 1983.

85. Fritzsche, W., Böhm, K., Unger, E., and Köhler, J.M. *Nanotechnology* **9**, 177-183, 1998.

86. Fritzsche, W., Koehler, J.M., Boehm, K., Unger, E., Wagner, T., Kirsch, R., Mertig M., and Pompe, W. *Nanotechnology* **10**, 331-335, 1999.

87. Fujita, Y. and Sakata, H. *J Neurophysiol* **25**, 209-222, 1962.

88. Fung, Y. C., *Biomechanics: Mechanical Properties of Living Tissues*, Springer, Berlin, 1993.

89. Fye, R.M., Martins, M.J., and Scalapino, D.J. *Phys. Rev.* **B45**, 7311-7314, 1992.

90. Fygenson, D.K., Braun, E., and Libchaber, A. *Phys. Rev. D* **50**, 1579-1588, 1994.

91. Gamblin, T.C., Nachmanoff, K., Halpain, S., and Williams, R.C. Jr. *Biochemistry* **35(38)**, 12576-12586, 1996.

92. Grad, J., Hernandez, G., and Mukamel, S. *Phys. Rev. A* **37**(10), 3835–3846, 1987.

93. Gittes, F., Mickey, E., and Nettleton, J. *J. Cell Biol.* **120**, 923-934, 1993.

94. Glanz, J. *Science* **276**, 678, 1997.

95. Gliksman, N.R., Skibbens, R V., and Salmon, E.D., *Molecular Biology and Chemistry* **4**, 1035-1050, 1993.

96. Goddard, G. and Whittier, J.E. *Proceedings of SPIE – Volume 6172*, Vijay K. Varadan, Editor, 617206, 2006 .

97. Gonzales-Billault, C., Owen, R., Gordon-Weeks, P.R., and Avila, J. *Brain Res.* **943**, 56-67, 2002.

98. Gaididei, Y. B., Rasmussen, K.Ø., and Christiansen, P.L. *Phys. Lett. A* **203**, 175–180, 1995.

99. Gaididei, Y. B., K.Ø. Rasmussen, and Christiansen, P.L. *Phys. Rev. E* **52**, 2951, 1995.

100. Greenough, W.T. *Am Sci.* **63(1)**, 37-46, 1975.

101. Griffith, L.M. and Pollard, T.D. *J. Biol. Chem.* **257(15)**, 9143-9151, 1982.

102. Gaididei, Y.B., Rasmussen, K.Ø., and Serikov, A.A. *Theor. Math. Phys.* **27**, 242–253, 1976.

103. Guillaud, L., Setou, M., and Hirokawa, N. *J. Neurosci.* **23**, 131-140, 2003.

104. Haken, H. *Quantum Field Theory*, North-Holland, Amsterdam, 1976.

105. Haken, H., *Synergetics: an Introduction*, 2nd edition, Springer, Berlin, 1990.

106. Hagan, S., Hameroff, S.R., and Tuszyński, J.A. *Phys. Rev. E* **65**, 061901-1 to -11, 2002.

107. Hempel, C.M. et al. *J. Neurophysiol.* **83**, 3031-3081, 2000.

108. Huth, G.C., Gutmann, F., and Vitiello, G. *Phys. Lett. A* **140**(6), 339–342, 1989.

109. Hille, B. *Sinauer Associates*, Sunderland, MA, 1992.

110. Hillson, N., Onuchi, J.N., and García A.E. *Proc. Natl. Acad. Sci. U.S.A.* **96**, 14848-14853, 1999.

111. Hinner B., Tempel M., Sackmann E., Kroy K., and Frey, E. *Phys. Rev. Lett.* **81**, 2614–2618, 1998.

112. Hirokawa, N. In *The Neuronal Cytoskeleton*, Wiley-Liss, New York, 5-74, 1991.

113. Horio T. and Hotani. H. *Nature, London* **321**, 605-607, 1986.

114. Horio, T. and Hotani, H., *Nature* **321**, 605, 1986.

115. Houchin, J. *J. Physiol.* **232**, 67-69, 1973.

116. Houchmandzadeh, B. and Vallade, M. *Phys. Rev. E* **6320**, 53, 1996.

117. Howard, J. *Mechanics of Motor Proteins and the Cytoskeleton*, Sinauer Associates, Sunderland, 2001.

118. Howard, W., and Timasheff, S. *Biochem.* **25**, 8292, 1986.

119. Hummer, G., Garca, A.E., and Garde, S. *Phys. Rev. Lett.* **85**, 2637-2640, 2000.

120. Hyman, J.M., McLaughlin, D.W., and Scott, A.C. *Physica D* **3**, 23-44, 1981.

121. Hyman, A.A., Salser, S., Dreschel, D.N., Unwin, N., and Mitchison, T.J. *Molec. Biol. Cell* **3**, 1155-1167, 1992.

122. Hyman, A.A., Chrétien, D., Arnal, I., and Wade, R.H. *J. Cell. Biol.* **128**, 117-125, 1995.

123. Inoue, T. *Thin Solid Films* **132**, 21–26, 1985.

124. Ingber, D.E. *J Cell Sci.* **104**, 613-627, 1993.

125. Ishijama, A., and Yanagida, T. a., *Science* **283**, 1667-1695, 2001; *Physics Today*, Oct. 2001, 46-51.

126. Ishijama, A. and Yanagida, T. b., *TIBS* **26**, 438-444, 2001.

127. Jelly, E.E. *Nature* **138**, 1009, 1936.

128. Jaffe, D. B. and Brown T. H. *J Neurophysiol* **72**, 471-474, 1994.

129. Janmey, P. A., Hvidt, S., Käs, J., Lerche, D., Maggs, A., Sackmann, E., Schliwa, M., and Stossel, T. P. *J. Biol. Chem.* **269**, 32503-32513, 1994.

130. Janmey, P. *Physio. Rev.* **78**, 763-781, 1998.

131. Jaslove, S.W. *Neuroscience* **47**, 495-519, 1992.

132. Jobs, E., Wolf, D.E. and Flyvbjerg, H., *Physical Review E* **79**, 519-522, 1997.

133. Jørgensen, M.F., Tuszyński, J.A., and Sept, D. *Phys. Lett. A*, 1996.

134. Juul Rasmussen, J. and Rypdal, K. *Physica Scripta,* **33**, 481–497, 1986.

135. Kaech S, Brinkhaus H, and Matus A. *Proc. Natl. Acad. Sci. U.S.A.* **96**, 10433-10437, 1999.

136. Käs, J., Strey, H., Tang, J. X., Finger, D., Ezzell, R., Sackmann, E. , and Janmey, P. A. *Biophys. J.* **70**, 609-625, 1996.

137. Kasumov, A.Y., Kociak, M., Gueron, S., Reulet, B., Volkov, V.T., Klinov, D.V., and Bouchiat, H. *Science* **291**, 280-282, 2001.

138. Katz, B. *J Physiol.* **11(3-4)**, 261-82, 1950.

139. Kay, A.R. and Wong, R.K. *The Journal of Physiology* **392(1)**, 603-616, 1987.

140. Khuchua, Z., Wozniak, D.F., Bardgett, M.E., Yue, Z., McDonald, M., Boero, J., Hartman, R.E., Sims, H., and Strauss, A.W. *Neuroscience.* **119**, 101-111, 2003.

141. Kiebler, M.A., and DesGroseillers, L. *Neuron* **25**, 19-28, 2000.

142. Kim, C.H. and Lisman, J.E. *J. Neurosci.* **21**, 4188-4194, 2001.

143. Kinosian, H. J., Selden, L.A., Estes, J. E. and Gershman, L. C. *Biochem.* **32**, 12353-12357, 1993.

144. Kitamura, K., Tokunaga, M., Iwane, A. H., and Yanagida, T. *Nature* **397**, 129-134, 1999.

145. Kirstein, S. and Möhwald, H. *Adv. Mater.* **7**, 460–463, 1995.

146. Knops, J., Kosik, K.S., Lee, G., Pardee, J.D., Cohen-Gould , L., and McConlogue, L. *The Journal of Cell Biology* **114**, 725-733, 1991.

147. Kobayasi, S., Asai, H., and Oosawa, F. *Biochim. Biophys. Acta* **88**, 528-540, 1964.

148. Koch, C. and Segev, I. *Nature Neuroscience* **3**, 1171-1177, 2000.

149. Kowalski, R.J. and Williams, R.C. Jr. *J. Biol. Chem* **268(13)**, 9847-9855, 1993.

150. Krebs, A., Goldie, K.N., and Hoenger, A. *J. Mol. Biol.* 2004 Jan 2;**335 (1)**, 139-53, 2004.

151. Kubo, R. *J. Phys. Soc. Japan* **12**, 570-586, 1957.

152. Kuhn, H. *Journal of Photochemistry* **10**, 111-132, 1979.

153. Kurzynski, M. *Biophys Chem.* **65(1)**, 1-28, 1997.

154. Kurzynski, M. *Acta Physica Polonica B* **28(8)**, 1853, 1997.

155. Kurzynski, M. *Progr. Biophys. Molec. Biol.* **69**, 23-82, 1998.

156. Lader, A., Woodward, H., Lin, E., and Cantiello, H. *METMBS'00 International Conference*, 77-82, 2000.

157. Lange, G., Mandelkow, E.M, Jagla, A. and Mandelkow, E., *European Journal Biochemistry* **178**, 61-69, 1988.

158. LeClerc, N., Kosik, K.S., Cowan, N., Pienkowski, T.P., and Baas, P.W. *Proceedings of the National Academy of Sciences* **90**, 6223-6227, 1993.

159. Letourneau, P. C. and Ressler, A. H. *J. Cell Biol* **97**, 963-973, 1983.

160. Lewis, A.K. and Bridgman, P.C. *The Journal of Cell Biology* **119**, 1219-1243, 1992.

161. Lieb, E.H. In *The Hubbard Model*, The Hubbard Model: some rigorous results and open problems, Plenum Press, New York, 1995.

162. Limbach, H.J., Sept, D., Bolterauer, H. and Tuszinsky, J.A. *Journal of Theoretical Biology* 1998, in press.

163. Lin, E. and Cantiello, H. *Biophys. J.* **65**, 1371-1378, 1993.

164. Llinas, R. and Nicholson, C. *Neurophysiol* **34**, 532-551, 1971.

165. Lomdahl, P.S. *Los Alamos Science* **10**, 27, 1984.

166. Lonngren, K. *Solitons in Action*, Observations of solitons on non-linear dispersive transmission lines, Academic Press, New York, 127-152, 1978.

167. Luby-Phelps, K. *Curr. Opin. Cell Biol.* **6**, 3-9, 1994.

168. Möbius, D. *Berichte der Bunsengesellschaft, International Journal of Phys. Chem.* **82**, 848-858, 1978.

169. Ma, Z., Wang, J., and Guo, H. *Phys. Rev. B* **59**, 7575, 1999.

170. Maass, W., Natschläger, T., and Markram, H. *Neural Computation* **14**, 2531-2560, 2002.

171. Magee J.C., and Johnston, D. *Science* **268(5208)**, 301-4, 1995.

172. Magee J.C., and Johnston, D. *The Journal of Physiology* **487(1)**, 67-90, 1995.

173. Mahan, G.D. *Many-Particle Physics*, Plenum Press, New York, 1981.

174. Maldague, P.F. *Phys. Rev.* **B16**, 2437-2446, 1977.

175. Malenka R.C., and Bear, M.F. *Neuron.* **44**, 5-21, 2004.

176. Malenka, R. C. *Neuron* **6**, 53-60, 1991.

177. Mandelkow, E. M., Lange, G., Jagla, A., Spann, U. and Mandelkow, E. *EMBO Journal* **7**, 357-65, 1988.

178. Mandelkow, E.M., Mandelkow, E., and Milligan, R. *J. Cell Biol.* **114**, 977-991, 1991.

179. Mandelkow E.-M. and Mandelkow, E. *Cell Motil. and Cytoskel.* **22**, 235-244, 1992.

180. Maniotis, A.J., Chen, C.S., and Ingber, D.E. *Proc. Natl. Sci. U.S.A.* **94**, 849, 1997.

181. Marechal, Y. *Journal of Molecular Liquids* **48(2-4)**, 253-260, 1991.

182. Martin, S.R., Schilstra, M.J. and Bayley, P.M., *Biophysics Journal* **65**, 578-596, 1993.

183. Marx, A., Jagla, A. and Mandelkow, E., "Microtubule assembly and oscillations induced by flash photolysis of caged-GTP", *European Biophysics Journal* **19**, 1-9, 1990.

184. Marx, A., and Mandelkow, E. *Eur. Biophys. J.* **22**, 405, 1994.

185. Mehrez, H., Walch, S., and Anantram, M. P. *Phys. Rev. B* **72**, 035441, 2005.

186. Mehta, A. D., Rief, M., Spudich, J. A., Smith, D. A., and Simmons, R. M. *Science* **283**, 1689-1695, 1999.

187. Mejillano, M.R., Tolo, E.T., Williams, R.C. Jr., and Himes, R.H., *Biochemistry* **31**, 3478-3483, 1992.

188. Melki, R., Carlier, M.-F., and Pantaloni, D., *EMBO Journal* **7**, 2653-2659, 1988.

189. Mershin, A., Kolomenski, A.A., Schuessler, H.A., and Nanopoulos, D.V. *Biosystems* **77**, 73-85, 2004a.

190. Mickey, B. and Howard, J. *J. Cell Biol.* **130**, 909-917, 1995.

191. Miller, S., Yasuda, M., Coats, J.K., Jones, Y., Martone, M.E., and Mayford, M. *Neuron* **36**, 507-519, 2002.

192. Minoura, I. and Muto, E. *Biophysical Journal* **90**, 3739-3748, 2006.

193. Mitchison, J.M. *Biology of the Cell Cycle.* Cambridge University Press, Cambridge, 1973.

194. Mitchison, T. and Kirschner, M. *Nature* **312**, 232-237, 1984.

195. Mitchison, T. and Kirschner, M. *Nature, London.* **312**, 237-242, 1984.

196. Miyakawa, H., Ross, W.N., Jaffe, D., Callaway, J.C., Lasser-Ross, N., Lisman, J.E., and Johnston, D. *Neuron* **9(6)**, 1163-73, 1992.

197. Möbius, D. and Kuhn, H. *J. Appl. Phys.* **64**(10), 5138–5141, 1988.

198. Mogul D.J., Adams M.E., and Fox A.P. *Neuron* **10(2)**, 327-34, 1993.

199. Morowitz, H. J. *AJP-Regulatory, Integrative and Comparative Physiology* **235(3)**, 99-114, 1978.

200. Moskowitz, P. and Oblinger, M. *J. Neurosci.* **15**, 1545, 1995.

201. Moser, F. and Urbach, F. *Phys. Rev.* **102**, 1519–23, 1956.

202. Murphy, D. B., Gray, R. O. , Grasser, W. A., and Pollard, T. D. *J. Cell Biol.* **106**, 1947-1954, 1988.

203. Nicklas, R.B. and Ward, S.C. *J. Cell Biol.* **126**, 1241, 1994.

204. Nicklas, R.B., Ward, S.C., and Gorbsky, G.J. *J. Cell Biol.* **130**, 929, 1995.

205. Nymeyer, H., Garca, A.E., and Onuchic, J.N. *Proc. Natl. Acad. Sci. U.S.A.* **95**, 5921-5928, 1998.

206. Odde, D.J., Cassimeris, L. and Buettner, H.M. *AICHE Journal* **42**, 1434-1442, 1996.

207. Oosawa, F. *Biopolymers* **9**, 677-688, 1970.

208. Oosawa, F. *Polyelectrolytes.* Marcel Dekker, Inc., New York, 1971.

209. Oosawa, F. and Asakura, S. *Thermodynamics of the Polymerization of Protein.* Academic Press, London; New York, 1975.

210. Ostroff, L.E., Fiala, J.C., Allwardt, B. and Harris, K.M. *Neuron* **35**, 535-45, 2002

211. Parodi, M., Bianco, B., and Chiabrera, A. *Cell Biophysics* **7**, 215-235, 1985.

212. Passafaro, M., Sala, C., Niethammer, M. and Sheng, M. *Nat. Neurosci.* **2**, 1063-1069, 1999.

213. Paul, R., Chatterjee, R., Tuszyński, J.A., and Fritz, O.G. *J. Theor. Biol.* **104**, 169, 1983.

214. Pedrotti B., and Islam K. *FEBS Lett.* **388(2-3)**, 131-3, 1996.

215. Pirollet, F., Job, D., Margolis, R.L. and Garel, J.R. *EMBO Journal* **6**, 3247-52, 1987.

216. Pollard, T.D. *J. Cell Biol.* **103**, 2747-2754, 1986.

217. Porath, D., Bezryadin, A., de Vries, S. and Dekker, C. *AIP Conference* **544**, 452-456, 2000.

218. Post, F., Doster, W., Karvounis, G., and Settles, M. *Biophys J.* **64(6)**, 1833-1842, 1993.

219. Priel, A.; Tuszyński, J.A. and Woolf, N. *European Biophysics Journal*, 2005.

220. Priel, A., Ramos, A.J., Tuszyński, J.A. and Cantiello, H.F. *Biophysical Journal* **90**, 4639-4643, 2006.

221. Prigogine, I. *From Being to Becoming*, W.H. Freeman, San Francisco, 1985.

222. Rall, W. *Science* **126(3271)**, 454, 1957.

223. Rall, W. *Exp Neurol.* **1**, 491-527, 1959.

224. Ramón-Moliner, E. *Structure and Function of the Nervous Tissue* (edited by Bourne, G. F.), Academic Press, New York, 205-267, 1968.

225. Ratner, M.A., and Shriver, D.F. *MRS Bulletin* **Sept 1989**, 39-52, 1989.

226. Regehr, W.G., Connor, J.A., and Tank, D.W. *Nature* **341**, 533-536, 1989.

227. Regehr, W.G. and Tank, D.W. *Nature* **345**, 807-810, 1990.

228. Regehr, W.G. and Tank, D.W. *Journal of Neuroscience* **12**, 4202-4223, 1992.

229. Rickard, J. E. and Sheterline, P. *J. Mol. Biol.* **201**, 675-681, 1988.

230. Rieder, C. L., and Salmon, E. D. *J. Cell Biology* **24**, 223-233, 1994.

231. Ripoll, C., Norris, V., and Thellier, M. *BioEssays* **26(5)**, 549-557, 2004.

232. Rypdal, K. and Juul Rasmussen, J. *Physica Scripta* **33**, 498–504, 1986.

233. Rodriguez, O.C., Schaefer, A.W., Mandato, C.A., Forscher, P., Bement, W.M., and Waterman-Storer, C.M. *Nat Cell Biol.* **5**, 599-609, 2003.

234. Roweis S., Winfree, E., Burgoyne, R., Chelyapov, N., Goodman, M., Rothemund, P. and Adleman, L.M. *Journal of Computational Biology* **5(4)**, 615-629, 1998.

235. Scheibe, G. *Augen. Chem.* **49**, 567, 1936.

236. Sackett, D.L. In *Subcellular Biochemistry-Proteins: Structure, Function and Engineering*. B.B. Biswas, and S. Roy, editors. Kluwer Academic Publishers, Dordrecht, 24:255-302, 1995.

237. Sackmann, B. and Naher, E., *Single-Channel Recording*, 2nd ed. Plenum, New York, 1995.

238. Salmon, W.C., Adams, M.C., and Waterman-Storer, C.M. *The Journal of Cell Biology* **158(1)**, 31-37, 2002.

239. Sanchez, C., Diaz-Nido, J., and Avila, J. *Prog. Neurobiol.* **61(2)**, 133-68, 2000.

240. Sansom, F.G., Ball, C.J., Kerry, R., McGee, R.L. and Usherwood, P.N.P., *Biophys. J.* **56**, 1229, 1989.

241. Santarella, R.A., Skiniotis, G., Goldie, K. N., Tittmann, P., Gross, H., Mandelkow, E. M., Mandelkow, E., and Hoenger, A. *Journal of Molecular Biology* **339**, 539-553, 2004.

242. Sattilaro, R.F. *Biochemistry* **25**, 2003-2009, 1986.

243. Schaefer, A.W., Kabir, N., and Forscher, P. *The Journal of Cell Biology* **158(1)**, 139-152, 2002.

244. Schwartzkroin, P.A., and Slawsky, M. *Brain Res.* **135(1)**, 157-61, 1977.

245. Scott, A.C., Chu, F.Y.F., and McLaughlin, D.W. *Proceedings of the IEEE* **61**, 1443, 1973.

246. Scott, A.C. *Physics Reports* **217**, 1-67, 1992.

247. Segel, L. and Parnas, H. *Biologically Inspired Physics*, Peliti, L., editor, Plennum Press, New York, Segev and London, 2000.

248. Segev, I. and London, M. *Science* **290(5492)**, 744-750, 2000.

249. Selden, S.C. and Pollard, T.D. *J. Biol. Chem.* **258**(11), 7064-7071, 1983.

250. Semënov, M.V. *J. Theor. Biol.* **179**, 91-117, 1996.

251. Sept D. *PhD thesis*, University of Alberta, 1997.

252. Sept D. *Physical Review E* **60**, 838-841, 1999.

253. Sept, D., Elcock, A. H., and McCammon, J. A. *Journal of Molecular Biology* **294**, 1181-1189, 1999.

254. Sept, D., Limbach, H.-J., Bolterauer, H., and Tuszyński, J.A. *J. Theor. Biol.* **197**, 77-88, 1999.

255. Skiniotis, G., Cochran, J.C., Muller, J., Mandelkow, E., Gilbert, S.P., and Hoenger, A. *EMBO J.* **23**, 989-999, 2004.

256. Softky W. *Neuroscience* **58(1)**, 13-41, 1994.

257. Spano, F.C., Kuklinski, J.R. and Mukamel, S. *Phys. Rev. Lett.* **65**(2), 211–214, 1990.

258. Spruston, N., Jonas, P., and Sakmann, B. *The Journal of Physiology* **482(2)** 325-352, 1995.

259. Spruston, N., Schiller, Y., Stuart, G., Sakmann, B. *Science* **268**, 297-300, 1995.

260. Stein, D.L. *Spin Glasses and Biology*, World Scientific, Singapore, 1992.

261. Steward, O. and Schuman, E.M. *Annu. Rev. Neurosci.* **24**, 299-325, 2001.

262. Stryer, L. *Biochemistry*, 4th Ed., Ch. 8, Freeman, New York, 1995.

263. Stuart, G.J. and Sakmann, B. *Nature* **367**, 69-72, 1994.

264. Szent-Gyorgyi, A. *Nature* **14**, 157-187, 1941.

265. Tabony J. and Job. D. *Nature, London* **346**, 448-451, 1990.

266. Tang D. and Goldberg D.J. *Molecular and Cellular Neuroscience* **15(3)**, 303-313, 2000.

267. Tang, J. and Janmey, P. *J. Biol. Chem.* **271**, 8556-8563, 1996.

268. Thorn, K.S., Ubersax, J.A., and Vale, R.D. *J. Cell Biol.* **151**, 1093-1100, 2000.

269. Tobacman, L. S. and Korn, E. D. *J. Biol. Chem.* **258**, 3207-3214, 1983.

270. Togel, M., Wiche, G., and Propst, F. *J. Cell Biol.* **143**, 695-707, 1998.

271. Tran, P.T. Walker, R.A., and Salmon. E.D. *J. Cell Biol.* **138**, 105-117, 1997.

272. Tuszyński, J.A., Brown, J.A., and Hawrylak, P. *Philos. Trans. R. Soc. London Ser. A.* **356**, 1897, 1998.

273. Tuszyński, J.A., Joergensen, M.F. and Möbius, D. *Physical Review E* **59**, 4374-4383, 1999.

274. Tuszyński, J. A. and Dixon, J. M. *Biomedical Applications of Introductory Physics.* John Wiley and Sons, New York, 2001.

275. Tuszyński, J.A., Portet, S., Dixon, J.M., Luxford, C., and Cantiello, H.F. *Biophysical Journal* **86**, 1890-1903, 2004.

276. Verde, F., Dogterom, M., Stelzer, E., Karsenti, E. and Leibler, S., *Journal Cell Biology* **118**, 1097-1108, 1992.

277. Wade, R.H., Pirollet, F., Margolis, R.L., Garel, J.R., and Job, D., *Biology of the Cell* **65**, 37-44, 1989.

278. Walker, R. A., O'Brien, E.T., Pryer, N.K., Soboeiro, M.F., Voter, W.A., Erickson, H.P., and Salmon, E.D., *Journal Cell Biology* **107**, 1437-1448, 1988.

279. Walker, R.A., Pryer N.K. and Salmon, E.D., *Journal Cell Biology* **114**, 73, 1991.

280. Wang, B., Zhao, X., and Guo, H. *Appl. Phys. Lett.* **74**, 2887, 1999.

281. Wegner, A. and Savko, P. *Biochemistry* **21**, 1909-1913, 1982.

282. Weisshaar, B., Doll, T., and Matus, A. *Development* **116**, 1151-1161, 1992.

283. Westenbroek, R.E., Ahlijanian, M.K., and Catterall, W.A. *Nature* **347**, 281-284, 1990.

284. Wong, R. K. S., Prince, D. A., and Basbaum, A. I. *PNAS* **76(2)**, 986-990, 1979.

285. Woolf, N.J., Zinnerman, M.D. and Johnson, G.V.W. *Brain Res.* **821**, 241-249, 1999.

286. Yamada, K.M., Spooner, B.S., and Wessell, N.K. *J. Cell Biology* **49 (3)**, 614, 1971.

287. Yu, W., Ling , C. and Baas, P.W. *Journal of Neurocytology* **30(11)** , 861-875, 2001.

288. Yu, X., Shacka, J.J., Eells, J.B., Suarez-Quian, C., Przygodzki, R.M., Beleslin-Cokic, B., Lin, C.-S., Nikodem, V.M., Hempstead, B., Flanders, K.C., Costantini, F., and Noguchi, C.T. *Development* **129**, 505-516, 2002.

289. Zakharov, V.E. and Shabat, A.B. *Sov. Phys. JETP* **34**, 62–69, 1972.

290. Zimm, B. *Coulombic Interactions in Macromolecular Systems*, Use of the Poisson-Boltzmann equation to predict ion condensation around polyelectrolytes, American Chemical Society, Washington D.C., 212-215, 1986.

Glossary

acetate Salts or esters of acetic acid in which the terminal hydrogen atom is replaced by a metal or a radical.

acetylcholine A neurotransmitter. In vertebrates, it is the major transmitter at neuromuscular junctions, autonomic ganglia, and at many sites in the central nervous system.

acid Chemical compounds which yield hydrogen ions or protons when dissolved in water, whose hydrogen can be replaced by metals or basic radicals, or which react with bases to form salts and water.

actins Filamentous proteins that are the main constituent of the thin filaments of muscle fibers. They can be dissociated into globular subunits, each of which is composed of a single polypeptide 375 amino acids long. In conjunction with myosins, actin is responsible for the contraction and relaxation of muscle.

action potentials A momentary change in electrical potential on the surface of a nerve or muscle cell that takes place when it is stimulated, especially by the transmission of a nerve impulse.

active transport The movement of materials across cell membranes and epithelial layers against an electrochemical gradient, requiring the expenditure of metabolic energy.

adenosine diphosphate (ADP) An adenine nucleotide containing two phosphate groups esterified to the sugar moiety at the 5'-position.

adenosine triphosphate (ATP) An adenine nucleotide containing three phosphate groups esterified to the sugar moiety at the 5'-position. In addition to its crucial roles in metabolism, adenosine triphosphate is a neurotransmitter.

adiabatic Without loss or gain of heat.

alcohols Alkyl compounds containing a hydroxyl group (-OH).

aldehydes Organic compounds containing a carbonyl group (-CHO).

aliphatic Having carbon compounds which are linked in open chains rather than aromatic rings.

allometry The variation in the relative rates of growth of various parts of the body, which helps shape the organism.

allosteric site A site on an enzyme, which upon binding of a modulator causes the protein to undergo a conformational change that may alter the catalytic or binding properties of the enzyme.

amines A group of compounds derived from ammonia by substituting organic radicals for the hydrogen atoms.

amino acid Organic compounds that generally contain an amino (-NH2) and a carboxyl (-COOH) group. There are twenty alpha amino acids that are the subunits polymerized to form proteins.

anaphase The third phase of cell division, in which the chromatids separate and migrate to opposite poles of the spindle.

anions Negatively charged atoms, radicals, or groups of atoms which travel to the anode or positive pole during electrolysis.

antibodies Immunoglobulin molecules that have a specific amino acid sequence such that they only interact with the antigen that induced their synthesis in cells of the lymphoid series (especially plasma cells), or with an antigen closely related to it.

antigens Substances that are recognized by the immune system and induce an immune reaction.

apoptosis One of the two mechanisms by which cell death occurs (the other being the pathological process of necrosis). It is the mechanism responsible for the physiological deletion of cells and appears to be intrinsically programmed. It balances mitosis in regulating the size of animal tissues and in mediating pathologic processes associated with tumor growth.

axon A nerve fibre that is capable of rapidly conducting impulses away from the neuron cell body.

bacteria Unicellular prokaryotic microorganisms which generally possess rigid cell walls, multiply by cell division, and exhibit three principal forms: round or coccal, rodlike or bacillary, and spiral or spirochetal.

bifurcation Separation into two parts or branches; a change in the stability or in the types of solutions which occurs as a parameter is varied in a dissipative dynamic system.

Brownian motion The random motion of small particles suspended in a gas or liquid.

calmodulin A calcium-binding protein which assists in controlling biochemical processes in cells.

capacitance The quantity of electric charge that may be stored per unit electric potential, expressed in farads.

capacitor A device used to store electrical energy by accumulating charge on conductors situated close to one another.

carbohydrate The largest class of organic compounds, including starches, glycogens, cellulose, gums, and simple sugars. Carbohydrates are composed of carbon, hydrogen, and oxygen in a ratio of $C_n(H_2O)_n$.

carbonic acid The hypothetical acid of carbon dioxide and water (H_2CO_3). It exists only in the form of its salts (carbonates), acid salts (hydrogen carbonates), amines (carbamic acid), and acid chlorides (carbonyl chloride).

catalyst A substance that accelerates a chemical reaction without being permanently changed by the reaction. Enzymes are an example of a biological catalyst.

cations Positively charged atoms, radicals, or groups of atoms that travel to the cathode or negative pole during electrolysis.

cellular automata A regular spatial lattice of "cells", each of which can have any one of a finite number of states. The states of all cells in the lattice are updated simultaneously and the state of the entire lattice advances in discrete time steps. The state of each cell in the lattice is updated according to a local rule which may depend on the state of the cell and its neighbors at the previous time step.

cellulose A polysaccharide of linked glucose units and the chief constituent of plant fibers.

centriole Self-replicating, short, fibrous, rod-shaped organelles. Each centriole is a short cylinder containing nine pairs of peripheral microtubules, arranged like blades of a fan to form the wall of the cylinder.

centromere The constricted portion of the chromosome at which the chromatids are joined and by which the chromosome is attached to the spindle during cell division.

chaos A property of some non-linear dynamic systems which exhibit sensitive dependence on initial conditions.

chlorophyll A porphyrin derivative containing magnesium that acts to convert light energy in photosynthetic organisms.

chloroplast Plant cell inclusion bodies that contain the photosynthetic pigment chlorophyll. Chloroplasts occur in cells of leaves and young stems of higher plants.

cholesterol The principal sterol (steroids with a hydroxyl group at C-3) of all higher animals. It is distributed in body tissues (especially the brain and spinal cord), as well as in animal fats and oils.

chromatid Either of the two longitudinally adjacent threads formed when a eukaryotic chromosome replicates prior to mitosis. The chromatids are held together at the centromere. Sister chromatids are derived from the same chromosome.

chromatin The material of chromosomes. It is a complex of DNA, histones, and non-histone proteins found within the nucleus of a cell.

chromosome In a prokaryotic cell or in the nucleus of a eukaryotic cell, a structure consisting of or containing DNA which carries the genetic information essential to the cell.

cilia Thin, motile appendages found covering the surface of cells. Each cilium arises from a basic granule in the superficial layer of cytoplasm. The movement of cilia propels ciliates through the liquid in which they live.

cis Having a pair of identical atoms or groups on the same side of the plane that passes through two carbon atoms linked by a double bond.

clathrate A "cage compound", where one type of molecule is completely contained within the lattice structure of a second type of molecule.

coenzyme Substances that are necessary for the action or enhancement of action of an enzyme. Many vitamins are coenzymes.

cofactor Any non-protein molecule or ion that is required for the proper functioning of an enzyme. Cofactors can be permanently bound to the active site or may bind loosely with the substrate during catalysis.

coherence The state in which two signals maintain a fixed phase relationship with each other or with a third signal that can serve as a reference for each.

collagen A polypeptide substance comprising about one third of the total protein in mammalian organisms. It is the main constituent of skin, connective tissue, and the organic substance of bones and teeth.

condensation The process by which a gas or vapor changes to a liquid; a chemical reaction in which water or another simple substance is released by the combination of two or more molecules.

conductance A measure of conductivity, given by the ratio of the current flowing through a conductor to the difference in potential between the ends of the conductor.

conduction The transfer of sound waves, heat, nervous impulses or electricity.

conductor A material that has sufficient free electrons in the conduction band to allow an electrical current to flow when a potential difference is applied. Conductors are usually metallic in nature, but not always (as in the case of graphite).

conservation law The maintenance of certain quantities unchanged during chemical reactions or physical transformations.

convection Heat transfer in a gas or liquid by the circulation of currents from one region to another.

covalent bond A type of strong chemical bond in which two atoms share one pair of electrons in a mutual valence shell.

current The quantity of charge per unit time, measured in amperes.

cytochromes Proteins whose characteristic mode of action involves the transfer of reducing equivalents, associated with a reversible change in the oxidation state of the prosthetic group.

cytokines Non-antibody proteins secreted by inflammatory leukocytes and some non-leukocytic cells. They are locally-acting intercellular mediators, and they differ from classical hormones in that they are produced by a number of tissue or cell types rather than by specialized glands.

cytokinesis The fission of a cell, occurring after the cell nucleus division is complete.

cytoplasm The part of a cell that contains the cytosol and small structures, excluding the cell nucleus, mitochondria, and large vacuoles.

cytoskeleton The network of filaments, tubules, and interconnecting filamentous bridges which give shape, structure, and organization to the cytoplasm.

dendrite Short, branched extensions of the nerve cell body that receive stimuli from other neurons.

density The mass per unit volume of a substance under specified conditions of pressure and temperature.

dialysis A process of selective diffusion through a membrane. It is usually used to separate low-molecular-weight solutes which diffuse through the membrane from the high-molecular-weight solutes which do not.

dielectric constant The ratio of the permittivity of a substance to the permittivity of free space. It is an expression of the extent to which a material concentrates electric flux.

diffraction Change in the directions and intensities of a group of waves after passing by an obstacle or through an aperture whose size is on the order of the wavelength of the waves.

diffusion The tendency of a gas or solute to pass from a point of higher pressure or concentration to a point of lower pressure or concentration and to distribute itself throughout the available space.

DNA A deoxyribonucleotide polymer that is the primary genetic material of all cells. Eukaryotic and prokaryotic organisms normally contain DNA in a double-stranded state, yet several important biological processes transiently involve single-stranded regions. It consists of a polysugar-phosphate backbone with projections of purines (adenine and guanine) and pyrimidines (thymine and cytosine). It forms a double helix that is held together by hydrogen bonds between these purines and pyrimidines (adenine to thymine and guanine to cytosine).

dwell time The period during which a dynamic process remains halted in order that another process may occur.

dynamics The branch of mechanics that is concerned with the effects of external forces on the motion of a body or system of bodies.

efficiency The ratio of the output to the input of a system.

electric field A region of space characterized by the existence of a force generated by electric charge.

electrode The medium used between an electric conductor and the object to which the current is to be applied.

electrolyte A substance that dissociates, to some extent, into two or more ions in water. Solutions of electrolytes thus conduct an electric current and can be decomposed by it.

electromotive force The energy per unit charge that is converted reversibly from chemical, mechanical, or other forms of energy into electrical energy.

electron pair Two electrons in an orbital (with opposing spins) which are responsible for a chemical bond.

enantiomers A pair of compounds (crystals or molecules) that are mirror images of each other but are not identical.

endocytosis Cellular uptake of extracellular materials within membrane-limited vacuoles or microvesicles.

endoplasmic reticulum A system of cisternae in the cytoplasm of many cells. It is continuous with the cell membrane or outer membrane of the nuclear envelope in certain places. If the outer surfaces of the endoplasmic reticulum membranes are coated with ribosomes, the endoplasmic reticulum is said to be rough-surfaced; otherwise it is said to be smooth-surfaced.

endothermic Characterized by or causing the absorption of heat.

energy The capacity of a physical system to do work.

enthalpy A thermodynamic function of a system, equivalent to the sum of the internal energy of the system plus the product of its volume multiplied by the pressure exerted on it by its surroundings.

entropy A measure of the disorder or unavailability of energy within a closed system. More entropy means less energy available for doing work.

enzymes Biological molecules that possess catalytic activity. They can occur naturally or be synthetically created. Enzymes are usually proteins; however catalytic RNA and catalytic DNA molecules have also been identified.

epithelial cells Cells that line the inner and outer surfaces of the body.

equilibrium The state of a body or physical system at rest or in unaccelerated motion, in which the resultant of all forces acting on it is zero and the sum of all torques about any axis is zero; the state of a chemical reaction in which its forward and reverse reactions occur at equal rates so that the concentration of the reactants and products does not change with time.

erythrocyte A red blood cell. Mature erythrocytes are non-nucleated, biconcave disks containing hemoglobin, whose function is to transport oxygen.

ester Any of a class of organic compounds corresponding to the inorganic salts and formed from an organic acid and an alcohol.

eukaryotic cells Cells of higher organisms, containing a true nucleus bounded by a nuclear membrane.

evolution A process of cumulative change over successive generations through which organisms acquire their distinguishing morphological and physiological characteristics.

exothermic A reaction that releases energy, usually in the form of heat.

fibroblasts Connective tissue cells which secrete an extracellular matrix rich in collagen and other macromolecules.

field Any macroscopic quantity that exists (and typically varies) throughout a region of space.

first-order A term describing the reaction rate of a chemical reaction in which the rate is proportional to the concentration of only one of the reactants.

flagella Whip-like motility appendage present on the surface of cells. Prokaryote flagella are composed of a protein called flagellin. Flagella have the same basic structure as cilia, but flagella are longer in proportion to the cell bearing them and present in much smaller numbers.

fluid A continuous, amorphous substance whose molecules move freely past one another and that has the tendency to assume the shape of its container.

flux The rate of flow of fluid, particles, or energy through a given surface.

force A vector quantity such as a push or pull on a body that tends to produce an acceleration of the body in the direction of its application.

fractal An entity that can be subdivided into parts, each of which is (at least approximately) a smaller copy of the whole. Fractals are generally self-similar, independent of scale, and possess a fractal dimension that differs from its topological dimension.

free energy A thermodynamic quantity that is the difference between the internal energy of a system and the product of its absolute temperature and entropy.

gas The state of matter distinguished from the solid and liquid states by: relatively low density and viscosity, relatively great expansion and contraction with changes in pressure and temperature, the ability to diffuse readily, and the spontaneous tendency to become distributed uniformly throughout any container.

genetics The branch of science concerned with the means and consequences of transmission and generation of the components of biological inheritance.

glucose A primary source of energy for living organisms. It is naturally occurring and is found in fruits and other parts of plants in its free state.

Golgi apparatus A stack of flattened vesicles that functions in posttranslational processing and sorting of proteins. The Golgi apparatus receives proteins from the rough endoplasmic reticulum and directs them to secretory vesicles, lysosomes, or the cell membrane.

gravitational force The natural force of attraction between any two massive bodies, which is directly proportional to the product of their masses and inversely proportional to the square of the distance between them.

ground state The state of least possible energy in a physical system, as of elementary particles.

heat The transfer of energy from one body to another as a result of a difference in temperature or a change in phase.

heat capacity The number of units of heat required to raise the temperature of a unit mass of the substance at that temperature one degree.

heterocycles Organic chemical structures which contain components that are not carbon.

heterocyclic Containing more than one kind of atom joined in a ring.

histones Small chromosomal proteins possessing an open, unfolded structure and attached to the DNA in cell nuclei by ionic linkages. They are classified into various types based on the relative amounts of arginine and lysine in each.

homeostasis The processes whereby the internal environment of an organism tends to remain balanced and stable.

homology When chromosomes bear genes for the same characters.

hormone A chemical substance having a specific regulatory effect on the activity of an organ. It may or may not be produced by an endocrine gland.

hydrogen bond A low-energy attractive force between hydrogen and another element. It plays a major role in determining the properties of water, proteins, and other compounds.

hydrolysis The process of cleaving a chemical compound by the addition of a molecule of water.

hydrophilic Having an affinity for water; readily absorbing or dissolving in water.

hydrophobic Repelling, tending not to combine with, or incapable of dissolving in water.

hysteresis A lagging or retardation of the effect when the forces acting upon a body are changed.

iceberg A "cage" of frozen water molecules which forms around non-polar molecules, similar to a regular iceberg but on a molecular scale.

indole An aromatic heterocyclic organic compound, composed of a benzene ring and a pyrrole ring (containing nitrogen).

insulator A substance or body that prevents the transfer of electricity or heat to or from it.

internal energy The total kinetic and potential energy associated with the motions and relative positions of the molecules of an object, excluding the kinetic or potential energy of the object as a whole. An increase in internal energy results in a rise in temperature or a change in phase.

interphase The interval between two successive cell divisions during which the chromosomes are not individually distinguishable and DNA replication occurs.

interstitial fluid The liquid found between the cells of the body, constituting much of the liquid environment of the body.

ion An atom or group of atoms that have a positive or negative electric charge due to a gain (negative charge) or loss (positive charge) of one or more electrons. Atoms with a positive charge are known as cations; those with a negative charge are anions.

ionization The formation of or separation into ions by heat, electrical discharge, radiation, or chemical reaction.

isomers Chemical compounds that have the same elemental composition but different structures.

isothermal Of, relating to, or indicating equal or constant temperatures.

isotopes Atoms having the same atomic number but different mass numbers.

kinase Any of various enzymes that catalyze the transfer of a phosphate group from a donor, such as ADP or ATP, to an acceptor.

kinesin A microtubule-associated protein that uses the energy of ATP hydrolysis to move organelles along microtubules toward the plus end of the microtubule.

kinetochore Large multiprotein complexes that bind the centromeres of the chromosomes to the microtubules of the mitotic spindle during metaphase in the cell cycle.

Krebs cycle A series of reactions involving oxidation of a two-carbon acetyl unit to carbon dioxide and water, with the production of high-energy phosphate bonds by means of tricarboxylic acid intermediate. Also called the citric acid cycle.

ligand A molecule that binds to another molecule, used especially to refer to a small molecule that binds specifically to a larger molecule. Ligands are also molecules that donate or accept a pair of electrons to form a coordinate covalent bond with the central metal atom of a coordination complex.

ligases A class of enzymes that catalyze the formation of a bond between two substrate molecules, coupled with the hydrolysis of a pyrophosphate bond in ATP or a similar energy donor.

luminescence The property of giving off light without emitting a corresponding degree of heat. It includes the luminescence of inorganic matter or the bioluminescence of human matter, invertebrates and other living organisms.

lymphocyte White blood cells formed in the body's lymphoid tissue. Their nucleus is round or ovoid with coarse, irregularly clumped chromatin while the cytoplasm is typically pale blue and may contain granules. Most lymphocytes can be classified as either T or B (with subpopulations of each); those with characteristics of neither major class are called null cells.

lysosome A class of morphologically heterogeneous cytoplasmic particles in animal and plant tissues characterized by their content of hydrolytic enzymes. A single unit membrane of the lysosome acts as a barrier between the enclosed enzymes and the external substrate, and this membrane must be ruptured for the enzymes to be active.

macromolecule A very large molecule, such as a polymer or protein, consisting of many smaller structural units linked together.

macrophage A relatively long-lived phagocytic cell of mammalian tissues, derived from blood monocytes.

magnetic field A condition found in the region around a magnet or an electric current, characterized by the existence of a detectable magnetic force at every point in the region and by the existence of magnetic poles.

magnetic moment The torque exerted on a magnet or dipole when it is placed in a magnetic field.

mechanics The science of the action of forces on all bodies. That part of mechanics which considers the action of forces in producing rest or equilibrium is called statics; that which relates to such action in producing motion is called dynamics.

meiosis A special method of cell division occurring during the maturation of germ cells, by means of which each daughter nucleus receives half the number of chromosomes.

membrane A thin layer of tissue which covers parts of the body, separate adjacent cavities, or connect adjacent structures.

metabolism The sum of chemical changes that occur within the tissues of an organism, consisting of anabolism and catabolism (the buildup and breakdown of molecules for utilization by the body).

metaphase The second phase of cell division, in which the chromosomes line up across the equatorial plane of the spindle prior to separation.

microtubule A slender, cylindrical filament composed of the protein tubulin and found in the cytoskeleton of plant and animal cells.

microtubule-associated proteins (MAPs) High molecular weight proteins found in the microtubules of the cytoskeletal system. Under certain conditions they are required for the assembly of tubulin into microtubules as well as for the stabilization of assembled microtubules.

microtubule-organizing centre An amorphous region of electron-dense material in the cytoplasm, from which microtubule polymerization is nucleated. An example is the pericentriolar region of the centrosome which surrounds the centrioles.

mitochondria Semiautonomous, self-reproducing organelles that occur in the cytoplasm of most eukaryotic cells. Each mitochondrion is surrounded by a double membrane and includes distinctive ribosomes, transfer RNAs, and elongation and termination factors. Mitochondria are the sites of the reactions of oxidative phosphorylation, which result in the formation of ATP.

mitosis A method of indirect cell division by means of which the two daughter nuclei normally receive identical complements of the number of chromosomes of the somatic cells of the species.

mitotic spindle apparatus An organelle consisting of three components: the astral microtubules, which form around each centrosome and extend to the periphery; the polar microtubules, which extend from one spindle pole to the equator; and the kinetochore microtubules, which connect the centromeres of the various chromosomes to either centrosome.

momentum A vector quantity that is the product of the mass and the velocity of an object or particle.

monosaccharide A simple sugar - a carbohydrate that cannot be decomposed by hydrolysis. It is typically a colorless crystalline substance with a sweet taste and the general formula $C_nH_{2n}O_n$.

myosins A diverse family of translocating proteins. They can bind to actins and hydrolyse MgATP. Myosins generally consist of heavy chains which are involved in locomotion, and light chains which are involved in regulation. The superfamily of myosins is organized into structural classes based upon the type and arrangement of the subunits they contain.

noise A disturbance, especially a random and persistent one, which obscures or reduces the clarity of a signal.

non-linearity A state in which the output of a system does not vary in direct proportion to its input.

nuclear envelope The membrane system of the cell nucleus that surrounds the nucleoplasm. It consists of two concentric membranes separated by the perinuclear space. The structures of the envelope where it opens to the cytoplasm are called the nuclear pores.

nucleolus Within most types of eukaryotic cell nucleus, a distinct region, not delimited by a membrane, in which some species of rRNA are synthesized and assembled into ribonucleoprotein subunits of ribosomes.

nucleoside A purine or pyrimidine base attached to a ribose or deoxyribose.

nucleotide The monomeric units from which DNA or RNA polymers are constructed. They consist of a purine or pyrimidine base, a pentose sugar, and a phosphate group.

orbitals An area of space where the electron is most likely to be found.

oscillator A system whose values vary continually between two extremes in a cyclic manner.

oxidation-reduction A chemical reaction in which an electron is transferred from one molecule to another. The electron-donating molecule is the reducing agent or reductant, while the electron-accepting molecule is the oxidizing agent or oxidant.

pH A measure of the acidity or alkalinity of a solution, equal to seven for neutral solutions, increasing up to fourteen with increasing alkalinity and decreasing to zero with increasing acidity.

pKa The negative log (to base 10) of the acidity constant K_a.

paramagnetic Relating to or being a substance in which an induced magnetic field is parallel and proportional to the intensity of the magnetizing field but is much weaker than in ferromagnetic materials.

passive transport The movement of a chemical substance across a cell membrane without expenditure of energy by the cell, as in diffusion.

percolation The theory of the formation and structure of clusters in a large lattice whose sites or bonds are present with a certain probability.

persistence length A measure of the stiffness of a macromolecule of a polymer, defined as the distance from one end of the chain one must move to bend it 90°.

phase transition A change from one state (solid or liquid or gas) to another.

phosphate The inorganic salt of phosphoric acid.

phosphorylation The introduction of a phosphoryl group into a compound through the formation of an ester bond between the compound and a phosphorus component.

potential The work required to move a unit of positive charge, a magnetic pole, or an amount of mass from a reference point to a designated point in a static electric, magnetic, or gravitational field.

potential energy The energy of a particle or system of particles derived from position, or condition, rather than motion. A raised weight, coiled spring, or charged battery has potential energy.

pressure The force applied to a unit area of surface, measured in pascals.

prokaryotic cells Cells that lack a nuclear membrane so that the nuclear material is either scattered in the cytoplasm or collected in a nucleoid region. Examples are bacteria and blue green algae.

prophase The first phase of cell division, in which the chromosomes become visible, the nucleus starts to lose its identity, the spindle appears, and the centrioles migrate toward opposite poles.

protein Polymers of amino acids linked by peptide bonds. The specific sequence of amino acids determines the shape and function of the protein.

purine A series of heterocyclic compounds that includes the constituents of nucleic acids adenine and guanine, as well as many alkaloids such as caffeine and theophylline. Uric acid is the metabolic end product of purine metabolism.

pyrimidine A single-ringed, crystalline organic base that forms uracil, cytosine, or thymine and is the parent compound of many drugs, including the barbiturates.

quantization The limitation of possible values of a quantity to a discrete set of values by quantum mechanical rules.

quantum theory The physical theory that certain properties occur only in discrete amounts (quanta). It accounts for the nature and behavior of matter and energy at the atomic and subatomic level.

radiation Emission and propagation of energy in the form of a stream of particles or electromagnetic waves.

radioactivity Spontaneous emission of radiation, either directly from unstable atomic nuclei or as a consequence of a nuclear reaction.

residue A single unit within a polymer, such as an amino acid within a polypeptide or protein.

resistance The opposition of a body or substance to current passing through it, resulting in a change of electrical energy into heat or another form of energy.

resonance The increase in amplitude of oscillation of an electric or mechanical system exposed to a periodic force whose frequency is equal or very close to the natural undamped frequency of the system.

retroviruses Viruses that have RNA (instead of DNA) as their genetic material. They are able to create a DNA version of their genes using an enzyme called reverse transcriptase

ribosome A class of multicomponent structures found in all cells, mitochondria, and chloroplasts. It has roles both in the genetic translation of transcripts and in the manufacture and secretion of proteins.

RNA A polynucleotide consisting essentially of chains with a repeating backbone of phosphate and ribose units to which nitrogenous bases are attached. RNA is unique among biological macromolecules in that it can encode genetic information, it serves as an abundant structural component of cells, and it also possesses catalytic activity.

RNA, messenger RNA sequences that serve as templates for protein synthesis. Bacterial mRNAs are generally primary transcripts in that they do not require post-transcriptional processing. Eukaryotic mRNA is synthesized in the nucleus and must be exported to the cytoplasm for translation.

RNA, transfer The small RNA molecules that function during translation to align amino acids at the ribosomes in a sequence determined by the mRNA. There are about thirty different transfer RNAs, and each recognizes a specific codon set on the mRNA through its own anitocodon.

salts Substances produced from the reaction between acids and bases; compounds consisting of a metal (positive) radical and non-metal (negative) radical.

self-organized criticality Refers to a statistical steady state that is produced by processes with an infinite separation of time scales and that exhibits scale invariance without further fine tuning.

self-similarity A set is called strictly self-similar if it can be broken into arbitrary small pieces, each of which is a small replica of the entire set.

semiconductor A material, typically crystalline, which allows current to flow under certain circumstances. Common semiconductors are silicon, germanium, and gallium arsenide. Semiconductors are used to make diodes, transistors, and other basic solid state electronic components.

soliton A pulse-like wave that can exist in non-linear systems, does not obey the superposition principle, and does not disperse.

solvent Liquids that dissolve other substances (solutes), which are generally solids, without any change in chemical composition.

sound wave A longitudinal pressure wave of audible or inaudible sound.

species A taxonomic group whose members can interbreed.

spectroscopy The science of measuring the emission and absorption of different wavelengths (spectra) of visible and non-visible light.

spin The intrinsic angular momentum of a subatomic particle.

statistical mechanics The branch of physics that makes theoretical predictions about the behavior of macroscopic systems on the basis of statistical laws governing its component particles.

steady state A stable condition that does not change over time or in which change in one direction is continually balanced by change in another.

strange attractor A set of physical properties toward which a system tends to evolve, regardless of the starting conditions of the system.

symmetry The exact correspondence of form on opposite sides of a dividing line or plane.

teleost fish A fish with a bony skeleton as opposed to one made from cartilage.

telophase The final phase of cell division, in which two daughter nuclei are formed, the cytoplasm divides, and the chromosomes lose their distinctness and are transformed into chromatin networks.

temperature The property of a body or region of space that determines whether or not there will be a net flow of heat into it or out of it from a neighboring body or region and in which direction (if any) the heat will flow.

tension A force that tends to stretch or elongate.

thermodynamics The branch of physics that deals with the relationships and conversions between heat and other forms of energy.

trans Having a pair of identical atoms on opposite sides of two atoms, linked by a double bond.

transcription The transfer of genetic information from DNA to messenger RNA by DNA-directed RNA polymerase.

translation The formation of peptides on ribosomes, directed by messenger RNA.

tubulin A microtubule protein subunit.

turbulence Violent macroscopic fluctuations which can develop under certain conditions in fluids and plasmas and which usually result in the rapid transfer of energy through the medium.

ubiquitin A highly conserved peptide universally found in eukaryotic cells that functions as a marker for intracellular protein transport and degradation. Ubiquitin becomes activated through a series of complicated steps and forms an isopeptide bond to lysine residues of specific proteins within the cell. These "ubiquitinated" proteins can be recognized and degraded by proteosomes or be transported to specific compartments within the cell.

viscosity A physical property of fluids that determines the internal resistance to shear forces.

voltage The electromotive force or potential difference, expressed in volts.

vortex A mass of fluid, especially of a liquid, having a whirling or circular motion, with the tendency to form a cavity in the center of the circle and to draw bodies inwards.

wave A vibration propagated from particle to particle through a body or elastic medium.

work The transfer of energy from one physical system to another through the application of a force that moves the body in the direction of the force.

zwitterionic Having acid and base groups within the same molecule.

Index

Index

Milton Keynes UK
Ingram Content Group UK Ltd.
UKHW021925071024
449327UK00022B/1704

9 780367 388485